Wastewater Engineering: Treatment and Reuse

Wastewater Engineering: Treatment and Reuse

Editor: Gabriel Craig

R CALLISTO REFERENCE

www.callistoreference.com

Callisto Reference,
118-35 Queens Blvd., Suite 400,
Forest Hills, NY 11375, USA

Visit us on the World Wide Web at:
www.callistoreference.com

ISBN: 978-1-64116-266-1 (Hardback)

Cataloging-in-Publication Data

Wastewater engineering : treatment and reuse / edited by Gabriel Craig.
    p. cm.
Includes bibliographical references and index.
ISBN 978-1-64116-266-1
1. Sewerage. 2. Sewage disposal. 3. Water reuse. 4. Water--Purification. 5. Sanitary engineering. I. Craig, Gabriel.
TD645 .W37 2020
628.3--dc23

Table of Contents

Preface

The world is advancing at a fast pace like never before. Therefore, the need is to keep up with the latest developments. This book was an idea that came to fruition when the specialists in the area realized the need to coordinate together and document essential themes in the subject. That's when I was requested to be the editor. Editing this book has been an honour as it brings together diverse authors researching on different streams of the field. The book collates essential materials contributed by veterans in the area which can be utilized by students and researchers alike.

Wastewater is a combination of water and water-transported wastes from domestic, commercial, industrial and agricultural sites. It also includes surface and storm water inflow, and groundwater infiltration that may enter the sewer system. On a global scale, nearly 80% of wastewater generated is discharged into the environment without treatment, leading to massive levels of water contamination. There are several ways of treating wastewater based on the type of contamination. A combination of physical, chemical and biological methods can be used to treat wastewater in wastewater treatment plants. Wastewater, after it has been treated, can be reused for the artificial recharge of aquifers, rehabilitation of natural ecosystems including wetlands, and industrial purposes. Certain processes such as ultrafiltration, forward osmosis, reverse osmosis, ozonation and advanced oxidation ensure that wastewater is made reusable. This book studies, analyzes and upholds the pillars of wastewater engineering and its utmost significance in modern times. It includes some of the vital pieces of work being conducted across the world, on various topics related to the treatment and reuse of wastewater. It is a vital tool for all researching or studying wastewater engineering as it gives incredible insights into emerging trends and concepts.

Each chapter is a sole-standing publication that reflects each author's interpretation. Thus, the book displays a multi-facetted picture of our current understanding of application, resources and aspects of the field. I would like to thank the contributors of this book and my family for their endless support.

Editor

Combination of ozonation with aerobic sequencing batch reactor for soft drink wastewater treatment: experiments and neural network modeling

Negar Amiri[1], Mojtaba Ahmadi[1,*], Meghdad Pirsaheb[2], Yasser Vasseghian[1], Pegah Amiri[1]

[1]Chemical Engineering Department, Faculty of Engineering, Razi University, Kermanshah, Iran.
[2]Department of Environmental Health Engineering-Kermanshah Health Research Center (KHRC), Kermanshah University of Medical Sciences, Iran.

ARTICLE INFO

Keywords:
Soft drink
Wastewater
SBR
Ozone
Artificial neural network

ABSTRACT

In this study, ozone combination with a sequencing batch reactor was tested in laboratory scale for treating a soft drink wastewater characterized by high concentrations of chemical oxygen demand (COD). A bench scale aerobic sequencing batch reactor (SBR) is carried out by two stages. The system was operated under three different mixed liquid suspended solids (MLSS) concentrations (3000, 4500, 6000 mg/l). The results show that the integrated ozonation with biological process was able to achieve high removal efficiencies for chemical oxygen demand (COD), with residual concentrations much lower than the current discharge limits. Also, the process was characterized by a very low MLSS concentration. Hence, the ratio between ozone dose and the COD removal was 0.72, indicating that the removed COD was higher than the dosed ozone. Artificial neural networks (ANN) was also employed to model the COD data obtained. A network consisting of two layers of five neurons in the hidden layer was considered. Regression coefficient between experimental data and data predicted by neural networks and root mean square error (R^2, RMSE) obtained 0.991, 80.36, respectively. Very low error in the network estimation confirmed validity of the obtained networks for further analysis and optimization.

1. Introduction

Wastewater, a liquid waste that is discharged from, industries, agricultural, commercial, hospital properties potentially release significant amounts of toxic and pathogenic contaminants into local treatment plants for processing. The activated sludge process uses a mass of microorganisms to aerobically treat wastewater. The microorganisms use organic pollutions in the wastewater as a food source, converting a portion of the carbon in it into new biomass and the remainder into carbon dioxide. Biological treatment is often a cheaper and more environmental salubrious, offer for the treatment of the organic pollutants from industrial wastewater (Hai et al. 2007; Jadhav et al. 2010). Unfortunately, biological treatment alone is not profitable to eliminate hardly biodegradable such as toxic, refractory compounds (Ledakowicz et al. 2001), hence, additional special treatment step such as advanced oxidation processes (AOPs) are necessary for optimization of treating wastewater (Di Iaconi. 2012). The AOPs, are characterized by the generation of highly reactive free radicals in aqueous, such as hydroxyl radicals (OH·). With a high oxidation potential of 2.80 V, OH· radical has a singular demolition power (Kurniawan et al. 2006):

$$OH^· + H + +e^- \rightarrow H_2O \qquad E_0 = 2.80\,V \qquad (1)$$

AOPs are effective in destruction organic chemicals; because they react rapidly and have no selectively with nearly all organic compound (Stasinakis. 2008). Through the ·OH radicals, the ozone (O_3) oxidation is a powerful oxidizing method with a high oxidation–reduction potential of 2.07 V. Ozone reacts with great number of with organic and inorganic compounds directly by molecular ozone or via hydroxyl radicals generated by ozone decomposition in water (Somensi et al. 2010). But,

advanced oxidation processes (AOPs) cannot meet the environmental discharge standard by itself alone (Somensi et al. 2010). Therefore, the combination of ozonation with biological treatment would be capable to remove refractory organic compounds, including color, coliform and virus (Sangave et al. 2007). Furthermore, no solid residues are produced during ozonation (Baig and Liechti. 2001; Arslan-Alaton. 2007). In pre-ozonation, ozone reacts with many biorefractory compounds, converting them into simpler and biodegradable molecules (Yasa et al. 2007; Cortez et al. 2011). Combination of ozone used as pre-treatment stage with activated-sludge aerobic process was efficacy to remove organic carbon, and nitrogen from the winery wastewater and increased ability of biological treatment. Also the use of ozone has improved settling property of the sludge (Oller et al. 2011). Cortez et al. discovered that ozone pretreatment could improve the biodegradability of recalcitrant organic compounds for subsequent biological treatment (Cheng et al. 2011; Cortez et al. 2011). On the other hand, post-ozonation also improved the quality of a secondary effluent by increasing the dissolved oxygen (Paraskeva and Graham. 2005). Variety research works have been conducted to evaluate the performance of ozonation in combination with biological treatment to remove chemical oxygen demand (COD) from wastewater. Treatment of paper mill wastewater with combination of ozone and algal treatment was examined by Balcıoğlu et al. (2007). The results showed that the BOD_5/COD ratio increased from 0.11 to 0.28 (Akmehmet Balcıoğlu et al. 2006; Balcıoğlu et al. 2007).

Treguer et al. (2010) studied effect of ozonation on natural organic matter in drinking water treatment in membrane bioreactor containing activated carbon (Treguer et al. 2010). The soft drink industry generates large volumes of wastewaters, characterized by high COD concentrations reflecting their high organic content. Typical wastewaters from soft drink industries are mainly composed of washing

Corresponding author Email: m_ahmadi@razi.ac.ir

waters from production and packing of syrup, bottling production runs, purification of process water, and equipment and conduit cleaning.

The use of artificial intelligence techniques such as fuzzy logic and artificial neural networks has been recommended in environmental pollution issues due to system complexity, validity and reliability (Mjalli et al. 2007; Ongen et al. 2013; Kundu et al. 2014). Also application of artificial intelligent model (MLP) and Adaptive Neuro fuzzy inference system (ANFIS) to predict the side weir discharge coefficient has been studied by Parsaie et al. (2014).

Prediction of dissolved oxygen (DO) in water reservoirs in Serbia by Rankovi et al. (2010) indicate that using ANN with training algorithm Levenberg–Marquardt (LM) was suitable. Which good agreement between experimental and predicted data was established (Ranković et al. 2010). Similarly, the ability of the neural network model to modeling urban wastewater quality characteristic was examined by Güçlü and Dursun (2010). Their findings showed that root mean square error (RMSE) for predicting SS, COD and MLSS were 5, 17.1 and 3.8 %, respectively, which indicated the ability of ANN to model the phenomenon.

Kundu et al. (2014) carried out a research work on ANN application for modeling in biological removal of COD and total Kjeldahl nitrogen (TKN) for the treatment of slaughterhouse wastewater. Thus, this study aimed to develop and evaluate the efficiency of the neural network model to investigate the performance of the hybrid system SBR and ozonation for COD removal of wastewater produced from the soft drink factory.

2. Materials and methods
2.1. Wastewater composition

Wastewater samples, with a pink color, were collected from a soft drink production plant located in Kermanshah-Iran. A sample was reaped in plastic containers from the effluent channel and transferred to the laboratory. The characterization of wastewater is presented in Table 1.

Table 1. Characterization of soft drink wastewater used.

Parameter	Value
COD	2500 mg/l
Biodegradability (BOD$_5$/COD)	0.43
pH	6.5-8

2.2. Biological process

Biological unit was based on sequencing batch reactor (SBR) technology. The SBR unit consisted of a cylindrical reactor (working volume: 1L). Oxygen was supplied by infusion at a flow rate of 2 L.h^{-1} using porous stones set at the bottom of the reactor in the liquid phase. All treatment process is performed at environment temperature which is about 25 ^{0}C. Activated sludge from a soft drink wastewater treatment plant was used as inoculums and nutrients (such as nitrogen and phosphorous) to obtain a COD: N: P ratio of about 100: 7: 1 in order to give a suitable growth of microorganisms. During the first 9 days of operation, the biomass was permitted to acclimatize to the influent substrate. The operation of a sequencing batch reactor (SBR) is based on five steps: fill, react, settle, decant, and idle (Fig. 1), which all steps are accomplished in a single tank.

2.3. Ozonation process

The ozone experiments were carried out in a bubble column reactor (working volume: 1L). Ozone generator consists of a stainless steel cylindrical vessel with a mercury vapor lamp inside the running length of the tube (Fig. 2). The lamp power was 300 W. The ozone stream was fed into the wastewater through a bubble gas.

2.4. Analytical methods

The ozone was bubbled into 2 % KI solution, where the potassium iodide solution reacted with the excess ozone resulting the following equation:

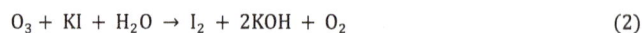

$$O_3 + KI + H_2O \rightarrow I_2 + 2KOH + O_2 \qquad (2)$$

The produced iodine was titrated using standard sodium thiosulfate, in the presence of starch as indicator. The quantities of unused ozone were determined, accordingly (Turhan and Turgut. 2009). Analytical procedures followed in this study for COD and MLSS determinations were those outlined in standard methods for the examination of water and wastewater (Awwa. 1998). For the MLSS measurements, 50 ml of sample was filtered and placed in oven at 103-105 ^{0}C for 1h and the biomass weight was found out by drying filter paper and its weight. The pH, DO and temperature in the reactor were measured by on-line monitors.

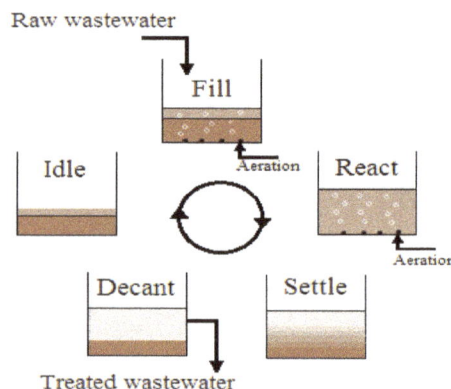

Fig. 1. Schematic of the SBR process.

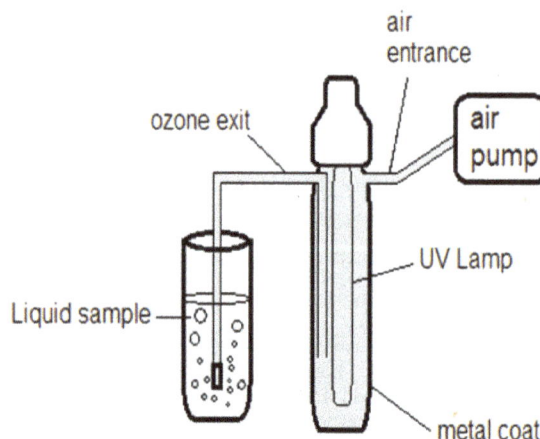

Fig. 2. Schematic representation of the ozone generator.

2.5. Artificial neural network (ANN) modeling

In this work, a feed forward neural network multilayer model was applied to predict the COD removal of refinery wastewater. The MATLAB software was employed for the ANN modelling. Fig. 3 shows the layout of the ANN architecture used in this work. In the ANN architecture, the numbers of input and output neurons were determined according to the problem definition. In this work, the input and output layers have four and one neurons, respectively. Theoretical methods for determining the appropriate number of hidden layers are not available. Moreover, the number of neurons in each hidden layer prior to training cannot be obtained theoretically. Therefore, the trial-and-error method is commonly used to design the ANN.

In the present study, the trial-and-error method was employed to attain the number of layers and neurons. For this purpose, a MATLAB script was written that creates and screens different structure several times, such as: one and two hidden layers as well as various numbers of neurons in each hidden layer. Because the initial weights may have a large effect on the convergence, runs were repeated 100 times with different randomly generated initial values. The results presented in this work are the best ones obtained from this procedure.

An increase in the number of hidden neurons may cause over fitting that usually occurs when a model is complicated. The complicated neural network means that there are a large number of neurons (with a large number of weights and biases) relative to the number of training data. In these networks, the optimal number of neurons should be determined.

In the each network, there are many types of transfer functions. In this work, the "tansig" transfer function was used for the hidden layers and "purelin" function was considered for the output layer. The training process of the ANN was carried out using the Levenberg–Marquardt back propagation algorithm. In this training method, connection weights w_{ij} and biases b_j are iteratively adjusted to minimize the output deviation

(predicted by ANN) from the target values according to Levenberg–Marquardt (Levenberg. 1944; Marquardt. 1963; Eslamloueyan et al. 2011) optimization method. The early stopping method was used to avoid memorizing the training data set. In this method, in case of a decrease in deviation of the predicted values without decrease in predicted values of the validation data set, the training will be stopped. Moreover, the error values should be calculated for different network structures with different numbers of neurons in hidden layers with one and two hidden layers. Table 2 gives the overall range of data points employed in this work for developing the ANN model.

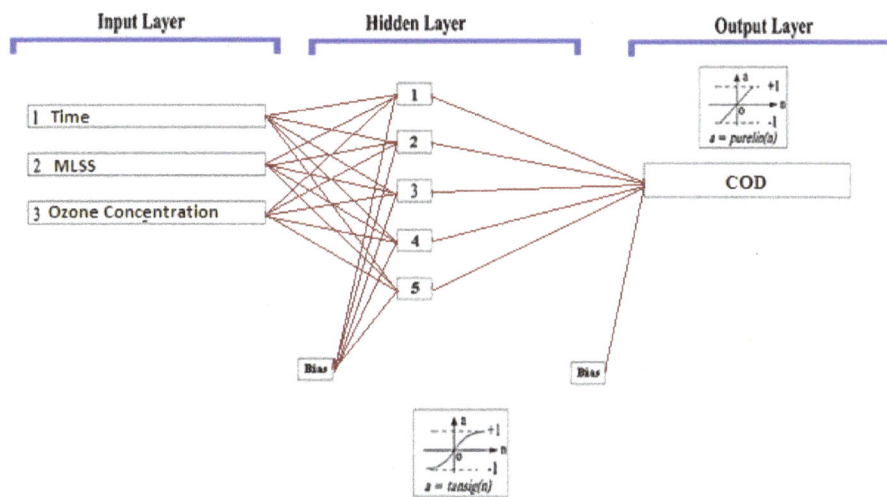

Fig. 3. The ANN architecture.

Table 2. The range of variables as employed data.

Variable	Range
Time (h)	0-24
Ozone concentration (mg/h)	0-500
MLSS (mg/L)	3000-6000

3. Results and discussion
3.1. One-stage SBR

The effect of aeration time on effluent chemical oxygen demand (CODe) during one stage SBR was examined at different MLSS concentrations. Therefore, the reactor was aeration for a period of 24 h. The system was operated at different MLSS concentrations (3000, 4500, 6000 mg/l). The COD removal was increasing with the increase in MLSS concentration. This clearly shows that the mass of substrate and bacterial can have an important influence on the COD removal. At the MLSS concentration of 6000 mg, COD removal was 94±1.4 %. When the MLSS value is equal to 3000 and 4500 mg/l, COD removal was 75±1.8 and 85.68±1.5 %, respectively. The optimum aeration time was 12 h. COD removal efficiency increased with rising aeration time up to 12 h (Fig. 4). No significant COD removal efficiency was observed when aeration was applied more than 12 h.

3.2. Two-stage SBR

Basically, the two-stage process is a combination of two independent SBR plants which work in series. The two-stage SBR was operated with selected HRT of 24 h. The effluent from the first-stage SBR after 12h is feeding the second SBR (Fig. 5). MLSS concentrations in two consecutive reactors are equal and changed to 3000, 4500 and 6000 mg/l.

Percentage of COD removal at this stage for the amounts of MLSS, 3000, 4500, 6000 mg/l, were 85.64±1.1, 94.28±2.12 and 96.67±0.79, respectively, which were higher than the single stage process used. Over time, aerobic biological treatment increased the reproduction of bacteria and the bacteria consumed organic matter in the wastewater. By providing sufficient oxygen to the wastewater, Microbial growth is somewhat accelerated by the time when the organic material in the wastewater is not enough. So, in the next stage, COD removal efficiency is severely less. The two-stage SBR process is a very good approach to resolve this problem. Due to the high ability of young microorganisms in consumption of the organic material; COD removal rate will be higher. The purpose of this two-stage process is to achieve the desired removal efficiency of COD with MLSS values less than that of a single-step process.

The soft drink wastewater treatment was studied by Kalyuzhnyi et al. (1997). The COD of wastewater was between 1.1 to 30.7 g/L. The soft drink wastewater was treated by using a UASB reactor and a hybrid reactor, an anaerobic filter combined with a sludge bed reactor. Results showed that the efficiency of removal COD for hybrid reactor and UASB reactor was 80 and 73 %, respectively (Kalyuzhnyi et al. 1997).

3.3. Two-stage SBR integrated with ozonation

Fig. 6 shows the sketch and operation of the plant based on SBR system integrated with ozonation used for treating the soft drink wastewater. The examination was accomplished at different MLSS concentrations in each process. The operation was based on the succession of 24.5 h treatment cycles each consisting of three consecutive stage: a preliminary biological degradation (for 12 h), ozonation degradation (for 0.5 h) and finally the ozonated wastewater came back to the SBR reactor for the final biological treatment. The effluent from the one-stage process is used as feed of the next stage. The samples are taken from the effluent of every stage in order to module the residual COD concentration.

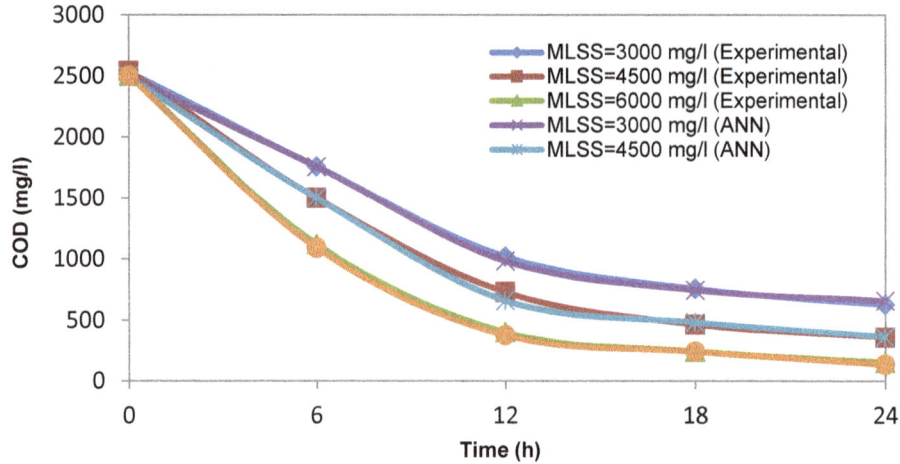

Fig. 4. Profile of effluent chemical oxygen demand (COD_e) during one-stage SBR.

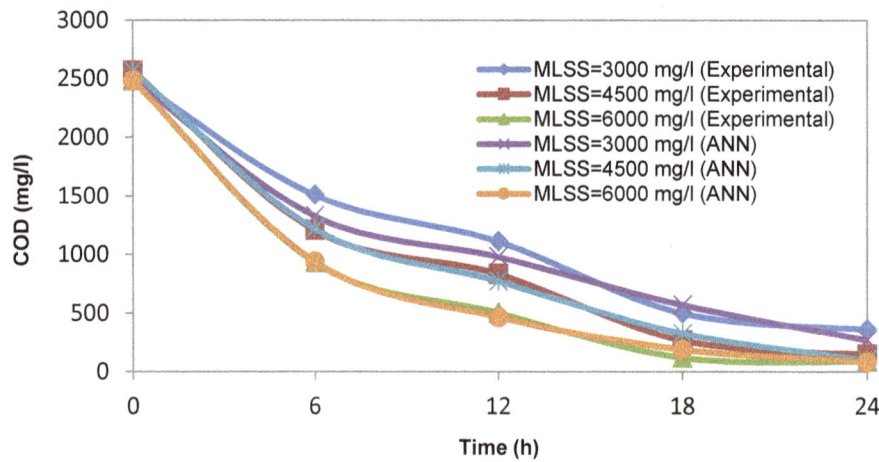

Fig. 5. Profile of effluent chemical oxygen demand (COD_e) during two-stage SBR.

COD concentration in the effluent was about 125±33, 65±55, 34±15 mg/l for MLSS concentrations of 3000, 4500 and 6000 mg/l, respectively. Hence, the ratio between ozone dose and the COD removal was 0.72. This ratio showed that the removed COD was higher than the dosed ozone. Combination of ozonation and biological treatment is considered as a new integrated treatment system. The results shows chemical oxidation can change the molecular structure of the slowly biodegradable compounds and breakdown them into smaller molecules (Giannis et al. 2007). These results indicated that the multi-stage ozonation-biological treatment gives higher COD removal efficiency than the conventional single-stage biological treatment.

Therefore, the treatment time and cost are substantially minimized. The COD concentration in effluent and removal efficiencies of the SBR and integrated treatments of soft drink wastewater reported in Table 3. The data reported in Table 2 show that biological treatment by one-stage SBR was able to reduce concentrations in the effluent as high as 625, 358, 150 mg/L for MLSS 3000, 4500, 6000 mg/l, respectively. The data obtained for the two-stage SBR (Table 3) show high removal efficiencies for COD, with residual concentrations about the discharge limits.

Table 3. Experimental values of final COD and % COD removal for SBR and combined treatment processes.

MLSS(mg/l)	Treatment	COD effluent (mg/L)	% COD removal
3000	One stage SBR	625±46	75.00±1.8
	Two stage SBR	359± 29	85.64 ± 1.1
	O_3+ biodegration	125 ± 33	95.00 ± 1.3
4500	One stage SBR	358±38	85.68 ± 1.5
	Two stage SBR	143±53	94.28 ± 2.12
	O_3+ biodegration	65 ± 55	97.4± 2.2
6000	One stage SBR	150± 35	94± 1.4
	Two stage SBR	84± 19	96.67± 0.79
	O_3+ biodegration	34± 15	98.64 ± 0.6

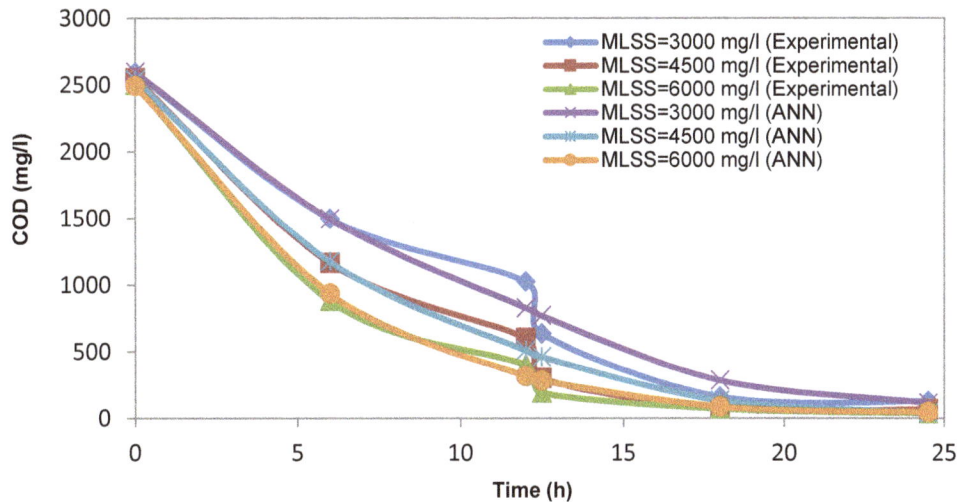

Fig. 6. Profile of effluent chemical oxygen demand (CODe) during two-stage SBR/O$_3$.

In the latter study, olive mill wastewater (OMW) treatment is perused experimentally in several combined processes of photo degradation by UV radiation, advanced oxidation with ozone (O$_3$), and an aerobic biodegradation by stirred tank reactor. For both single stage treatment of O$_3$ and two-stage treatment of O$_3$/UV, COD remains quite high. But, an integrated system of biological and UV/O$_3$ process for the olive mill wastewater treatment seems to be an efficient alternative in the reduction of the COD. In particular, biodegradation of UV/O$_3$ pretreated OMW found to have the highest removal levels; the percent of COD removal reached about 91 % (Lafi et al. 2009; Shannak et al. 2009).

3.4. Modeling results with ANN

In the present study, the ANN was developed to COD removal of wastewater emanated from the soft drink factory using some variables as ANN input data. These data are accessible by experimental studies. In this research, the input variables were selected according to the experimental studies.

In this work, the application of ANN models with one hidden layer was investigated. The ANN with different architectures lead to different outcomes in each run, so there is more possibility to reach the best answer with more runs. Fig. 7 shows the lowest obtained RMSE after several runs in terms of increase in the number of neurons in one hidden layer for COD concentration. Here, the ANN with a structure of 3-5-1 (five neurons in hidden layer) was chosen as optimum topologies.

Table 4 depicts the ANN network performances for COD using different number of neurons in the hidden layer and the Levenberg-Marquardt (LM) algorithm. The optimum number of neurons in the hidden layer was obtained 5 neurons.

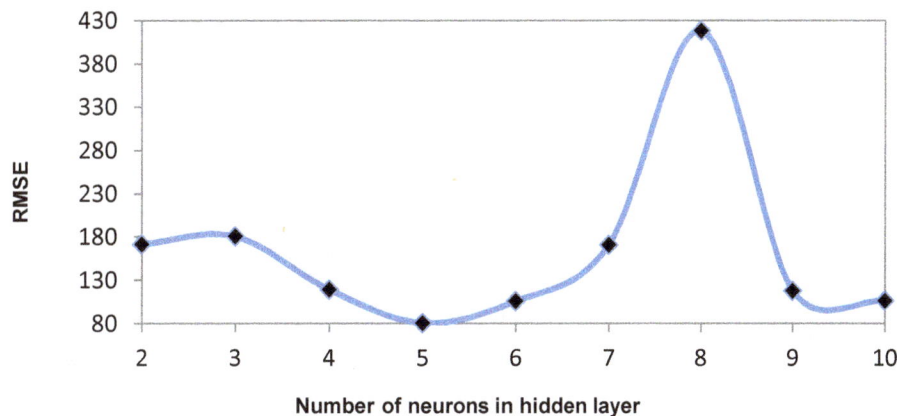

Fig. 7. Effect of number of neurons in hidden layer on ANN performance.

Table 4. Effect of number of neurons in the hidden layer on the performance of neural networks for COD removal.

Number of neurons in the hidden layer	RSME	R^2	Eq.
2	171.20	0.960	Y=0.939X+74.08
3	180.50	0.958	Y=0.974X+27.81
4	119.10	0.981	Y=0.968X+12.90
5	80.36	0.991	Y=0.982X+74.08
6	105.70	0.985	Y=0.974X+22.88
7	170.60	0.960	Y=0.944X+83.22
8	418.00	0.816	Y= X+10.23
9	117.70	0.983	Y=0.991X-2.50
10	106.00	0.986	Y=0.990X+5.07

In addition to the Levenberg–Marquardt algorithm (trainlm), other algorithms such as traincgb (conjugate gradient back propagation with Powell–Beale restarts), trainb (batch training with weight and bias learning rules), trainbr (bayesian regularization back-propagation), traingdm (gradient descent with momentum back-propagation), and traingda (gradient descent with adaptive learning rate back-propagation), were used to train the network. Table 5 shows the mean square error (RMSE) and correlation coefficient (R^2) obtained for each algorithm.

The reported errors in the table illustrate that using the Levenberg–Marquardt algorithm leads to the lowest error. Fig. 8 indicates a comparison between the predicted results from the ANN and the experimental values.

Table 5. Best RMSE and R^2 values of different training algorithm for ANN with 3-5-1 configuration.

Neuron No.	algorithm	Transfer fun.		RMSE	R^2
		Hidden lay.	Output lay.		
1	BFG	Tansig	Purelin	158.60	0.967
2	CGB	Tansig	Purelin	364.60	0.838
3	CGF	Tansig	Purelin	208.60	0.945
4	CGP	Tansig	Purelin	321.90	0.853
5	GDA	Tansig	Purelin	186.40	0.950
6	LM	Tansig	Purelin	80.36	0.991
7	OSS	Tansig	Purelin	155.80	0.967
8	RP	Tansig	Purelin	230.20	0.919
9	SCG	Tansig	Purelin	200.90	0.944

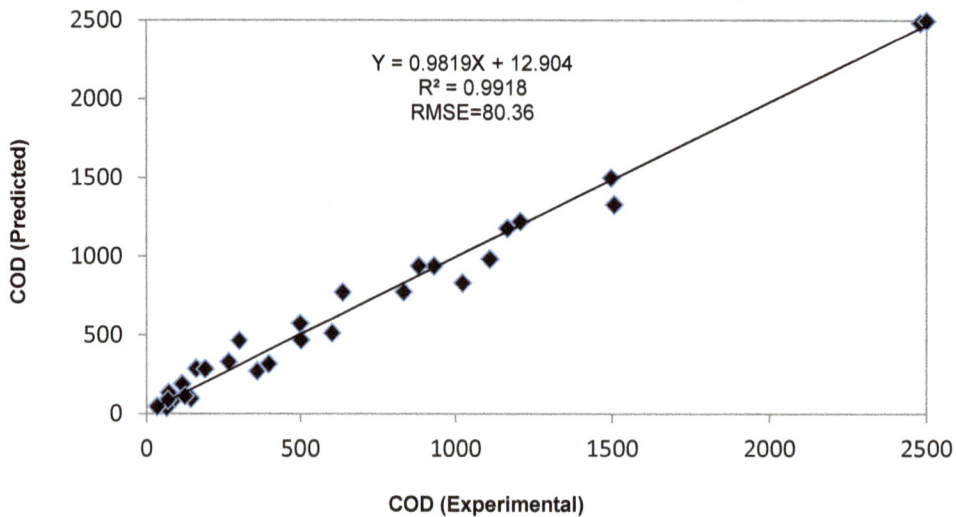

Fig. 8. Cross-correlation between measurement values of COD and predicted values of COD by ANN.

This figure shows that the ANN predicted values for all data points are quite close to the experimental data values. The developed two-layered ANN (3-5-1) in this work provides the weights, which are listed in Table 6. Using the parameters (W, b) presented in Table 6, the COD can be calculated from the following equation:

$$D = F_p \left\{ \sum_{j=1}^{7} W_{kj} \left[F_t \left(\sum_{i=1}^{4} W_{ji} X_i + b_j \right) \right] + b_k \right\} \quad (1)$$

where, X is the input value of the network, W is the weight, b is the bias, 'i', 'j' and 'k' refer to the input, hidden, and output layer, respectively. F is the transfer function that is used to get the normalized output values from the neurons. In this study, the "tansig" transfer function was considered for hidden layers and the "purelin" transfer function was used for the output layer.

The presented ANN in this work introduces a technique to avoid curve fitting of a large number of polynomials, any of which can be useful just for one system. Here, the developed ANN works similar to a box that contains many polynomial equations. One can enter input and find the COD value with a high precision.

Table 6. Connection weights and biases.

Neuron No.	W_1			b_1	b_2= 586.12
	Time	MLSS	O_3 Concentration		W_2
1	-1.4501	3.3237	0.5166	1.4686	0.0177
2	-2.0461	-0.3550	1.1135	-1.4180	0.4253
3	2.4355	6.2937	1.7096	1.0733	-0.0399
4	1.3671	0.3792	0.7346	0.7078	- 0.4830
5	4.4350	-0.2905	0.0715	4.3544	-0.6541

4. Conclusion

This work represents treatment of soft drink industrial wastewater by a two-stage system, combination of SBR and ozonation. The effect of several parameters on the efficiency of COD removal e.g. time, MLSS, and O_3 concentration were examined experimentally, and the following results are obtained:

1. In one-stage SBR, notable COD removal efficiency was observed when aeration was applied more than 12 h.

2. In two-stage SBR, COD removal efficiency could be improved in comparison to the one-stage treatment.

3. The role of ozonation was to breakdown the big molecule to small molecules that is easily biodegraded in biological treatment.

The presented ANN in this work introduces a technique to predict COD of the wastewater with a high precision by introducing input data component instead of sophisticated polynomial fitting.

References

Akmehmet Balcıoğlu I., Sarac C., Kıvılcımdan C., Tarlan E., Application of ozonation and biotreatment for forest industry wastewater, Ozone: Science and Engineering 28 (2006) 431-436.

Arslan-Alaton I., Degradation of a commercial textile biocide with advanced oxidation processes and ozone, Journal of Environmental Management 82 (2007) 145-154.

Awwa A., Standard methods for the examination of water and wastewater, Washington, DC Standard Methods for the Examination of Water and Wastewater 20 (1998).

Baig S., Liechti P., Ozone treatment for biorefractory COD removal, Water Science & Technology 43 (2001) 197-204.

Balcıoğlu I. A., Tarlan E., Kıvılcımdan C., Türker Saçan M., Merits of ozonation and catalytic ozonation pre-treatment in the algal treatment of pulp and paper mill effluents, Journal of Environmental Management 85 (2007) 918-926.

Cheng L., Bi X., Jiang T., Liu C., Effect of Ozone Enhanced Flocculation on the Treatment of Secondary Effluent, Procedia Environmental Sciences 10 (2011) 555-560.

Cortez S., Teixeira P., Oliveira R., Mota M., Evaluation of Fenton and ozone-based advanced oxidation processes as mature landfill leachate pre-treatments, Journal of Environmental Management 92 (2011) 749-755.

Di Iaconi C., Biological treatment and ozone oxidation: Integration or coupling?, Bioresource Technology 106 (2012) 63-68.

Eslamloueyan R., Khademi M. H., Mazinani S., Using a Multilayer Perceptron Network for Thermal Conductivity Prediction of Aqueous Electrolyte Solutions, Industrial & Engineering Chemistry Research 50 (2011) 4050-4056.

Giannis A., Kalaitzakis M., Diamadopoulos E., Electrochemical treatment of olive mill wastewater, Journal of Chemical Technology and Biotechnology 82 (2007) 663-671.

Güçlü D., Dursun Ş., Artificial neural network modelling of a large-scale wastewater treatment plant operation, Bioprocess and biosystems engineering 33 (2010) 1051-1058.

Hai F.I., Yamamoto K., Fukushi K., Hybrid Treatment Systems for Dye Wastewater, Critical Reviews in Environmental Science and Technology 37 (2007) 315-377.

Jadhav J.P., Kalyani D.C., Telke A. A., Phugare S. S., Govindwar S. P., Evaluation of the efficacy of a bacterial consortium for the removal of color, reduction of heavy metals, and toxicity from textile dye effluent, Bioresource Technology 101 (2010) 165-173.

Kalyuzhnyi S., Saucedo J.V., Martinez J. R., The anaerobic treatment of soft drink wastewater in UASB and hybrid reactors, Applied biochemistry and biotechnology 66 (1997) 291-301.

Kurniawan T. A., Lo W.-H., Chan G.Y. S., Degradation of recalcitrant compounds from stabilized landfill leachate using a combination of ozone-GAC adsorption treatment, Journal of Hazardous Materials 137 (2006) 443-455.

Kundu P., Debsarkar A., Mukherjee S., Kumar S., Artificial neural network modelling in biological removal of organic carbon and nitrogen for the treatment of slaughterhouse wastewater in a batch reactor, Environmental Technology 35 (2014) 1296-1306.

Lafi W. K., Shannak B., Al-Shannag M., Al-Anber Z., Al-Hasan M., Treatment of olive mill wastewater by combined advanced oxidation and biodegradation, Separation and Purification Technology 70 (2009) 141-146.

Ledakowicz S., Solecka M., Zylla R., Biodegradation, decolourisation and detoxification of textile wastewater enhanced by advanced oxidation processes, Journal of Biotechnology 89 (2001) 175-184.

Levenberg K., A method for the solution of certain problems in least squares, Quarterly of applied mathematics 2 (1944) 164-168.

Marquardt D.W., An algorithm for least-squares estimation of nonlinear parameters, Journal of the Society for Industrial & Applied Mathematics 11 (1963) 431-441.

Mjalli F. S., Al-Asheh S., Alfadala.H. E., Use of artificial neural network black-box modeling for the prediction of wastewater treatment plants performance, Journal of Environmental Management 83 (2007) 329-338.

Oller I., Malato S., Sánchez-Pérez J. A., Combination of Advanced Oxidation Processes and biological treatments for wastewater decontamination-A review, Science of The Total Environment 409 (2011) 4141-4166.

Ongen A., Kurtulus Ozcan H., Arayıcı S., An evaluation of tannery industry wastewater treatment sludge gasification by artificial neural network modeling, Journal of Hazardous Materials 263 (2013) 361-366.

Paraskeva P., Graham N. J. D., Treatment of a secondary municipal effluent by ozone, UV and microfiltration: microbial reduction and effect on effluent quality, Desalination 186 (2005) 47-56.

Parsaie A., Haghiabi A. H., Assessment of some famous empirical equation and artificial intelligent model (MLP, ANFIS) to predicting the side weir discharge coefficient, Journal of Applied Research in Water and Wastewater 2 (2014) 75-79.

Radetski C. M., Use of ozone in a pilot-scale plant for textile wastewater pre-treatment: Physico-chemical efficiency, degradation by-products identification and environmental toxicity of treated wastewater, Journal of Hazardous Materials 175 (2010) 235-240.

Ranković V., Radulović J., Radojević I., Ostojić A., Čomić L., Neural network modeling of dissolved oxygen in the Gruža reservoir, Serbia, Ecological Modelling 221 (2010) 1239-1244.

Sangave P. C., Gogate P. R., Pandit A. B., Combination of ozonation with conventional aerobic oxidation for distillery wastewater treatment, Chemosphere 68 (2007) 32-41.

Somensi C. A., Simionatto E. L., Bertoli S. L., Wisniewski Jr A., Fahmi C. Z., Rosmady N., Multi stage ozonation and biological treatment for removal of azo dyes industrial effluent, International Journal of environmental science and development 1 (2010) 193-198.

Stasinakis A., Use of selected advanced oxidation processes (AOPs) for wastewater treatment—a mini review, Global NEST Journal 10 (2008) 376-385.

Treguer R., Tatin R., Couvert A., Wolbert D., Tazi-Pain A., Ozonation effect on natural organic matter adsorption and biodegradation – Application to a membrane bioreactor containing activated carbon for drinking water production, Water Research 44 (2010) 781-788.

Turhan K., Turgut Z., Decolorization of direct dye in textile wastewater by ozonization in a semi-batch bubble column reactor, Desalination 242 (2009) 256-263.

Yasar A., Ahmad N., Chaudhry M., Rehman M., Khan A., Ozone for color and COD removal of raw and anaerobically biotreated combined industrial wastewater, Polish Journal of Environmental Studies 16 (2007) 289.

Removal of eosin Y and eosin B dyes from polluted water through biosorption using Saccharomyces cerevisiae: Isotherm, kinetic and thermodynamic studies

Nader Bahramifar[1,]*, Maryam Tavasolli[2], Habibollah Younesi[1]

[1]Department of Environmental Science, Faculty of Natural Resources, Tarbiat Modares University, Imam Reza Street, Noor, P.O. Box 46414-356, Iran.
[2]Department of chemistry, Payame Noor University (PNU), POBOX 1939-3697, Tehran, Iran.

ARTICLE INFO

Keywords:
Biosorption
Eosin Y
Eosin B
Saccharomyces cerevisiae,
Kinetics

ABSTRACT

Biosorption of two anionic dyes, eosin Y and eosin B, from aqueous solution using Saccharomyces cerevisiae was investigated in a batch mode. The influence process parameters such as contact time, initial dye concentration, sorbent dosage, pH and temperature of aqueous solution were studied. The maximum adsorption capacities were found to be at 200 and 100 mg g^{-1} for eosin Y and 1 eosin B, respectively. The Langmuir and Temkin model were found to be appropriate for the description of biosorption process of eosin Y and eosin B, respectively. The pseudo-second order kinetic model fitted well in correlation to the experimental results for both dyes. Thermodynamic parameters such as enthalpy change ($\Delta H°$), entropy change ($\Delta S°$) and free energy change ($\Delta G°$) were also investigated. Thermodynamic studies indicated that biosorption of both dyes onto S. cerevisiae was an endothermic process. The negative values of free energy change showed that the biosorption of both dyes was spontaneous at the temperatures under investigation. These results indicate that biomass S. cerevisiae particles with clean surface and high porosity are an interesting alternative for dye removal from the wastewater effluents.

1. Introduction

Many industries, such as textile, paper, plastics and dyestuffs, consume substantial volume of water, and also use chemicals during manufacturing and dyes to color their products. As a result, they generate a considerable amount of polluted wastewater (Crini and Badot. 2008). Color is a visible pollutant and presence of even very minute amount of coloring substance makes it undesirable due to its appearance (Kumara and Porkodi. 2007). Generally, dyes are stable to light, heat and oxidizing agents, and are usually biologically non-degradable (Crini and Badot. 2008). Eosin Y and eosin B coal tar xanthene dye, which used extensively in the printing and dyeing industries was chosen as the model anionic dye to avoid environmental hazards during investigation, as this dye is not specifically listed as toxic by different health agencies (Chatterjee et al. 2005).

The removal of color arising from the presence of the water-soluble reactive dyes is a major problem due to the difficulty in treating such wastewaters by conventional treatment methods (Ghouti et al. 2009). Dye wastewater is usually treated by physical or chemical treatment processes. These include flocculation combined with flotation, electroflocculation, membrane filtration, electrokinetic coagulation, electrochemical destruction, ion-exchange, irradiation, precipitation, ozonation, and katox treatment method involving the use of activated carbon and air mixtures (Srinivasan and Viraraghavan. 2010). Although they can remove dyes partially, their initial investment and operational costs are so high that they can be widely used in dyeing and finishing industries, especially in developing countries (Aksu and Tezer. 2000). Adsorption is one of the processes, which besides being widely used for dye removal also has wide applicability in wastewater treatment (Gupta and Suhas. 2009). Studies have shown that activated carbons are good materials for the removal of different types of dyes in general but their use is sometimes restricted in view of higher cost. Also, the activated carbons after their use (treatment of wastewater) become exhausted and are no longer capable of further adsorbing the dyes (Gupta and Suhas. 2009). However, because of the cost involved search for alternative adsorbent that could provide an economical solution is very important in developing countries (Ponnusami. 2007). Many studies have been undertaken to investigate the use of low-cost adsorbents such as biosorbent (Aksu. 2003; Crini and Badot. 2008; Fu and Viraraghavan. 2002a; Fu and Viraraghavan. 2002b), fly ash (Gupta. 2000), rice husk (Ponnusami. 2007), tea waste (Hameed. 2009), palm ash (Ahmad. 2007), agricultural waste (Hameed and Daud. 2008). The accumulation and concentration of pollutants from aqueous solutions by the use of biological materials is termed biosorption. In this instance, biological materials, such as chitin, chitosan, peat, yeasts, fungi or bacterial biomass, are used as chelating and complexing sorbents in order to concentrate and to remove dyes from solutions. These biosorbents and their derivatives contain a variety of functional groups which can complex dyes. The biosorbents are often much more selective than traditional ion-exchange resins and commercial activated carbons, and can reduce dye concentration to µg l^{-1} levels. Biosorption is a novel approach, competitive, effective and cheap (Crini. 2006). Among the promising biosorbents for heavy metal and dyes removal which have been researched during the past decades, Saccharomyces cerevisiae has received increasing attention due to the unique nature in spite of its mediocre capacity for metal and dye uptake. Compared with other fungi, S. cerevisiae is widely used in beverage production, is easily cultivated using cheap media, is also a by-product in large quantity as a waste of the fermentation industry, and is easily manipulated at molecular level (Wang and Chen. 2006).

In the present study, biosorption techniques are employed for removal of eosin Y and B by S. cerevisiae yeast from aqueous solution. The effects of temperature, pH, initial dye concentration and sorbent dosage on biosorption were investigated. Moreover, the biosorption isotherms, kinetics and thermodynamic were also explored.

2. Materials and methods

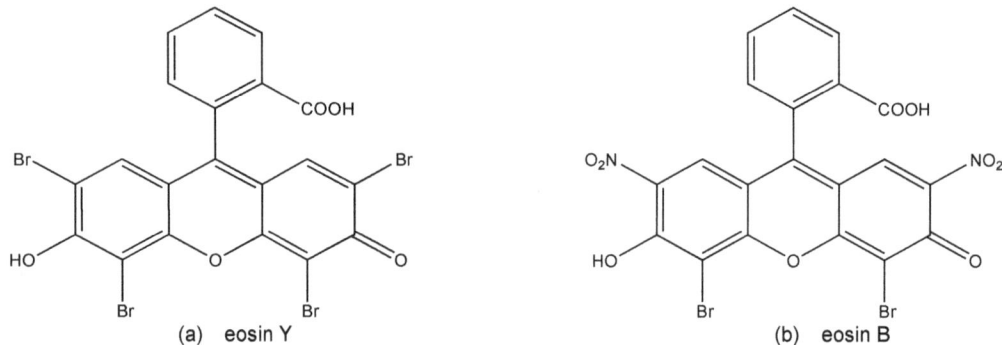

Fig.1. Structure of (a) eosin Y (b) eosin B.

2.2. Preparation of the biosorbent

S. cerevisiae was provided from Research and Technology Department of Ministry of Sciences (Persian Type Culture Collection) in the form of freeze dry, and then cultured in sterilized medium. The composition of growth medium was (grams per liter): glucose, 15; $(NH4)_2SO_4$, 9; $MgSO_4$, 2.5; yeast extract, 1; KH_2PO_4, 1; K_2HPO_4, 0.2. The medium was sterilized by autoclaving at a pressure of 1.5 atm and temperature of 121 °C for 20 min. The yeast cells were grown for 16 h and then filtered. Yeast biomass was deactivated by heating in an oven at 80 °C for 24 h. The dried yeast was ground and screened through a sieve with 100 mesh. The pretreatment of the biosorbent was carried out with nonviable yeast cells in 700 g L^{-1} ethanol solution 20 min at room temperature. Then, it was centrifuged at 3600 rpm for 10 min and the ethanol solution was discarded. The ethanol washed biomass was rinsed several times with de-ionized water to remove excess ethanol and adsorbed nutrient ions. The rinsed yeast was again centrifuged and remaining biomass was dried at 70 °C for 12 h. The dried cells were ground and screened as mentioned above. The purpose of grinding dried yeast was to make a homogenized yeast biomass in order to destroy biomass aggregates and increase uptake capacity (Ghorbani. 2008). The ground biomass was stocked in the refrigerator for use in biosorption studies.

2.3. Biosorption studies

Batch biosorption experiments were carried out in 250 ml glass stoppered and Erlenmeyer flasks with 150 ml of dye solution. Necessary amount of adsorbent was then added to the solution. The flasks were agitated at a constant speed of 170 rpm for 3 h in an incubator shaker (JalTajhiz,TSL20, Iran) for different time intervals at room temperature (25 °C).The influence of pH (2.0, 3.0, 4.0, 5.0, 6.0, 7.0, 8.0 and 9.0), contact time (10, 15, 25, 35, 60, 90, 120, 150 and 180 min), initial dye concentration (50, 100, 150, 200 and 250 mg L^{-1}), biosorbent dosage (0.30, 0.50, 1.00, 2.00 and 3.00 g L^{-1} for eosin Y and 0.07, 0.10, 0.30, 0.50,0.75 and 1.00 g L^{-1} for eosin B) and temperature (15, 25, 35 and 45 °C) was evaluated during the present study. Samples were collected from the flasks at predetermined time intervals for analyzing the residual dye concentration in the solution. Then the supernatant was centrifuged at 5000 rpm for 5 min to remove particulates. The residual amount of dye in each flask was investigated using UV-Vis spectrophotometer (GBC, Cintra 20, Australia) at an absorbance wavelength of 510 nm and 505 nm for eosin Y and eosin B, respectively. The pH of the solution was adjusted with HCl and NaOH solutions and measured using a pH meter (Jenway-3510). The amount of dye sorbed onto unit weight of biosorbent and color removal were calculated using the following equations 1 and 2, respectively:

2.1. Materials

The anionic dyes used in this study were eosin Y and eosin B that molecule structure of the dye is shown in Fig. 1. The dye eosin Y and eosin B were obtained from Merck, Germany. Stock dye solutions were prepared by dissolving 1.00 g of dyes in 1 L of double distilled water. All working solutions were prepared from the stock solutions by further dilution. The NaOH pellets and HCl solution used for adjusting of pH were obtained from Merck, Germany.

$$q_e = \frac{(C_o - C_e)V}{W} \tag{1}$$

$$R = \frac{C_o - C_e}{C_o} \times 100 \tag{2}$$

where, q_e is the amount of dye adsorbed onto the unit amount of the biomass (mg g^{-1}), C_o and C_e (mg L^{-1}) are the liquid phase concentrations of dye at initial and equilibrium solution, respectively, V (L) is the volume of the solution and W (g) is the mass of dry sorbent used.

2.4. Biosorption equilibrium

Equilibrium experiments were carried out by taking known amount of yeast in 250 ml flasks containing 150 ml of different initial concentrations (50–250 mg L^{-1}) of dye solution in optimum pH and temperature of 25 °C. The mixture was shaken in an incubator shaker at 170 rpm for 180 min, which is sufficient time to reach the equilibrium. The samples were then centrifuged and analysis was performed as previously mentioned.

2.5. Biosorption kinetics

Sorption kinetics experiments were carried out at different initial dye concentrations. Kinetic experiments were performed by mixing 2.00 g L^{-1} of sorbent for eosin Y and 0.10 g L^{-1} for eosin B with 150 ml dye solution (100 mg L^{-1}). The suspensions were shaken for 180 min with constant temperature of 25°C. The samples were taken at different time intervals, centrifuged and analyzed for remaining dye concentrations as described before.

3. Results and discussion
3.1. Effect of pH on dye adsorption

The ionic forms of the dye in solution and the surface electrical charge of the biomass depend on solution pH. Therefore, solution pH influences both the fungal biomass surface dye binding sites and the dye chemistry in the medium (Srinivasan and Viraraghavan. 2010). The effect of pH on the percentage both of dyes adsorption by S. cerevisiae was shown in Fig. 2. For an initial dye concentration of 100 mg L^{-1} the biosorption uptake of S. cerevisiae particles decreased with the increase of the solution pH. The maximum removal percentage for eosin Y and eosin B dyes were found to be 91.72 and 97.11, at the pH of 4.0 and 2.0, respectively. In fact, the eosin is a dipolar molecule at low pH, as shown in Fig.1 with a decrease in the pH of the dye solution, more dye molecules are protonated and are suitable to interact with negatively charged groups in biomass. On the other hand, the outer layer of the cell wall of S. cerevisiae consists on a coat protein that can develop a charge by dissociation of ionizable side groups of the constituent amino acids. The ionic state of ligands such as carboxyl, phosphate, imidazole and amino groups will be to promote reaction with the positively dye ions. At lower pH, cell wall ligands were closely associated with the hydronium ions [H_3O^+] and restricted the approach of dye cations as a result of the repulsive force. Electrostatic attraction to negatively charged functional groups may be one of the specific biosorption mechanisms (Özer and Özer. 2003; Wang and Chen. 2006).

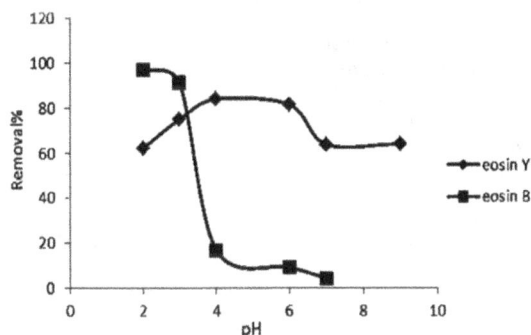

Fig. 2. Effect of pH on the percentage biosorption of the eosin Y and eosin B dye by S. cerevisiae. (T = 25 ° C, biosorbent dosage = 2.0 for eosin Y and 0.1 g L^{-1} for eosin B, C$_0$ = 100 mg L^{-1}, t = 3h).

3.2. Effect of biosorbent dosage on dye adsorption

Biosorbent dosage is a significant factor to be considered for effective pollutant removal as it determines sorbent–sorbate equilibrium of the system (Iftikhara. 2009). The effect of biosorbent dosage on the removal both of dyes were studied for an initial dye concentration of 100 mg L^{-1}, by varying the dosage, keeping all other parameters constant. The effect of biosorbent dosage on equilibrium uptake, q$_e$ (mg g^{-1}) and % removal against biomass dosage (g L^{-1}) was shown in Fig. 3. It was observed that the amount of dye adsorbed onto unit weight of biosorbent gets decreased with increasing biomass concentration for both of dyes. For eosin Y dye uptake decreased from 275.10 to 27.84 mg g^{-1} for an increase in biomass dosage from 0.30 to 3.00 g L^{-1} and for eosin B dye uptake decreased from 908.58 to 58.23 mg g^{-1} for an increase in biomass concentration from 0.07 to 1.00 gL-1. Higher uptake was obtained when the dosage was low. This is due to the fact that the active sites could be effectively utilized when the dosage was low. When the adsorbent dosage is higher, biosorption sites remain unsaturated during adsorption reaction (Ponnusami et al. 2009). Whereas at equilibrium time the removal increases from 85.1 to 94.7% and 71.26 to 96.22% for an increase in biosorbent dosage from 0.3 to 3.0 g L^{-1} and 0.07 to 1.0 g L^{-1} for eosin Y and eosin B, respectively. The effective amounts of biomass were found to be 2.00 and 0.10 g L^{-1} for eosin Y and eosin B, respectively. The increase in color removal was due to the increase of the available sorption surface and availability of more adsorption sites (Bennani et al. 2009).

3.3. Effect of initial dye concentration and contact time

Dye concentration also affects the efficiency of color removal. Initial concentration provides an important driving force to overcome all mass transfer resistances of the dye between the aqueous and solid phases. Hence a higher initial concentration of dye may enhance the adsorption process (Aksu. 2005). The experimental results of adsorption of eosin Y and eosin B onto S. cerevisiae at various concentrations with different contact time were shown in Fig. 4. It can be seen that the actual amount adsorbed per unit mass of S. cerevisiae increased with the increase for both of dyes concentration. The amount of eosin Y and eosin B adsorbed at equilibrium (q$_e$) increased from 18.67 to 82.92 mg g^{-1} and 451.90 to 622.40 mg g^{-1},respectively as the initial concentration was increased from 50 to 200 mg L^{-1}.This indicates that the initial dye concentration plays an

important role in the adsorption capacity of dye (Gurses et al. 2006).Moreover, the initial rate of adsorption was greater for higher initial dye concentration because the resistance to the dye uptake decreased as the mass transfer driving force increased (Karima et al. 2009).The percentage of dye removal decreases with the increase in the initial dye concentration. This may be due to the saturation of the sorption sites on the biosorbent as the concentration of the dye increases (Farah et al., 2007). It can be seen from Fig. 4, the process was found to be initially very rapid, and a large fraction of the total amount of dye was removed within approximately 15 min for eosin Y and 30 min for eosin B. The rapid uptake of the dye indicates that the sorption process could be ionic in nature where the dye molecules bind to the various negatively charged organic functional groups present on the surface of the biomass (Farah et al. 2007). From the results, it was obvious that the adsorption occurred quickly and reached equilibrium within 3 h.

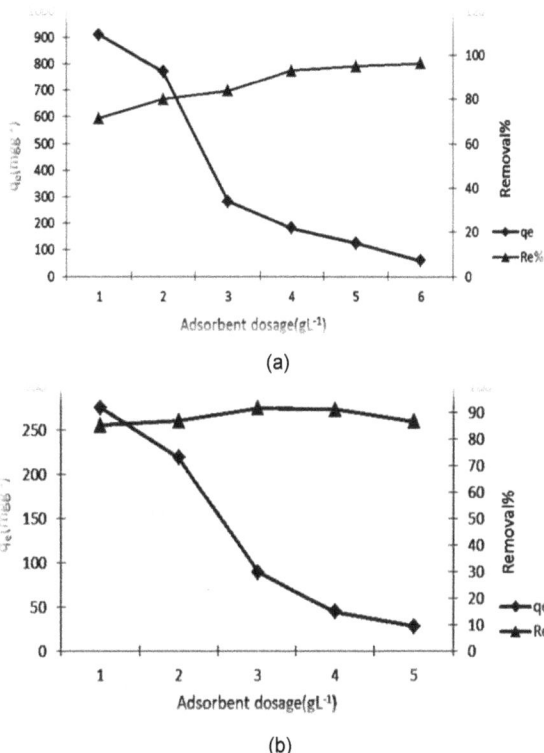

(a)

(b)

Fig.3. Effect of biosorbent dosage on the biosorption of (a) eosin Y and (b) eosin B dye by S. cerevisiae. (T = 25 °C, pH = 4 and 2 for eosin Y and eosin B, respectively, C$_o$ = 100 mg L^{-1}, t = 3h).

3.4. Effect of temperature

Most textile and other dye effluents are produced at relatively high temperatures and hence temperature will be an important factor in real application of biosorption in future (Srinivasan and Viraraghavan, 2010). Increasing the temperature is known to increase the diffusion rate of the adsorbate molecules across the external boundary layer and in the internal pores of the adsorbent particles, owing to the decrease in the viscosity of the solution. In addition, the change temperature affects the equilibrium uptake of the adsorbent for a particular adsorbate (Farah et al. 2007). The effect of temperature on the dyes biosorption experiments was investigated with initial concentration of 200 mg L^{-1} at temperatures 15, 25, 35 and 45 °C. As seen from Fig. 5, the maximum equilibrium uptakes were found to be at 45 °C. The equilibrium biosorption uptake of both dyes increased sharply with increase in temperature. The increase in equilibrium biosorption uptake indicates that higher temperature favor eosin Y and eosin B removal therefore this system is endothermic. This may be due to a higher temperature would lead to higher affinity of sites for dye or binding sites on the yeast. The energy of the system facilitates dyes attachment on the cell surface to some extent (Wang and Chen. 2006). It may also be due to the fact that at higher temperatures, an increase in free volume occurs due to increased movement of the solute (Ho and McKay. 2003). Thus, an increasing the temperature of the reaction from 15 to 45 °C, the equilibrium uptake of the dye increased from 142.40 to 202.84 mg g^{-1} for eosin Y and 451.80 to 1454.52 mg g^{-1} for eosin B.

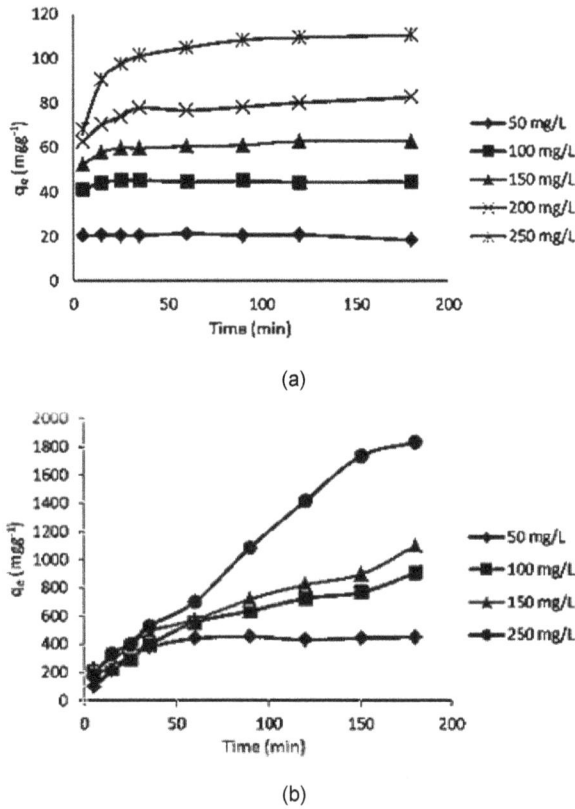

(a)

(b)

Fig.4. Effect of the initial concentration and contact time for the biosorption of (a) eosin Y and (b) eosin B by S. cerevisiae. (T = 25 °C, pH = 4 and 2, biosorbent dosage = 2.0 and 0.1 g L^{-1} for eosin Y and eosin B, respectively, t = 3h).

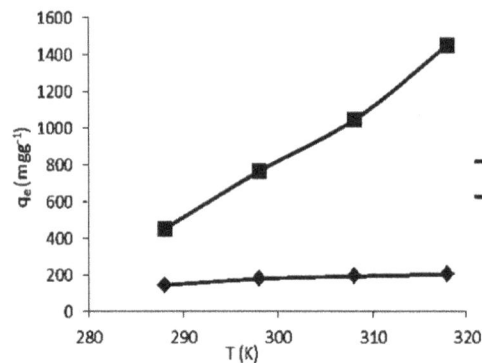

Fig. 5. Effect of temperature on the biosorptin uptake of eosin Y and eosin B by S. cerevisiae. (pH = 4 and 2, biosorbent dosage = 2.0 and 0.1 g L^{-1} for eosin Y and eosin B, respectively, C$_o$= 200 mg L^{-1},t = 3h).

3.4.1. Thermodynamic parameters

Thermodynamic parameters such as changes in free energy (ΔG°), enthalpy (ΔH°) and entropy (ΔS°) for this biosorption process have been determined using the following equations 3 and 4:

$$\Delta G^o = \Delta H^o - \Delta S^o \qquad (3)$$

$$\log\left(\frac{q_e}{C_e}\right) = \frac{\Delta S^o}{2.303R} + \frac{-\Delta H^o}{2.303RT} \qquad (4)$$

where q$_e$ is the maximum amount of dye adsorbed per unit weight of the S. cerevisiae (mg g^{-1}), C$_e$ is equilibrium concentration (mg L^{-1}) and T is temperature in Kelvin and R is the universal gas constant, 8.314 J mol^{-1} K^{-1}. The values of ΔH° and ΔS° were determined from the slope and intercept of the linear plot of log (q$_e$/C$_e$) versus 1/T, respectively (Fig. 6). The value of Gibbs free energy (ΔG°) is then calculated from Eq. (3). The obtained values of ΔH°, ΔS, and ΔG° are given in Table 1 for the initial dye concentration of 200 mg L^{-1}. The positive value of ΔH° for eosin Y and eosin B removal with S. cerevisiae biomass indicated that the dye biosorption process was endothermic in nature and adsorption process is favorable at higher temperatures and possible strong bonding between dye and each sorbent. The negative values of ΔG° confirm the feasibility of the processes and spontaneous nature of adsorptions at different temperature with a high degree of affinity of the dye molecules for each sorbent surface (Aksu et al. 2008). The positive value of ΔS° suggests increased randomness during biosorption at the solid–solution interface during the adsorption of dye onto biomass (Karima et al. 2009). Generally, the value of ΔG$_{ads}$ for chemical adsorption is more than −4.7 Kcal mol^{-1}. The value of ΔG$_{ads}$ for this case is less than −4.7 K calmol^{-1} suggesting that the process is controlled by chemical adsorption (Özer and Özer. 2003). It is a chemisorption mechanism where there is an increase in the number of molecules acquiring sufficient energy to undergo chemical reaction with increasing temperature (Farah et al. 2007).

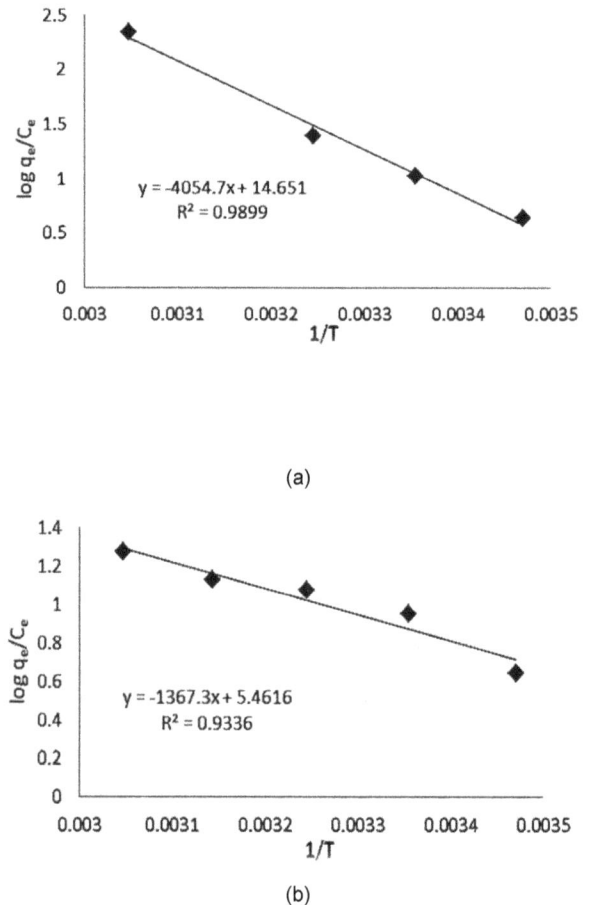

(a)

(b)

Fig. 6. Van't Hoff plot for the biosorption of (a) eosin Y and (b) eosin B dye by S. cerevisiae.

3.5. Biosorption isotherm

Analysis of equilibrium data is important to develop an equation which accurately represents the results and could be used for the design of biosorption systems used for the removal of organic pollutants. The empirical models for single solute systems used to describe the biosorption equilibrium are Langmuir, Freundlich and Temkin models. These models can provide information of dye-uptake capacities and differences in dye uptake between various species.

3.5.1. Langmuir isotherm

The most widely used isotherm equation for modeling equilibrium is the Langmuir equation, based on the assumption that there is a finite number of binding sites which are homogeneously distributed over the adsorbent surface, these binding sites have the same affinity for adsorption of a single molecular layer and there is no interaction between adsorbed molecules is given by the following equation (Langmuir. 1918):

$$q_e = \frac{q_e b C_e}{1 + b C_e} \tag{5}$$

where, C_e is the equilibrium concentration (mg L^{-1}), q_e is the amount of metal ion adsorbed (mg g^{-1}), q_m and b are the Langmuir constants related to maximum biosorption capacity describing a complete mono-layer adsorption (mg g^{-1}) and bonding energy of adsorption (L mg^{-1}), respectively.

Table 1. Thermodynamic parameters for biosorption of eosin Y and eosin B dye by S. cerevisiae.

Type of dye	$\Delta H°$, kJ mol^{-1}	$\Delta S°$, J mol^{-1}K^{-1}	$-\Delta G°$, kJ mol^{-1}				
			288	298	308	318	328
Eosin Y	26.174	104.56					
Eosin B	77.62	280.51	39.55	50.01	60.26	70.62	81.34
			73.02	75.83	78.64	81.44	-

The values of various Langmuir constants were calculated from this isotherm and their values are listed in Table 2.

The essential characteristics of the Langmuir isotherm can be expressed in terms of a dimensionless constant separation factor RL that is given by Eq. (4) (Hall. 1966):

$$R_L = \frac{1}{1 + b C_o} \tag{6}$$

where, C_o is the critical concentration (mg L^{-1}) and b is the Langmuir constant. The value of R_L indicates the shape of the isotherm to be either unfavorable ($R_L > 1$), linear ($R_L = 1$), favorable ($0 < R_L < 1$) or irreversible ($R_L = 0$). The R_L values between 0 and 1 indicate favorable adsorption. In this study, the value of R_L was obtained to be in the range of 0–1, telling that the biosorption process is favorable for the both of dyes (Hameed. 2009).

Table 2. Langmuir, Freundlich and Temkin biosorption isotherm constant for the biosorption of eosin Y and eosin B by S. cerevisiae.

Type of dye	Langmuir				Freundlich			Temkin		
	q_m, mg g^{-1}	b, L mg^{-1}	R_L	R_2	K_F, (mg g^{-1})(mg L^{-1})n	n	R_2	A, L mg^{-1}	B, J mol^{-1}	R_2
Eosin Y	200	0.0203	0.330	0.271	2.616	0.754	0.896	4.01	71.17	0.959
Eosin B	1000	0.166	0.057	0.982	418.79	3.436	0.635	2.85	277.2	0.555

3.5.2. Freundlich isotherm

The Freundlich isotherm (Freundlich. 1906) is an empirical equation assuming that the adsorption process takes place on heterogeneous surfaces, where the sorption energy distribution decreases exponentially. This equation is also applicable to multilayer adsorption (Padmavathy. 2008). The Freundlich equation has the general form:

$$q_e = K_F C_e^{\frac{1}{n}} \tag{7}$$

where, K_F and n are the Freundlich's constants related to the adsorption capacity and adsorption intensity of the sorbent characteristics of the system, respectively. The value of n indicates whether the biosorption process is favorable or not and the high values of K_F showed ready uptake of the dye from wastewater with high adsorptive capacities of these biosorbents (Aksu and Karabayır. 2008). The value of K_F, correlation coefficient and n of both dyes are presented in Table 2.

3.5.3. Temkin isotherm

The Temkin model considered the effects of some indirect adsorbent/adsorbate interactions on adsorption isotherms. Its main assumption is the uniformity in the distribution of binding energies up to some maximum binding energy and also because of these interactions the heat of adsorption of all the molecules in the layer would decreases linearly with coverage. The linear form of Temkin isotherm can be written as (Temkin. 1941):

$$q_e = B \ln A + B \ln C_e \tag{8}$$

Where B=RT/b, T is the absolute temperature in Kelvin and R is the universal gas constant (8.314 J mol^{-1} K^{-1}). A is the equilibrium binding constant and B is related to the heat of adsorption. The values of the parameters are given in Table 2. Based on the correlation coefficients, the applicability of the isotherms was compared (Table 2). The experimental results indicate that sorption of eosin B onto S. cerevisiae follows the Langmuir model. The fact that the Langmuir isotherm fits the experimental data very well may be due to homogeneous distribution of active sites onto adsorbent surface. On the other hand, according to Table 2, it was found that R_2 of Temkin model were close to 1.0 for eosin Y. These strongly suggest that Temkin isotherm model was slightly better for describing the biosorption equilibrium than Langmuir and Freundlich model. The results indicated that the surface of S. cerevisiae is heterogeneous in nature and did possess equal distribution of binding energies on the available binding sites.

3.6. Biosorption kinetics

An ideal adsorbent for wastewater pollution control must not only have a large adsorbate capacity but also a fast adsorption rate. Therefore, the adsorption rate is another important factor for the selection of the material and adsorption kinetics must be taken into account since they explain how fast the chemical reaction occurs and also provides information on the factors affecting the reaction rate (Chatterjee et al. 2005). Thus, in the removal of dyes from wastewater, it is necessary to know the rate of adsorption for process design, operation control and adsorbent evaluation. In order to investigate the mechanism of adsorption at different initial concentrations characteristic constants of adsorption rate were determined by using a pseudo first-order equation of Lagergren based on solid capacity, and pseudo second-order equation based on solid phase adsorption (Aksu. 2005).

3.6.1. Pseudo-first-order kinetic model

The pseudo-first-order kinetic model considers the rate of occupation of the adsorption sites to be proportional to the number of unoccupied sites. The pseudo- first order rate expression based on solid capacity is generally expressed as follows (Aksu and Donmez. 2003):

$$\log(q_e - q_t) = \log q_e - \frac{k_1 t}{2.303} \qquad (9)$$

where, q_e and q_t are the amounts of eosin dyes (mg g^{-1}) adsorbed on the sorbent at equilibrium, and time t, respectively, and k_1 is the rate constant (min^{-1}). The plot of log (q_e -q_t) versus t of Eq. (7) should give a linear relationship from which q_e and k_1 can be determined from the slope and intercept of the plot, respectively. The values of rate constant k_1, q_e were calculated. q_e experimental and R_2 of eosin Y and eosin B are presented in Table 3. This table showed that the q_e calculated is not equal to q_e experimental. Mostly, the first-order kinetic model is not fitted well for whole data range of contact time and can be applied for preliminary stage of biosorption mechanism (Safa and Bhatti. 2011). However, the biosorption of eosin Y and eosin B is not likely to follow the pseudo-first-order kinetic model.

3.6.2. Pseudo-second-order kinetic model

The pseudo-second order kinetic model assumes that the rate limiting step may be biosorption involving valence forces through sharing or exchange of electrons between the biosorbent and sorbate (Aksu et al. 2008; Iftikhara et al. 2009). The pseudo-order equation is also based on the sorption capacity of the solid phase and on the assumption that the sorption process involves chemisorption mechanism and is expressed as:

$$\frac{t}{q_t} = \frac{1}{k_2 q_e^2} + \frac{t}{q_e} \qquad (10)$$

where, k_2 (g mg^{-1}min^{-1}) is pseudo-second order biosorption rate constant, and q_e values were determined from the slope and intercept of the plot of t/q_t versus t (Fig. 6). The values of the rate constants are shown in Table 3 for both of dyes. As can be seen from Table 3, for the both of dye the correlation coefficients for the second-order kinetic model were close to 1.0 for all concentrations studied and the theoretical values of q_e also agreed well with the experimental values. These results indicate that the adsorption system studied belongs to the second-order kinetic model.

Table 3. Comparison study of kinetic parameters for the biosorption of eosin Y and eosin B by S. cerevisiae.

dye	C_0, mgL-1	q_e exp., mg g^{-1}	Pseudo-first-order kinetic model			Pseudo-second-order kinetic model		
			q_e, mg g^{-1}	K_1, min^{-1}	R_2	q_e, mg g^{-1}	K_2 x 10^{-5}, g mg^{-1}min^{-1}	R_2
Eosin Y	50	19.31	-	-	-	20.83	0.043	0.999
	100	44.81	0.52	0.046	0.102	45.45	0.097	0.999
	150	63.1	57.54	0.001	0.385	62.5	0.015	0.999
	200	82.92	16.71	0.016	0.946	83.3	0.051	0.999
Eosin B	50	451.9	212.32	0.023	0.744	500	0.138	0.992
	100	907.28	758.57	0.011	0.941	1000	2.300	0.946
	150	897.5	724.43	0.011	0.965	1000	3.125	0.976
	200	622.4	539.51	0.011	0.885	1000	1.750	0.916

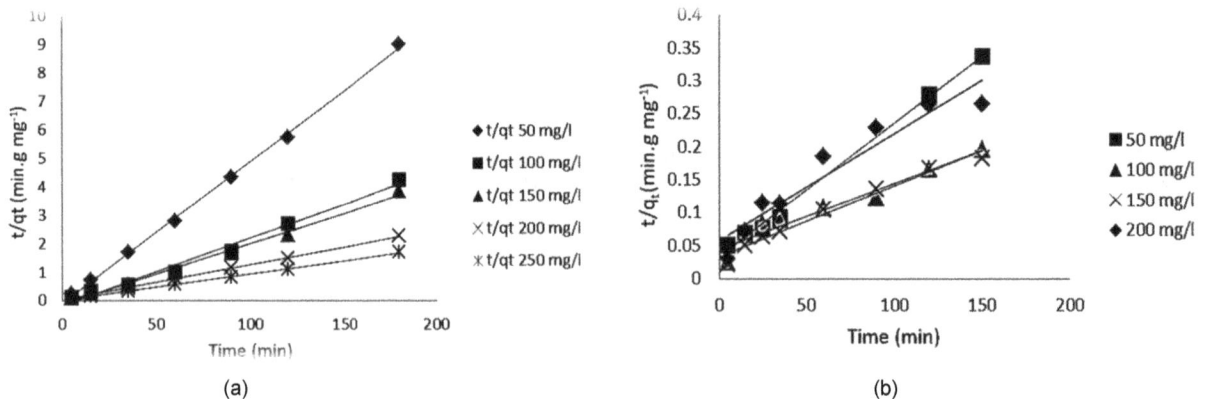

(a)　　　　　　　　　　　　　　　　　　(b)

Fig. 7. Pseudo second-order kinetic model data for the biosortion of eosin Y and eosin B on S. cerevisiae.

4. Conclusion

In the present investigation it has been clearly shown that S. cerevisiae could be effectively used as a low cost adsorbent for the removal dyes from aqueous effluents. The results showed that the biosorption uptake was optimal under acidic conditions. Increase in concentration and decrease in adsorbent dosage resulted in increase in equilibrium dye uptake. The equilibrium data were fitted to non-linear models of Langmuir, Freundlich and Temkin, and the equilibrium data were best described by the Langmuir isotherm model for eosin B dye, with maximum monolayer adsorption capacity of 1000 mg g^{-1} of S. cerevisiae. The Temkin isotherm was found to be the most suitable for eosin Y adsorption by S. cerevisiae. Furthermore, the biosorption followed pseudo-second order adsorption kinetics,

suggesting that the adsorption process was controlled by chemisorption. Thermodynamic parameters such as change in free energy, enthalpy, and entropy were also determined for each sorbent–dye system indicating the spontaneous, endothermic and irregular nature of sorption in each case. Therefore, the adsorbent is expected to be economically feasible for removal of dye from aqueous solutions.

Acknowledgements

The authors wish to thank the Ministry of Science and the Department of Chemistry of the Payame Noor University (PNU) for their financial support, which funding and a research grant made this study possible.

Reference

Ahmad A.A., Hameed B.H., Aziz N., Adsorption of direct dye on palm ash: kinetic and equilibrium modeling, J. Hazard. Mater 141 (2007) 70–76.

Ahmad A.A., Hameed B.H., Aziz N., Adsorption of direct dye on palm ash: kinetic and equilibrium modeling, J. Hazard. Mater 141 (2007) 70–76.

Aksu Z., Tezer S., Equilibrium and kinetic modelling of biosorption of Remazol Black B by Rhizopusarrhizus in a batch system: effect of temperature, Process. Biochem 36 (2000) 431–439.

Aksu Z., Reactive dye bioaccumulation by Saccharomyces cerevisiae, Process. Biochem 38 (2003) 1437–1444.

Aksu Z., Application of biosorption for the removal of organic pollutants: A review, Process. Biochem 40 (2005) 997–1026.

Aksu Z., Tatl A.S., Tunc O., Comparative adsorption/biosorption study of Acid Blue 161: Effect of temperature on equilibrium and kinetic parameters, Chem. Eng. J. 142 (2008) 23–39.

Aksu Z., Donmez G., A comparative study on the biosorption characteristics of some yeast for Remazol Blue reactive dye, Chemosphere 50 (2003) 1075–1083.

Aksu Z., Karabayır G., Comparison of biosorption properties of different kinds of fungi for the removal of Gryfalan Black RL metal-complex dye, Bioresour.Technol 99 (2008) 7730–7741.

Al-Ghouti M.A., Yehya S., Al-Degs Y.S. ,Khraisheh M.A.M., Ahmad M.N., Allen S.J., Mechanisms and chemistry of dye adsorption on manganese oxides-modified diatomite, J. Env. Manage 90 (2009) 3520–3527.

Baysal Z., Cinar E., Bulut Y., Alkan H., Dogru M., Equilibrium and thermodynamic studies on biosorption of Pb (II) onto Candida albicans biomass, J. Hazard.Mater 161 (2009) 62–67.

Bennani K.A., Mounir B., Hachkara M., Bakassec M., Yaacoubib A., Removal of Basic Red 46 dye from aqueous solution by adsorption onto Moroccan clay. J. Hazard. Mater 168 (2009) 304–309.

Chatterjee S., Chatterjee S., Chatterjee B.P., Das A.R., Guha A.K., Adsorption of a model anionic dye, eosin Y, from aqueous solution by chitosan hydrobeads, J. Colloid Interface Sci 288 (2005) 30–35.

Crini G., Non-conventional low-cost adsorbents for dye removal: A review, Bioresour.Technol. 97 (2006) 1061-1085.

Crini G., Badot P.M., Application of chitosan, a natural amino polysaccharide, for dye removal from aqueous solutions by adsorption process using batch studies: A review of recent literature, Prog.Polym. Sci 33 (2008) 399-447.

Farah J.Y., El-Gendy N., Farahat L.A., Biosorption of Astrazone Blue basic dye from an aqueous solution using dried biomass of Baker's yeast, J. Hazard. Mater 148 (2007) 402–408.

Freundlich H., Adsorption in solution, Phys. Chem. Soc 40 (1906) 1361–8.

Fu, Y., Viraraghavan T., Removal of Congo Red from an aqueous solution by fungus Aspergillus niger, Adv. Environ. Res 7 (2002a) 239-247.

Fu Y., Viraraghavan T., Dye biosorption sites in Aspergillus niger, Bioresour. Technol 82 (2002b) 139-145.

Ghorbani F., Younesi H., Ghasempouri S.M., Zinatizadeh A.A., Amini M., Daneshi A., Application of response surface methodology for optimization of cadmium biosorption in an aqueous solution by Saccharomyces cerevisiae, Chem. Eng. J 145 (2008) 267-275.

Gupta V.K., Mohan D., Sharma S., Sharma M., Removal of basic dye (Rhodamine B and Methylene blue) from aqueous solutions using bagasse fly ash, Sep. Sci. Technol 35 (2000) 2097–2113.

Gupta V.K., Suhas., Application of low-cost adsorbents for dye removal - A review, J. Environ. Manage 90 (2009) 2313-2342.

Gurses A., Dogar C., Yalcin M., Acikyildiz M., Bayrak R., Karaca S., The adsorption kinetics of the cationic dye, methylene blue, onto clay, J. Hazard. Mater. B 131 (2006) 217–228.

Hall K.R., Eagleton L.C., Acrivos A., Vermeulen T., Pore and solid-diffusion kinetics in fixed-bed adsorption under constant-pattern conditions, I&EC Fundam 5 (1966) 212–223.

Hameed B.H., Spent tea leaves: A new non-conventional and low-cost adsorbent for removal of basic dye from aqueous solutions, J. Hazard. Mater 161 (2009) 753–759.

Hameed B.H., Daud F.B.M., Adsorption studies of basic dye on activated carbon derived from agricultural waste: Heveabrasiliensis seed coat, Chem. Eng. J 139 (2008) 48–55.

Ho Y.S., McKay G., Sorption of dyes and copper ions onto biosorbents, Process Biochem 38 (2003) 1047-1061.

Iftikhara A.R., Bhattia H.N., Hanif M.A., Nadeema R., Kinetic and thermodynamic aspects of Cu (II) and Cr (III) removal from aqueous solutions using rose waste biomass, J. Hazard. Mater 161 (2009) 941–947.

Karima A. B., Mounira, B., Hachkara M., Bakassec M., Yaacoubi A., Removal of Basic Red 46 dye from aqueous solution by adsorption onto Moroccan clay, J. Hazard. Mater 168 (2009) 304–309.

Kumara K.V., Porkodi K., Mass transfer, kinetics and equilibrium studies for the biosorption of methylene blue using Paspalumnotatum, J. Hazard. Mater 146 (2007) 214–226.

Langmuir I., The adsorption of gases on plane surfaces of glass, mica and platinum, J. Am. Chem. Soc 40 (1918) 1361–1368.

Özer A., Özer D., Comparative study of the biosorption of Pb(II), Ni(II) and Cr(VI) ions onto S. cerevisiae: determination of biosorption heats, J. Hazard. Mater. B 100 (2003) 219–229.

Padmavathy V., Biosorption of nickel (II) ions by baker's yeast: Kinetic, thermodynamic and desorption studies, Bioresour.Technol 99 (2008) 3100 –3109.

Ponnusami V., Krithika V., Madhuram R., Srivastava S.N., Biosorption of reactive dye using acid-treated rice husk: Factorial design analysis, J. Hazard. Materi 142 (2007) 397–403.

Ponnusami V., Gunasekar V., Srivastava S.N., Kinetics of methylene blue removal from aqueous solution using gulmohar (Delonixregia) plant leaf powder: Multivariate regression analysis, J. Hazard. Materi 169(2009) 119-127.

Safa Y., Bhatti H. N., Kinetic and thermodynamic modeling for the removal of Direct Red-31 and Direct Orange-26 dyes from aqueous solutions by rice husk, Desalination 272 (2011) 313–322.

Srinivasan A., Viraraghavan T., Decolorization of dye waste waters by biosorbents: A review, J. Envir. Manage. 91 (2010) 1915-29.

Temkin M.I., Adsorption equilibrium and the kinetics of processes on non homogeneous surfaces and in the interaction between adsorbed molecules, Zh. Fiz. Chim 15 (1941) 296–332.

Wang J., Chen C., Biosorption of heavy metals by Saccharomyces cerevisiae, Biotech. Advanc 24 (2006) 427-451.

3

Preparation and characterization of a high antibiofouling ultrafiltration PES membrane using OCMCS-Fe₃O₄ for application in MBR treating wastewater

Zahra Rahimi, Ali Akbar Zinatizadeh*, Sirus Zinadini

Water and Wastewater Research Center (WWRC), Department of Applied Chemistry, Faculty of Chemistry, Razi University, Kermanshah, Iran.

ARTICLE INFO	ABSTRACT
Keywords: Membrane bioreactor Antifouling Ultrafiltration Carboxymethyl chitosan-Fe₃O₄	An innovative method based on the membrane bioreactor (MBR) technology was developed as a potential remedy for the water shortage. MBRs attracted much attention in the field of wastewater treatment and reuse. It is reported from many researchers that membrane bioreactor technology is feasible and an efficient method for the treatment of wastewater. However, MBRs are faced to membrane fouling which lead to short membrane lifetime and increase operating costs. Here we were modified polyethersulfone (PES) ultrafiltration membrane by blending of O-carboxymethyl chitosan/ Fe_3O_4 nanoparticles in a PES solution (14% polymer weight) and casted by a phase inversion process. Membranes with four different weight percentage of O-Carboxymethyl chitosan bound Fe_3O_4 magnetic nanoparticles (OCMCs-Fe_3O_4) to PES of 0.05, 0.10, and 1 wt. % were tested. The OCMCS-Fe_3O_4 nanoparticles were prepared by the binding of carboxymethyl chitosan (CC) onto the surface of Fe_3O_4 magnetic nanoparticles, which were prepared by co-precipitating method. The synthesized nanoparticles were characterized by the Fourier transform infrared (FTIR) technique. Moreover, OCMCS-Fe_3O_4 nanoparticales blend membranes were also characterized using scanning electron microscopy (SEM), and permeation tests. Antifouling performance was studied using activated sludge as a biological suspension and measuring the pure water flux recovery ratio (FRR). The 0.1 wt. % OCMCS-Fe_3O_4-PES membrane revealed the highest FRR value (89%). The results exhibited that addition of OCMCS-Fe_3O_4 nanoparticales lead to membranes with high pure water flux compared to the unmodified PES membrane.

1. Introduction

Due to increasing the water source shortage concerns, recently technology of membrane bioreactors (MBRs) is become an attractive option for the treatment and reuse of municipal and industrial wastewaters because of many favorable features that it offers: high quality of processed water, reduction in excess sludge, controllability of solids and hydraulic retention time, and minimization possible in required footprint (Le-Clech et al. 2010; Kraume et al. 2010; Judd. 2008). However, a major issue associated with MBR is membrane fouling that the bioreactor suffers from it. Complicated interactions between membrane material and various components of activated sludge mixed liquor result in biofouling of the membrane. Unlike physical or chemical fouling, biofouling can irreversibly damage membrane surfaces and often causes permanent permeability loss, which makes the MBRs for wastewater treatment costly (Lee et al., 2013; Flemming et al. 1991; Liu et al. 2010). Therefore, it is highly desirable to have a membrane with antifouling capability, or anti-biofouling membrane.

A membrane material among synthetic polymers that has widely been used in membrane processes is polyethersulfone (PES), because of having many good physico-chemical characteristics such as desirable thermal and mechanical properties as well as chemical stability and easy processing (Marchese et al. 2003). However, the natural hydrophobicity of PES due to its structure is caused it is not immune from the biofouling problem and provides a low membrane flux and poor anti- fouling properties (Akar et al. 2013). Therefore, various approaches have been taken to increase the hydrophilic properties of PES either by chemical or physical modifications including blending [Yi et al. 2010; Teli et al. 2012), coating (He et al., 2008) and grafting (Yune et al. 2011; Deng et al. 2009). Akar and coworker (2013), prepared polyethersulfone ultrafiltration membranes with selenium and copper nanoparticles and investigated the morphology, performance and anti-fouling properties of membranes. It was shown that the blending membranes with nanoparticles are considered to be suitable for the prevention of biofouling. Yu et al. (2013) were prepared SiO₂@N-Halamine/polyethersulfone (PES) ultrafiltration membranes by phase inversion method. It was reported that hybrid membranes showed good antifouling and antibacterial properties, which might expand the usage of PES in water treatment and also could make some potential contributions to membrane antifouling.

In the another study by Huang et al. (2012), mesoporous silica (MS) particles was synthesized as inorganic fillers, and blended with polyethersulfone (PES) to achieve nanocomposite membranes with antifouling properties. The results indicated that the nanocomposite membrane with 2 % MS exhibited excellent hydrophilicity, water permeability and good antifouling performance.

*Corresponding author E-mail: zinatizadeh@razi.ac.ir

Chitosan (CS) widely researched for several applications such as commercial, industrial, environmental and biomedical applications due to is non-toxic, biodegradable and biocompatible biopolymer and also due to economic reasons, since it derives from chitin, which is very abundant in nature (Kumar. 2000). Chitosan has also have been a large capacity to fix molecules such as proteins and a high perm-selectivity for water (Huang et al. 2001). However, unmodified chitosan has few bottlenecks such as acidic solubility, low thermal and mechanical stability (Mansourpanah et al. 2013). To overcome these problems especially its limited solubility in aqueous media, many derivatives of CS have been synthetic. One of the hydrophilic derivatives which has a backbone structure like CS but hydroxyl group (–OH) is substituted by carboxyl group (COOH) is O-carboxymethyl chitosan (OCMCS).

Recently, intensive attention has been paid to Fe_3O_4 nanoparticles due to unique properties of them such as high specific surface area, magnetic, catalytic activities once used with powerful oxidants and/or UV/VIS light, easy to be synthesized, ecofriendly (Chen et al. 2009; lida et al. 2007). In addition to, magnetic nanoparticles and polymer composites not only can be tailored to reveal some new properties such as good film forming and processing properties, besides electrical, magnetic and optical properties, also do not destroy the polymer intrinsic properties (Guo et al. 2009). Despite those advantages, the magnetite (Fe_3O_4) nanoparticles have to be encapsulated to inhibit the agglomeration or to make them mono disperse in suspension. Afterward, it is important to modify the Fe_3O_4 nanoparticles to improve the stability.

In this work, a novel ultrafiltration membrane was prepared by blending of O-carboxymethyl chitosan coated Fe_3O_4 nanoparticles in PES matrix using the phase-inversion method. SEM and FTIR analyses together with permeation measurements were carried out to characterize the membranes containing different amounts of the Fe_3O_4/OCMCS nanoparticles. Investigation of fouling alleviation properties of Fe_3O_4/OCMCS entrapped PES membranes during the activated sludge filtration were studied.

2. Materials and methods
2.1. Materials

Polyethersulfone (Ultrason E 6020P, M_w=58,000 g/mol) and dimethylacetamide (DMAc) as solvent were supplied from BASF Co., Germany. Polyvinylpyrrolidone (PVP) with a molecular weight of 25,000 g/mol, sulfuric acid (H_2SO_4) (98 wt. %), $FeCl_3 \cdot 6H_2O$, $FeCl_2 \cdot 4H_2O$, and ammonium hydroxide (28 %) were purchased from Merck. O-carboxymethyl chitosan was purchased from Chung-mu Industrial Corp (Korea). All chemicals used in the experiments were of reagent grade and used as supplied.

2.2. Preparation of O-carboxymethyl chitosan–Fe_3O_4

Firstly, Fe_3O_4 nano particles was synthesized by co-precipitation from Fe^{2+} and Fe^{3+} ions by ammonia solution in N_2 condition at 70 °C, based on the method published previously (Mai et al. 2012). Briefly, ferric and ferrous chlorides were added into three-neck flask (molar ratio 2:1). Chemical precipitation was obtained under vigorous stirring by adding 2 M NH_4OH solution. During the reaction process, the temperature and the pH were maintained about 70 °C and 10, respectively. The precipitates were washed several times with distilled water and ethanol and resultant precipitates were dried for 24 h in vacuum condition at 50 °C. To prepare OCMCS modified Fe_3O_4 nanoparticles according to the method of Mai et al. (2012), 20 mg of Fe_3O_4 powder was dissolved in 10 ml of 2 mg ml^{-1} CC solution (pH= 7). Then, the mixture was stirred for 24 h. Fe_3O_4-OCMCS nanocomposite were finally separated from the reaction mixture by magnetic bar and then were washed with water and ethanol.

2.3. Preparation of PES/Fe_3O_4-OCMCS blended membranes

Phase inversion method has been used for fabrication of membranes. At first step, different precise amount of the Fe_3O_4-OCMCS nanoparticles was dissolved into DMAC and sonicated for 30 min for good dispersion. The composition of the casting solutions is detailed in Table 1. After dispersing nanoparticles in the solvent, the first PVP and next PES polymer were added to the dope solution and dispersed uniformly by using continuous stirring for 24 h at around 25°C. After that, the solution was left overnight to release the dissolved bubbles. The resulted polymer/nanofiller solution was again

sonicated 10 min to remove air bubbles. In the next step, the solution was cast in a glass plate by using casting knife with the same thickness of 150 μm and was put into the water immediately. The fabricated membrane was separated from the glass plate. After primary phase separation, the membrane was formed. The prepared membranes were stored in coagulation bath for 24 h for separation of residual solvent and then kept between filter papers at room temperature to dry for another 24 h.

Table 1. Specifications of the prepared membranes.

Membrane code	PES (wt. %)	OCMCS\-Fe_3O_4 (wt. %)	PVP (wt. %)
M_1	14	-	2
M_2	14	0.05	2
M_3	14	0.1	2
M_4	14	1	2

2.4. Characterization of the nanocomposite UF membranes

The Fourier transform infrared (FTIR) spectra of unmodified and OCMCS modified Fe_3O_4 nanoparticles for functional group determination was recorded on a Bruker FTIR spectrometer (Model: TENSOR 27). Membranes morphology studies were performed based on Philips-X130 and Cambridge scanning electron micro-scopes (SEM). Briefly, the membrane pieces were frozen in liquid nitrogen for 60–90 s, then broken and air dried. After sputtering with gold, they were viewed with the microscope at 26 kV.

2.5. Permeation experiments

The performance of the prepared ultrafiltration membranes was characterized by measuring pure water flux and activated sludge fouling tests by using a dead-end stirred cell filtration system connected with a nitrogen gas line. The schematic of the dead end system is depicted in Fig. 1 (Vatanpour et al. 2012). Each membrane was primarily immersed in distilled water for 30 min and next pressurized with distilled water at 3 bar for 30 min to compaction. After that, the pressure was reduced to the operating pressure value of 2 bar. The flux (J) via the cake and the membrane may be described as following:

$$j_{w,1} = \frac{M}{A \, \Delta t} \qquad (1)$$

where M is the weight of the permeate pure water (kg), A is the membrane effective area (m^2) and Δt is the permeation time (h).

2.6. Membrane antifouling property

After pure water flux measurement ($J_{w,1}$), the solution reservoir was rapidly refilled with 1000 ppm activated sludge suspension as the feed and the flux (J_p) based on the water quantity permeating the membranes at 2 bar for 6 min was attained. The flux was calculated according to the Eq. (1).

After filtration of activated sludge suspension, the membranes were washed with distilled water for 10 min and the water flux of cleaned membranes was measured ($J_{w,2}$). In order to evaluate the membrane antifouling property of the membrane, the flux recovery ratio (FRR) was calculated the following equation:

$$FRR \, (\%) = \left(\frac{J_{w,2}}{J_{w,1}} \right) \times 100 \qquad (2)$$

3. Results and discussion
3.1. Characterization of Fe_3O_4-OCMCS nanoparticles

For showing the successful coating of O-carboxymethyl chitosan on the Fe_3O_4 surface, spectroscopic analysis performed. Fig. 2 shows the FTIR spectra of unmodified and OCMCS surface modified magnetic particles. The band of 581 cm^{-1} is the characteristic peak of Fe–O–Fe in the unmodified Fe_3O_4. Presence of the peaks at 1633 cm^{-1} and 1391 cm^{-1} shows the C=O stretches of Fe_3O_4/OCMCS, indicating the carboxylate anion interacting with the FeO surface resulting in the formation of the iron carboxylate. The signal at 1057 cm^{-1} could be attributed to the C-O/C-N stretching vibration. Also, a broad band due to the stretching of O-H and N-H bond seemed at 3429 cm^{-1}. These results prove that O-carboxymethyl chitosan was adsorbed on the magnetic particles (Barroso et al. 2011; Shi et al. 2011; Vatanpour et al. 2012; Xu et al. 2004).

3.2. Morphology of the prepared membranes

To evaluate the changes induced in the skin-layer and sub-layer of the membranes, cross-sectional SEM images of the unmodified PES membrane and nanoparticle blended membranes are displayed in Fig. 3. A typical asymmetry structure composed of a thin skin-layer and a porous bulk with a finger-like structure can be clearly seen from the SEM images. According to the SEM images (Fig. 3), addition of the Fe_3O_4/OCMCS nanoparticles to the casting solutions causes the increased of number of finger-like channels, which led to increase of the porosity. However, the addition of more than 0.1 wt. % of the Fe_3O_4/OCMCS results in denser skin-layers with increased thickness and sub-layers with lower porosities. This was may be due to the increased viscosity of the Fe_3O_4/OCMCS /PES blend solution and agglomeration of the nanoparticle on the surface of the membranes. Increase of the viscosity usually delays the exchange of solvent and non-solvent. This suppresses the formation of large pore radius and reduces the porosity of membrane.

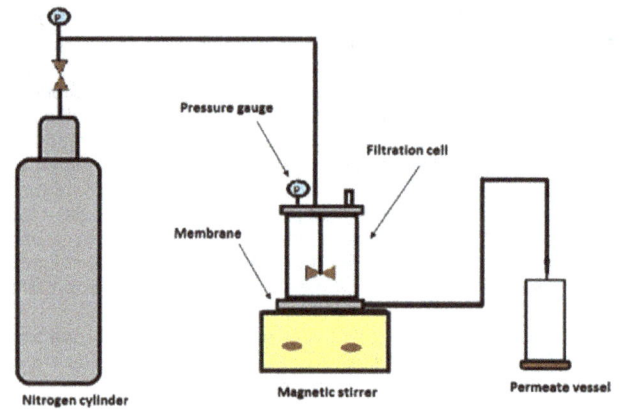

Fig. 1. Schematic of dead end system.

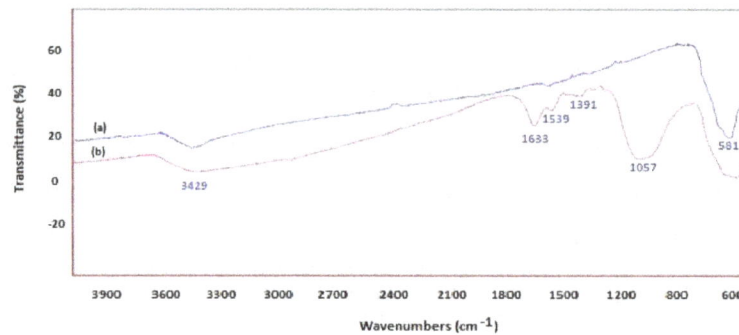

Fig. 2. FTIR spectra of unmodified (a) and CC modified Fe_3O_4 particles (b).

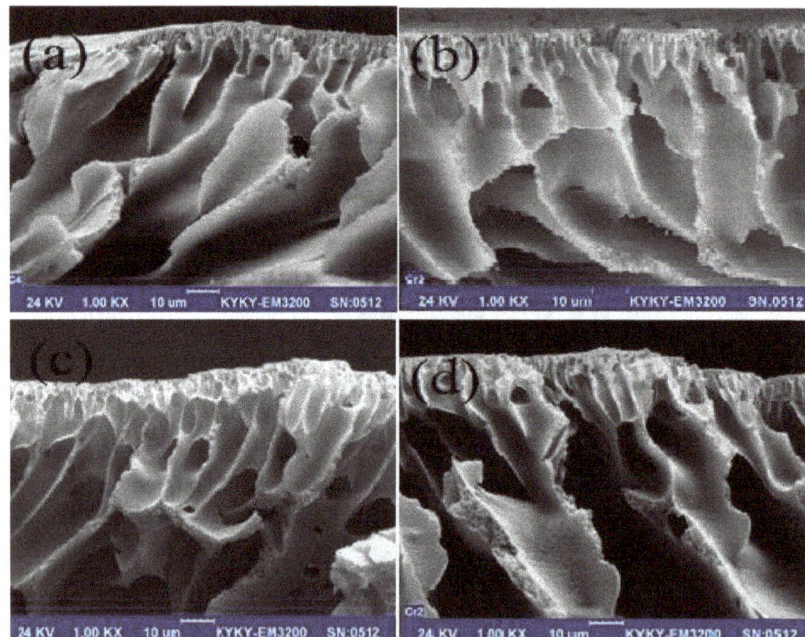

Fig. 3. Cross-sectional SEM images of the prepared membranes (a) M_1: Unfilled PES, (b) M_2: Fe_3O_4/OCMCS 0.05 wt. %, (c) M_3: Fe_3O_4/OCMCS 0.1 wt. % and (d) M_4: Fe_3O_4/OCMCS 1 wt. %.

3.3. Pure water flux

The pure water flux (PWF) of different types of Fe_3O_4/OCMCS embedded PES membranes as a function of the nanoparticle concentration shown in Fig. 4. As shown, the pure water flux of the membranes increased from 77.5 to 277 kg/m^2 h as the additive quantity increased from 0 to 0.1 wt. %. Although, the 1 wt. % Fe_3O_4/OCMCS content membranes have the highest percent of nanoparticles, but its pure water flux was the lowest. This contrary behavior can be ascribed to agglomeration of nanoparticles which cause to clogging of membranes pores (Vatanpour et al. 2012) as well as dense skin-layer of this membrane as shown in the SEM images (Fig. 3d).

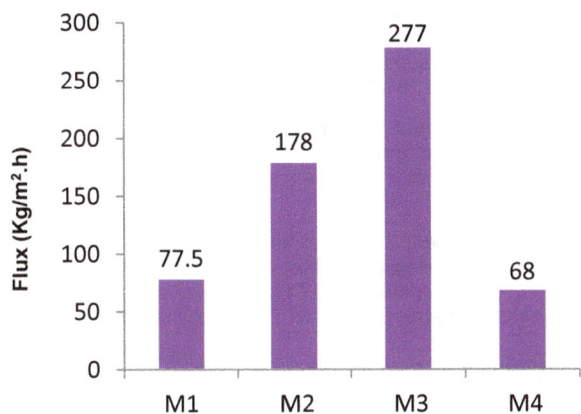

Fig. 4. Pure water flux of the prepared membranes.

3.4. Antifouling properties of the membranes

Fouling is the major obstacle of PES ultrafiltration membrane due to the hydrophobic interaction between foulant and membrane surface. The fouling is due to the adsorption and deposition of proteins on membrane surface and introducing of proteins within the membrane pores. Fouling mainly caused flux decline and shortening of membrane life time (Vatanpour et al. 2011). In order to increase membrane permeability and antifouling property, many efforts have been carried out for enhance membrane hydrophilicity (Kumar et al. 2013; Rana et al. 2010), which among these approaches, blending with hydrophilic particles has been considered as an effective and convenient approach for preparation of antifouling membranes (Ng et al. 2013; Zinadini et al. 2014).

The antifouling performance of the unfilled and the filled PES ultrafiltration membranes with nanoparticles was examined by measuring of water flux recovery after fouling by activated sludge suspension, which presented in Fig. 5. The highest FRR indicated that the membranes have more fouling resistant. It could be seen that the FRR values increased when the content of the nanoparticles in the casting solution increased, and the maximum FRR value reached 89

% for the M_3 blend membrane and then this was dropped to 78 %. This may be attributed to the agglomeration of nanofillers in the pores.

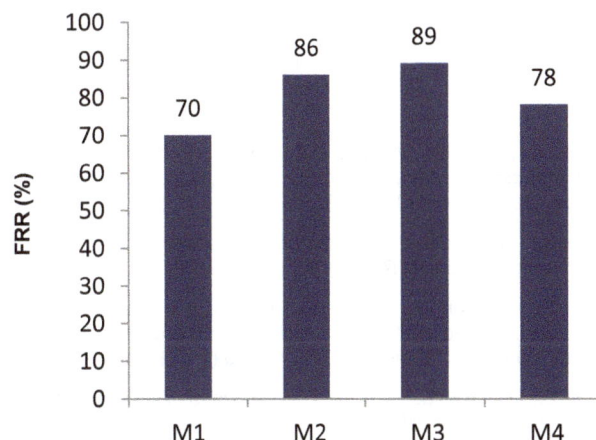

Fig. 5. Water flux recovery of the Fe_3O_4/OCMCS nanoparticles blended PES membranes after activated sludge suspension fouling. (Average of three replicates was reported).

4. Conclusions

The high performance PES nanocomposite membranes were prepared by addition of synthesized Fe_3O_4-OCMCS nanoparticles in the casting solution at different amounts. The water flux of the modified membrane improved by addition of nanoparticales from 0 to 0.1 wt. % and this was followed by a sharp decrease with further addition of Fe_3O_4-OCMCS nanoparticles. Similar trend to the pure water flux was observed in this study for antifouling properties and flux recovery ratio of modified membranes. The SEM images of membranes indicated that addition of the Fe_3O_4-OCMCS from 0 to 0.1 wt. % led to formation of membranes with the more number of finger-like channels.

References

Akar N., Asar B., Dizge N., Koyuncu I., Investigation of characterization and biofouling properties of PES membrane containing selenium and copper nanoparticles, Journal of Membrane Science 437 (2013) 216–226.

Barroso T., Temtem M., Casimiro T., Aguiar A., Antifouling performance of poly(acrylonitrile)-based membranes: From green synthesis to application, Journal of Supercritical Fluids 56 (2011) 312–321.

Deng B., Yang X.X., Xie L.D., Li J.Y., Hou Z.C., Yao S., Liang G.M., Sheng K.L., Huang Q., Microfiltration membranes with pH dependent property prepared from poly(methacrylicacid) grafted polyethersulfone powder, Journal of Membrane Science 330 (2009) 363–368.

Flemming H.C., Biofouling in Water Treatment, Geesey, G.G. (Eds.), Biofouling and Biocorrosion in Industrial Water Systems Springer, Hidelberg, (1991), pp. 4780.

Guo J.Y., Ye X.Y., Liu W., Wu Q., Shen H.Y., Shu K.Y., Preparation and characterization of poly(acrylonitrile-co-acrylic acid) nanofibrous composites with Fe_3O_4 magnetic nanoparticles, Material Letter 63 (2009) 1326-1328.

He T., Frank M., Mulder M.H.V., Wessling M., Preparation and characterization of nanofiltration membranes by coating polyethersulfone hollow fibers with sulfonated poly (etherether ketone) (SPEEK), Journal of Membrane Science 307 (2008) 62–72.

Huang J., Zhang K., Wang K., Xie Z., Ladewig B., Wang H., Fabrication of polyethersulfone-mesoporous silica nanocomposite ultrafiltration membranes with antifouling properties, Journal of Membrane Science 424 (2012) 362‾370.

Huang R.Y.M., Moon G.Y., Pal R., Chitosan/anionic surfactant complex membranes for the pervaporation of methanol/MTBE and characterization of the polymer/surfactant system, Journal of Membrane Science 184 (2001) 1–15.

Iida H., Takayanagi K., Nakanishi T., Osaka T., Synthesis of Fe_3O_4 nanoparticles with various sizes and magnetic properties by controlled hydrolysis, Journal of Colloid Interface Science 314 (2007) 274-280.

Judd S., The status of membrane bioreactor technology, Trends Biotechnology 26 (2008) 109 – 116.

Kraume M., Drews A., Membrane bioreactors in wastewater treatment-status and trends, Chemical Engineering Technology 33 (2010) 1251–1259.

Kumar R., Isloor A.M., Ismail A.F., Rashid S.A., Ahmed A. Al, Permeation, antifouling and desalination performance of TiO_2 nanotube incorporated PSf/CS blend membranes, Desalination 316 (2013) 76–84.

Le-Clech P., Membrane bioreactors and their uses in wastewater treatments, Applied Microbiological Biotechnology 88 (2010) 1253–1260.

Lee J., Chae H.-R., Won Y.J., Lee K., Lee C.-H., Lee H.H., Kim I.-C., Lee J.-m., Graphene oxide nanoplatelets composite membrane with hydrophilic and antifouling properties for wastewater treatment, Journal of Membrane Science 448 (2013) 223–230.

Liu C.X., Zhang D.R., He Y., Zhao X.S., Bai R., Modification of membrane surface for anti-biofouling performance: effect of anti-adhesion and anti-bacteria approaches, Journal of Membrane Science 346 (2010) 121–130.

Mansourpanah Y., Soltani H., Alizadeh K., Tabatabaei M., Enhancing the performance and antifouling properties of nanoporous PES membranes using microwave-assisted grafting of chitosan, Desalination 322 (2013) 60–68.

Marchese J., Ponce M., Ochoa N.A., Pradanos P., Palacio L., Hernandez A., Fouling behavior of polyethersulfone UF membranes made with different PVP, Journal of Membrane Science 211 (2003) 1–11.

Ng L.Y., Mohammad A.W., Leo C.P., Hilal N., Polymeric membranes incorporated with metal/metal oxide nanoparticles: A comprehensive review, Desalination 308 (2013) 15–33.

Rana D., Matsuura V, Surface modifications for antifouling membranes, Chemical Review 110 (2010) 2448–2471.

Shi Q., Su Y., Chen W., Peng J., Nie L., Zhang L., Jiang Z., Grafting short-chain amino acids onto membrane surfaces to resist protein fouling, Journal of Membrane Science 366 (2011) 398-404.

Teli S.B., Molina S., Calvo E.G., Lozano A.E., Abajo J.de., Preparation, characterization and antifouling property of polyethersulfone–PANI/PMA ultrafiltration membranes, Desalination 299 (2012) 113–122.

Trang Mai T.T., Ha P.T., Pham H.N., Huong Le T.T., Pham H.L., HoaPhan T.B., Tran D.L. Nguyen X.P., Chitosan and O-carboxymethyl chitosan modified Fe_3O_4 for hyperthermic treatment, Advance in Natural Scieince: Nanoscience and Nanotechnolgy 3 (2012) 1-5.

Vatanpour V., Madaeni S.S., Moradian R., Zinadini S., Astinchap B., Novel antibifoulingnano filtration polyether sulf one membrane fabricated from embedding TiO_2 coated multiwalled carbon nanotubes, Separation and Purification Technology 90 (2012) 69–82.

Vatanpour V., Madaeni S.S., Moradian R., Zinadini S., Astinchap B., Fabrication and characterization of novel antifouling nanofiltration membrane prepared from oxidized multiwalled carbon nanotube/polyethersulfonenanocomposite, Journal of Membrane Science 375 (2011) 284–294.

Vatanpour V., Madaeni S.S., Rajabi L., Zinadini S., Derakhshan A.A., Boehmite nanoparticles as a newnanofiller for preparation of antifouling mixed matrix membranes, Journal of Membrane Science 401-402 (2012) 132–143.

Vatanpour V., Madaeni S.S., Moradian R., Zinadini S., Astinchap B., Novel antibifoulingnanofiltrationpolyethersulfone membrane fabricated from embedding TiO_2 coated multiwalled carbon nanotubes, Separation Purification Technology 90 (2012) 69–82.

Xu v, Shen H., Xu J.R., Li X.J., Aqueous-based magnetite magnetic fluids stabilized by surface small micelles of oleolysarcosine, Applied Surface Science 221(2004) 430-436.

Yi Z., Zhu L.P., Xu Y.Y., Zhao Y.F., Ma X.T., Zhu B.K., Polysulfone-based amphiphilic polymer for hydrophilicity and fouling-resistant modification of polyethersulfone membranes, Journal of Membrane Science 365 (2010) 25–33.

Yu H., Zhang X., Zhang Y., Liu J., Zhang H., Development of a hydrophilic PES ultrafiltration membrane containing SiO_2@N-Halamine nanoparticles with both organic antifouling and antibacterial properties, Desalination 326 (2013) 69–76.

Yune P.S., Kilduff J.E., Belfort G., Fouling-resistant properties of a surface-modified poly (ether sulfone) ultrafiltration membrane grafted with poly (ethylene glycol)-amide binary monomers, Journal of Membrane Science 377 (2011) 159–166.

Zinadini S., Zinatizadeh A.A., Rahimi M., Vatanpour V., Zangeneh H., Preparation of a novel antifouling mixed matrix PES membrane by embedding graphene oxide nanoplates, Journal of Membrane Science 453 (2014) 292–301.

Nitrification of activated sludge effluent in low depth nitrifying trickling filter (LDNTF)

Mehraban Sadeghi[1], Akram Najafi Chaleshtori[1], Neda Masoudipour[1,*], Behnam Zamanzad[2]

[1]Department of Environmental Health Engineering, School of Public Health, Shahrekord University of Medical Sciences, Shahrekord, Iran.
[2]Department of Microbiology, School of Health, Shahrekord University of Medical Sciences, Shahrekord, Iran.

ARTICLE INFO	ABSTRACT
Keywords: Nitrifying trickling filter Nitrification Activated sludge Sewage treatment plant	Nitrogen in treatment plants effluent causing problems such as oxygen depletion, toxic impacts on aquatic organisms, eutrophication, and negative impacts on public health. The aim of present study was to determine the performance of integrated system activated sludge/nitrifying trickling filter to improve nitrification in the wastewater treatment plant of Isfahan. In this applied experimental study, an integrated activated sludge (AS) process (in full scale) was used with a trickling filter (TF) (in semi-industrial scale). The diameter and height of TF were 1.8 m and 3 m of steel, respectively. The volume of polypropylene media was 8 m^3 and surface area of 240 m^2/m^3. The hydraulic loading rate during the startup period was 2.4 m^3/h which was raised to 7.2 m^3/h in the operation period. Flow rate, BOD_5, COD, pH, TKN, $N-NH_3$, $N-NO_2^-$, $N-NO_3^-$, alkalinity and temperature were measured weekly according to standard methods during the operation period. The effect of filter depth on nitrification was studied in 3.6, 4.2, 5.4 and 6 m and HRT of 3.6 m/h. The samples were analyzed by SPSS. The results showed that the best hydraulic and ammonia loading rate achieved here were 3.6-4.2 m/h and 2-2.5 g N/m^2d, respectively. The AS/TF system efficiency were 86 % COD removal, 94 % BOD5 removal, 70 % turbidity removal, 94.4 % TSS removal, 55.5-75.5 % TKN removal and 85 % nitrification, respectively. The highest efficiency to reduce of wastewater pollution and nitrification was occurred in depth 4.5 m. Integration of the activated sludge and trickling filter processes, especially in old wastewater treatment plants is a good way to reduce the amounts of nitrogen in treatment plants effluent.

1. Introduction

As freshwater systems provide multiple environmental services such as supplying drinking water and irrigation, assimilating wastes through biotic/abiotic cycling, and supporting numerous species; water resource protection is critical and essential (Jackson et al. 2001; Naiman et al. 2000). Linkages between aquatic systems and terrestrial (Meyer et al. 1988) create a critical condition in freshwater systems that result from population growth and land use modifications. Estimates indicate that 60 % of the world will reside in urban areas by 2030, with the effect of increasing water demand creating larger volumes of wastewater in populous areas (Postel et al. 1996).

Nutrient content is a characteristic of WWTP effluent that often impacts receiving water bodies. Although one may assume that WWTPs are regulated (i.e. requiring a discharge permit) and therefore without a significant impact in pollutant loads, this is not necessarily true for nutrients (Hager et al. 1992; Andersen et al. 2004; Gibson et al. 2007).

Unfortunately, at present time, the wastewater treatment plants that are often used as activated sludge process for several reasons including overloaded, absence of suitable operation due to low qualification of operators or shortage of investment costs; cannot remove nutrients and high concentrations of these compounds can be seen in the effluent treatment plant (Sudol et al. 2014). In order to reduce these pollutants; developing of treatment methods and selecting of the best management techniques is necessary.

During recent years, biofilm processes with attached growth have been under special attention for biological treatment of municipal and industrial wastewater due to the low investment and operation costs, needless to return sludge system, simplicity of the operation, and high removal of nutrients (Wang et al. 2006; Yang et al. 2013). But in these systems there are also issues of high hydraulic losses, short circuiting problem, clogging and the need for periodic backwashing. That's why the idea development of designing integrated systems that have the benefit of the both suspended growth and attached growth system (Lee et al. 2006; Grady et al. 2012).

The use of trickling filter (TF) for upgrading the WWTPs is due to low operating costs and ease of operation (Aguilera et al. 2000). Application of plastic medium trickling filter systems for nitrification, with or without concurrent organic matter removal, has received considerable attention since the late 1960 (Huang et al. 1989).

Use of a TF system for nitrification of secondary effluent reduces capital cost because there is no need for final clarifiers (Almstrand et al. 2011).

Nourmohammadi et al., have reported the reduction of ammonia concentration at treatment plant of north of Tehran from 26.8 mg/l to 0.29 mg/l in an integrated process of trickling filter and activated sludge (Nourmohammadi et al. 2013). In the removal of inert organic fraction from municipal wastewater using the integrated activated sludge and trickling filter system showed that the integrated process can remove refractory organics two times higher in compared to single process (Sadeghi et al. 2014). Hannah et al., have also reported the removal of organics by using the activated sludge and trickling filter processes (Hannah et al. 1988). Chung et al. (2007), have done the removal of nitrogen in hybrid biofilm reactors/suspended growth. Daigger et al. (1993), founded that in compound process of trickling filter/activated sludge, nitrification occurs when loading of trickling filter is about 1kg BOD/m³d.

In Iran, many wastewater treatment plants discharge secondary effluent to the receiving waters without nutrient removal. There is growing interest in using methods to improve quality of the secondary effluent as organic materials, clearness, and nutrients.

This study examines the feasibility of using a XF (X-Flow) plastic media TF system with low depth for nitrification of activated sludge effluent. The aim of this work was nitrification of activated sludge effluent in the low depth nitrifying trickling filter.

2. Materials and methods
2.1. Bioreactor configuration

Considering a part of the experimental work was done on the AS process and was conducted in full scale, hence we tried to apply changes mostly on the second process (NTF) to examine and observe the reaction of the integrated process. Therefore, one TF tank in the semi-industrial scale was constructed using a 2.5 mm thickness steel sheet with two epoxy layers for avoiding the corrosion. Diameter and depth of NTF tank were 180 cm and 300 cm, respectively. Subtracting the depth of the flow distributor and pier media supporting system, there was an effective volume of 5.787 m³ in the TF tank (it is volume of applied media). Polypropylene XF media with a surface area of 240 m²/m³ was used as the biomass support material. The TF tank was coupled with an activated sludge process (Fig. 1).

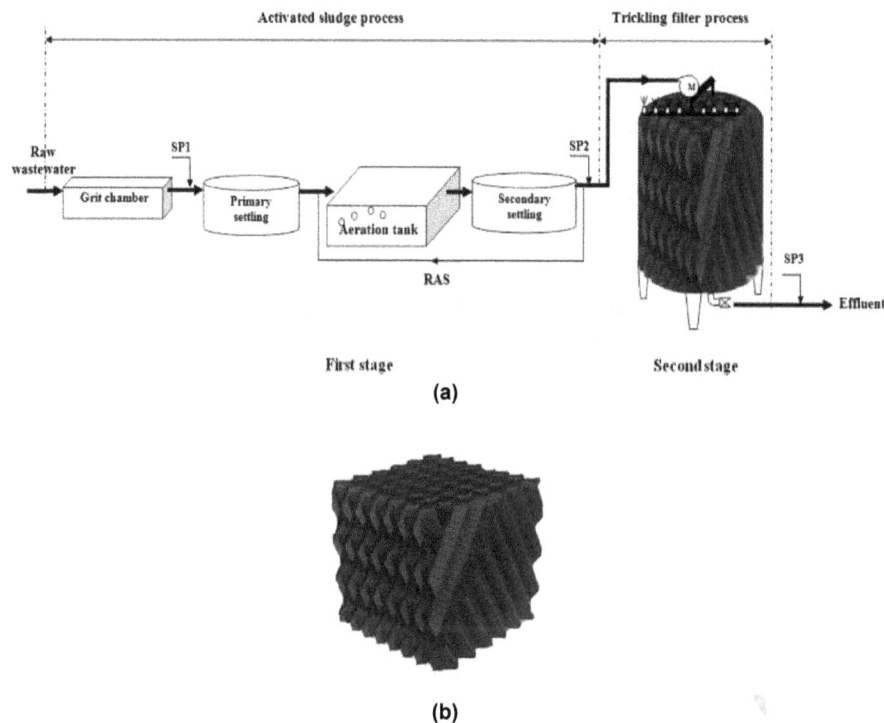

(a)

(b)

Fig. 1. Schematic of integrated activated sludge/nitrifying trickling filter (AS/NTF) process (a) and 2 HX-media (b).

Final effluent from the activated sludge in full scale, that was applied to treat a large region in Isfahan, was used as TF influent by tow installed pumps (1+1) on a channel between the clarifier basins and the chlorination unit (Fig. 2).

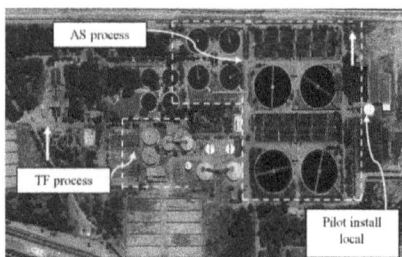

Fig. 2. A plan view from the wastewater treatment plant of Isfahan.

Flow rate was monitored by using a Rotameter for liquids with sp.gr. 1.0. Inlet flow to the TF tank was sprinkled on surface of the media by tow rotary arms equipped with 9 nozzles on each one and a rotor coupled to a gear box with 2.5 rpm in head of the tank. Hydraulic capacity minimum of the pilot in during the start-up phase was 3.6 m³/h. The pilot was operated about 3 months until set-up completed (i.e. fluctuations of control parameters in the pilot received below ±5 % in set-up period). Total surface of circle and square grooves applied in order to aeration in the TF tank was 0.0314-0.22 m² (Fig. 3).

Hydraulic Loading Rate (HLR) of the pilot in this study was 2.4-7.2 m/h in operation period (Fig. 4). Flow rate, BOD_5, COD, $N-NH_3$, $N-NO_2$, $N-NO_3^-$, Total Kjeldahl Nitrogen (TKN), Total Suspended Solids (TSS), alkalinity, pH and temperature were measured according to standard methods daily during the set-up and weekly during the operation period.

3. Results and discussion

Effluent from secondary treatment of wastewater (AS process) with hydraulic capacity of 2.4, 3.12, 3.6, 4.2, 5.4, 6 and 7.2 m/h were loaded on the NTF in separate stages and controlling parameters were measured during the operation periods. Changes occurred in the concentration of organic carbon compounds in various hydraulic loading rates is given in Table 1.

Fig. 3. Circle and square grooves applied in order to aeration in the NTF tank.

Fig. 4. Rotary sprinkle system on the pilot head.

Table 1. Average values of organic carbon compounds in AS/NTF system.

HLR (m/h)	2.4	3.12	3.6	4.2	5.4	6	7.2
COD_{raw} (mg/L)	365±76.02	385±61.94	460±11.13	447±22.0	485±15.0	530±32.78	565±60.0
BOD_{5raw} (mg/L)	220±46.63	244±24.78	218±37.04	230±20.29	262±19.07	251±31.04	265±8.88
$COD_{AS\text{-}outlet}$ (mg/L)	184±67	187.2±41.3	263.3±91.52	208.7±7.93	216±6.0	259±61.0	204±41.57
$BOD_{5AS\text{-}outlet}$ (mg/L)	113±7.0	101±6.0	72±2.0	82±0.71	80±2.82	140±5.65	87±9.89
$COD_{NTF\text{-}outlet}$ (mg/L)	131±61.05	100.8±24.4	62±9.16	78±6.0	90±18.0	113±11.26	134±3.11
$BOD_{5NTF\text{-}outlet}$ (mg/L)	82±6.0	41±5.0	15±2.0	18±3.0	18±2.0	50±5.65	70±2.82

In the present study, usefulness of an integrated system of the activated sludge/nitrifying trickling filter was investigated to upgrade the nitrification level. Fig. 5 shows the removal efficiency of organic materials versus the HLR in the NTF and the AS/NTF.

Study indicated that the average of numbers related to organic carbon compounds (COD and BOD_5) in activated sludge effluent exceeds the minimum necessary to be. The reason of this is the full-scale usage of activated sludge system and considering overload capacity of the wastewater treatment plant due to population growth, the wastewater provided by the system does not meet the standards of secondary treatment. The high concentration of organics (as BOD_5 and COD) in the inlet of NTF was one of the main nitrification limits in this study.

It can be concluded that the HLR within the range of 3.6 to 5.4 m/h, are the best conditions in which the removal efficiency of 86 % for COD and 94 % for BOD_5 have been achieved. Meanwhile, the maximum removal of TSS and nitrification was obtained in 4.2 m/h hydraulic load.

Aguilera et al. (1999) ascertained the concentration of COD and BOD_5 that are supposed to pass the nitrification stage should be respectively lower than 60 mg/L and 20 mg/L (Parker et al. 1986). However, in this study the average concentration at the NTF inlet have been respectively 216 mg/L and 96.5 mg/L. Moreover high concentrations of TSS entering the NTF through the outlet current of the activated sludge have been highly effective on role of this process in nitrification.

Table 2 shows that the average concentration of TSS entering NTF is 109 mg/L which may be led to reduce of nitrification rate occurred in the study (with average concentration of 15.8 mg/L nitrate formed in NTF effluent) (Fig. 6).

Table 2. Average concentration and the removal efficiency of TSS in NTF and AS/NTF in different loading rates.

HLR m/h	2.4	3.12	3.6	4.2	5.4	6	7.2
TSS_{raw} (mg/L)	178±36.91	183±37.33	170±5.0	178±30.61	220±32.78	308±48.28	311±25.06
$TSS_{AS\text{-}outlet}$ (mg/L)	70±28.47	118±6.63	102.7±64.29	96±19.69	112±24.0	140±15.10	125±30.02
$TSS_{NTF\text{-}outlet}$ (mg/L)	64±31.15	30±5.29	29±5.0	10±4.0	22±5.29	62±8.71	33±13.0
NTF. Eff. (%)	8.5	74.5	72	89.6	80	55.7	73.6
Overall Eff. (%)	64.04	83.61	82.94	94.38	90	79.87	89.39

(a)

(b)

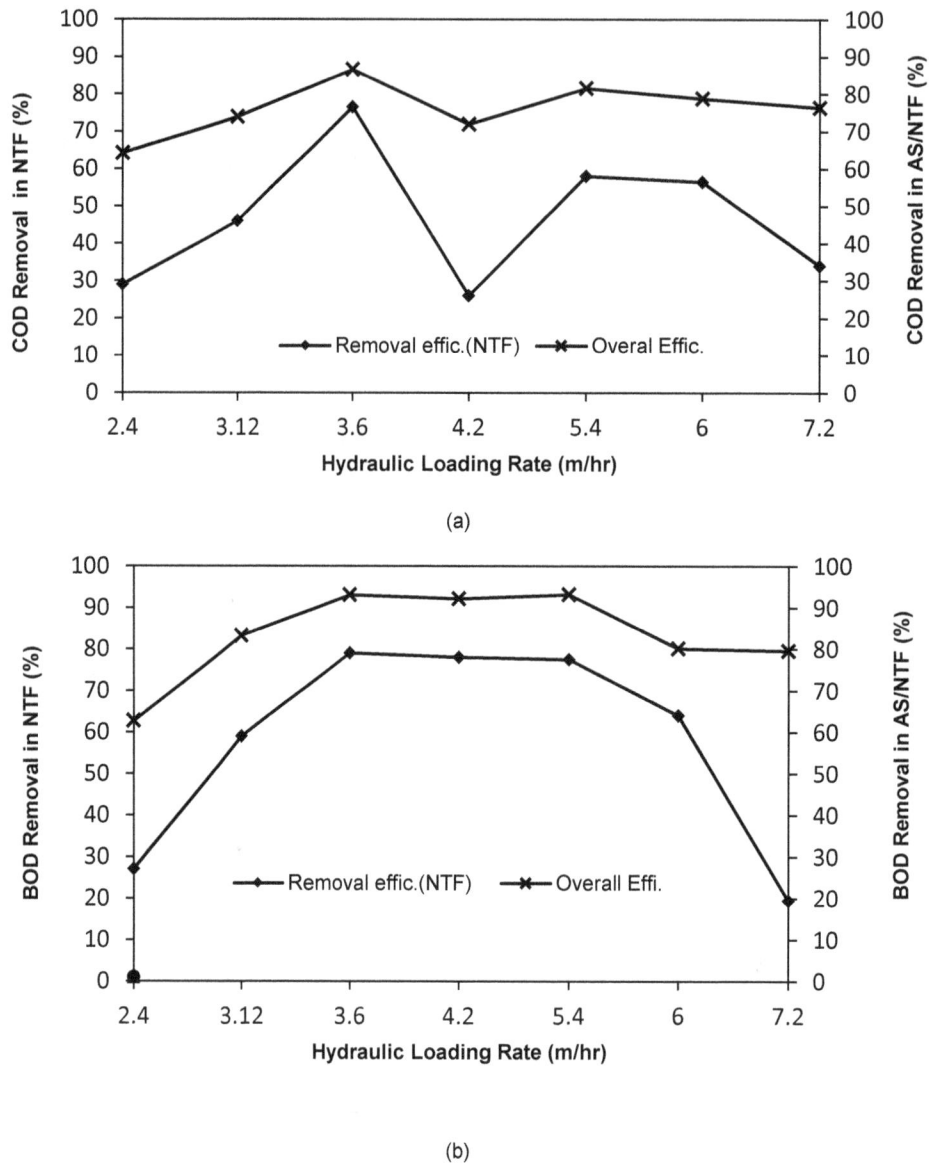

Fig. 5. Removal efficiency of organic compounds in NTF and AS/NTF Versus HLR, (a) COD and (b) BOD$_5$.

The occurred nitrification rate (actual nitrification) and the expected rate (based on alkalinity consumption) in different hydraulic loads are shown in Fig. 6.

Alkalinity reduction rate in the NTF effluent and the percentage of ammonia removal along with pH reduction rate compared to different hydraulic loads is shown in Table 3.

Table 3. Percentage of reduction in alkalinity, ammonia and pH in NTF effluent and changes in the concentration of added nitrate versus various hydraulic loads.

HLR m/h	2.4	3.12	3.6	4.2	5.4	6	7.2
Alkalinity reduction (%)	36.8	42	54.55	42.86	45.45	33.33	36.38
Ammonia reduction (%)	55.5	47.11	75.56	69.21	61.55	61.26	58.06
pH reduction (%)	2	0.63	0.9	2.26	3.65	0.4	0.8
Added nitrate concentration (mg/L)	16.3	14.54	15.65	16.13	17.77	17.12	13.02

Considering that the highest percentage of alkalinity reduction for the NTF reactor happened to be in the hydraulic load of 3.6 m/h, also given the acceptance of 7.14 mg/L CaCO$_3$ alkalinity consumption per 1 mg ammonia nitrogen nitrified. The theoretical rate of nitrogen nitrified in this hydraulic load is obtained nearly at 22.41 mg N/l. Nevertheless, in this hydraulic load the actual amount of ammonia removal has been the maximum and equal to 28.53 mg/L N-NH$_3$. Hence using this method, the actual amount of nitrified nitrogen with 27 % deviation from its theoretical amount represents a fair balance. In addition, the significant reduction

of pH in hydraulic loading range of 3.6-5.4 m/h proves the significant rate of nitrification. On the other hand, ammonia removal efficiency started in low amount 55.5 % at the hydraulic load of 2.4 m/h and increased to 75.56 % in hydraulic load of 3.6 m/h however from this point onwards despite the increase in hydraulic load the efficiency has been decreased. The reason of this could be that as hydraulic load gets higher in the system, the amount of inlet ammonia to the media will be increased as well and it leads to more nitrification. Moreover, in low hydraulic loading the media hydraulic wetting low and consequently the dearth of available ammonia leads to less nitrification. The efficiency obtained in this study for the ammonia removal is similar to the work done by G. Aguilera Soriano et al (Aguilera et al. 1999b) on an integrated AS/TF system for C and N removal in which they were able to achieve the 50-70 % of the removal efficiency.

This efficiency rate is low compared to 82 % ammonia removal observed in a study done by Bernard et al (Bernard et al. 1998) in which a form of third type of nitrification (TF filled with a different plastic Sessil) was implemented with external loading of ammonia 1.12 g $N-NH_3/m^2$.d. However, with increase in ammonia loading rate and growth of the biofilm on the media, effective area of media is reduced and acts as a limiting factor in nitrification process and is likely lead to a slight reduction in ammonia removal efficiency. Nevertheless, if increase in ammonia loading rate is due to increase in flow rate passed through the filter (increase in the HLR), in practice leads to increase in shear forces

on the media as well as reduction of biofilm containing nitrifiers and consequently less nitrification will occur.

As it can be seen, in the range of hydraulic loads 3.6 and 4.2 m/h, curves are closer to each other and it indicates the better balance of process in this range. The higher nitrification efficiency achieved in this research compared to similar studies perhaps is due to more ammonia loading (2-2.5 g $N-NH_3/m^2d$).

The nitrification rate occurred in the NTF at different temperatures of wastewater is shown in Fig. 7. The temperature measurements showed that in the coldest days (December) the average temperature of wastewater is not less than 16 $°C$, while the highest temperature average of wastewater has been 26.5 $°C$ which is registered in July has been equal to 26.5 $°C$. The nitrification rate in cold seasons compared to the best condition of this process (nitrification at 22 $°C$) shows the 57.5 % reduction ratio.

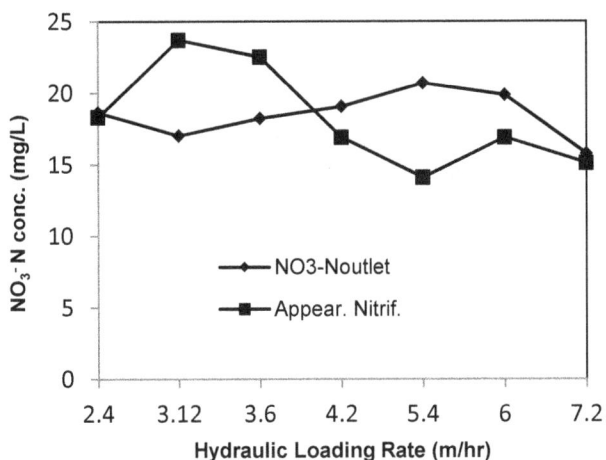

Fig. 6. Changes in the occurred nitrification rate in comparison to the expected rate versus hydraulic load rate.

Fig. 7. The nitrification rate occurred in NTF versus wastewater temperature (HLR=3.6 m/h).

The evaluation of TKN/COD and BOD_5/COD in NTF effluent is one of the most important parameters in this process. If the process could reduce these ratios, it can be concluded that the quality of effluent has been improved.

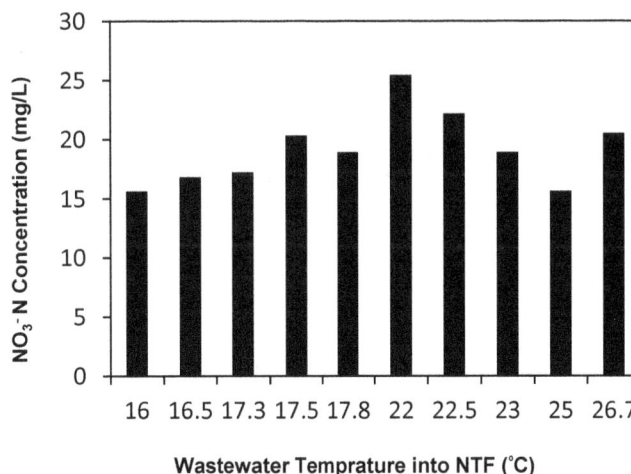

In Table 4, it is shown that the highest reduction of (TKN/COD)$_{out}$ and (BOD_5/COD)$_{out}$ has occurred in the hydraulic load of 4.2 m/hr.

Table 4. The ratio of TKN/COD and BOD_5/COD in NTF effluent.

HLR m/h	2.4	3.12	3.6	4.2	5.4	6	7.2
(TKN/COD)$_{out}$	0.21	0.23	0.26	0.17	0.25	0.25	0.29
(BOD_5/COD)$_{out}$	0.62	0.41	0.24	0.20	0.23	0.44	0.52

4. Conclusion

In general, the laboratory experiments involving analysis of various parameters in the integrated process of activated sludge/trickling filter (AS/NTF) included the following results:
- The best hydraulic and ammonia loading rate achieved here were 3.6-4.2 m/h and 2-2.5 g N/m^2. d, respectively.
- The AS/NTF system efficiency based on the research conditions has been equal to 86 % COD removal, 94 % BOD_5 removal, 70 % turbidity removal, 94.4 % TSS removal, 55.5-75.5 % TKN removal and 85 % nitrification, respectively.

- The nitrification rate in the depth of 3 m was not enough to reach an acceptable level of nitrified nitrogen in NTF effluent.
-Integration of activated sludge and trickling filter processes, especially in old wastewater treatment plants in which equipment depreciation and removal efficiency reduction of pollution parameters happen due to overload of individual process is a way to fix the problem and is able to withstand the hydraulic pressure and organic loading exerted to it.

Acknowledgements

This research was fully supported by the Water & Wastewater Company of Isfahan Province, Iran.

References

Aguilera S.G., Audic J.M., McCarley S., Ekama G.A., Wentzel MC. Option for the upgrading of WWTP's by means of trickling filters, Proceedings of 6th biennial water institute of Southern Africa; Suncity, South Africa 2000.

Aguilera S.G., Harduin H., Virloget F., Carbon removal by means of a combined trickling filter/activated sludge process, Project report France 1999b.

Almstrand R., Lydmark P., Sörensson F., Hermansson M., Nitrification potential and population dynamics of nitrifying bacterial biofilms in response to controlled shifts of ammonium concentrations in wastewater trickling filters, Bioresource Technology 102 (2011) 7685-7691.

Andersen B.C., Lewis G.P., Sargent K.A., Influence of wastewater-treatment effluent on concentrations and fluxes of solutes in the Bush River, South Carolina, during extreme drought conditions, Journal of Environmental Geosciences 11 (2004) 28-41.

Bernard J.F., Audoin L., Riotte M., Krier J., Etude de la nitrification tertiairesurle bioréacteurà ruissellement SESSIL, Techniques sciences methods 3 (1998) 35-43.

Chung J., Bae W., Lee Y.W., Rittmann B.E., Shortcut biological nitrogen removal in hybrid biofilm/suspended growth reactors, journal of Process Biochemistry 42 (2007) 320-328.

Daigger G.T., Norton L.E., Watson R.S., Crawford D., Sieger R.B., Process and Kinetic Analysis of Nitrification in Coupled Trickling Filter/Activated Sludge Processes, Journal of Water Environment Research 65 (1993) 750-758.

Gibson C.A., Meyer J.L., Nutrient Uptake in a Large Urban River1, Journal of the American Water Resources Association 2007 (43) 576-87.

Grady J.r., Daigger G.T., Love N.G., Filipe C.D.M., Biological wastewater treatment Taylor and Francis Group: LLC; 2012.

Hager S.W., Schemel L.E., Sources of nitrogen and phosphorus to Northern San Francisco Bay, Estuaries 15 (1992) 40-52.

Hannah S.A., Austern B.M., Eralp A.E., Dobbs R.A., Removal of organic toxic pollutants by trickling filter and activated sludge, Journal of Water Pollution Control Federation 60 (1988) 1281-1283.

Huang J.M., Hao O.J., Wu Y.C., Molof A.H., Nitrification of activated sludge effluent in a cross-flow medium trickling filter system, Journal (Water Pollution Control Federation) 61 (1989) 461-469.

Jackson R.B., Carpenter S.R., Dahm C.N., McKnight D.M., Naiman R.J., Postel S.L., Running S.W., Water in A Changing World, Ecological Applications 11 (2001) 1027-45.

Lee W.N., Kangm I.J., Lee C.H., Factors affecting filtration characteristics in membrane-coupled moving bed biofilm reactor, Journal of Water Reserch 40 (2006) 1827-1835.

Meyer J.L., McDowell W.H., Bott T.L., Elwood J.W., Ishizaki C., Melack J.M., Elemental dynamics in streams, Journal of the North American Benthological Society 7(1988) 410-432.

Naiman R.J., Turner M.G., A Future Perspective on North America's Freshwater Ecosystems. Ecological Applications, 10 (2000) 958-70.

Nourmohammadi D., Esmaeeli M.B., Akbarian H.G., hasemian M., Nitrogen Removal in a Full-Scale Domestic Wastewater Treatment Plant with Activated Sludge and Trickling Filter, Journal of Environment and Public Health 2013 (2013) 2-6.

Parker D., Richards T., Nitrification in trickling filters, Journal of Journal (Water Pollution Control Federation), 58 (1986) 896-902.

Postel S.L., Daily G.C., Ehrlich P.R., Human appropriation of renewable fresh water, Science 271 (1996) 785-788.

Sadeghi M., Fallahizadeh S., Mirzaei M., Removal of the Inert Organic Fraction of Municipal Wastewater Using the Integrated Activated Sludge/Trickling Filter System, Journal of Water and wastewater 6 (2014) 106-113.

Sudol M.Z., Walczak J., Effects of mechanical disintegration of activated sludge on the activity of nitrifying and denitrifying bacteria and phosphorus accumulating organisms, Journal of Water Reserch. 61 (2014) 200–209.

Wang X.J., Xia S.Q., Chen L., Zhao J.F., Renault N.J., Chovelon J.M., Nutrient removal from monicipal wastewater by chemical precipitation in a moving bed reactor, Journal Process Biochemistry 41 (2006) 824-828.

Yang Y., Yang X., Li W., Guo X., Enhanced nitrogen removal from reject water of municipal wastewater treatment plants using a novel excess activated sludge nased nitrification and denitrification process, Environmental Engineering and Management Journal 12 (2013) 1367-1373.

5

Three novel methods for removing inorganic species from contaminated industrial stormwater at a Smelter site in London

Lee Fergusson

Principal Consultant, Prana World Consulting, PO Box 1620, Oxenford, Queensland 4210, Australia.

ARTICLE INFO	ABSTRACT
Keywords: Heavy metals Stormwater Filtration Alumina refinery residue	Stormwater represents one of the least researched forms of wastewater in environmental science. Contaminated industrial stormwater, that is stormwater generated by runoff from industrial sites such as refineries, smelters and mine sites, is even less well understood. However, contaminated industrial stormwater can have damaging environmental impacts because it generally occurs in sudden bursts of high velocity and can result in significant downstream contamination. Flows of hundreds of thousands of litres of industrial stormwater are not uncommon in heavy rain events, and even when reduced through dilution, infiltration, co-mingling and by subsequent rain events, contaminants in stormwater can pose a risk to healthy urban and industrial environments. For these reasons, more research on contaminated industrial stormwater is desirable. This study considered two laboratory-scale experiments and an on-site field trial to assess three novel approaches to the treatment of heavy-metal contaminated stormwater at a smelter site in London. The approaches included the direct addition of a reagent derived from alumina refinery residue (ARR) and two filtration applications through laboratory and on-site reactive systems, both of which contained a form of pelletised media manufactured from alumina refinery residue. These three approaches resulted in the removal of inorganic contaminants from industrial stormwater, including cadmium from 0.08 mg/L to 0.0008 mg/L and copper from 0.7 mg/L to 0.0 mg/L by direct addition and arsenic from 0.34 mg/L to below the detection limit and antimony from 9.3 mg/L to 0.3 mg/L by filtration, with all post-treatment concentrations below the allowable limits for discharge. Although preliminary in nature, this study confirms other findings associated with the reuse of modified alumina refinery residue as a viable chemical raw material in industrial wastewater and solids treatment applications throughout the world, and the use of filtration of stormwater rather than the more common direct addition approach deserves further consideration.

1. Introduction

Stormwater is one of the most under-researched subjects in environmental science and wastewater treatment studies. Perhaps this is due to the fact that stormwater is a complex topic for analysis, with a variety of technical disciplines bearing directly upon it, including chemistry, civil engineering, hydrology, hydraulics, geomorphology, ecology and hydrogeology, among other fields of specialisation, making coherent and sustained investigation challenging.

Similarly, while the potential for complex organic and inorganic chemistry makes the study of contaminated stormwater appealing, its transient, high-volume/low-contaminant character and its apparently lower environmental polluting impacts also mean that stormwater, particularly industrial stormwater, is often the "forgotten wastewater". Typically coming under the purview of municipal councils and government water agencies but generally not of great interest to academic research (Erickson, et al. 2013; Washington State Department of Transport. 2014) the profound impact of stormwater on municipal sewerage systems, means stormwater management and mitigation are often the responsibility of local councils or state governments (City of Whittlesea. 2012) making it one step removed

from being of primary industrial concern or being part of a sustained research effort.

Moreover, because stormwater is generally associated with topics such as non-point source erosion and runoff in agriculture, pesticide dispersion and contamination, and inundation and infiltration impacts on impervious urban surfaces (such as along highways and from residential and other built-up areas like shopping malls and airports), industrial stormwater (i.e., contaminated stormwater generated at hard stand industrial sites) appears to be of even less interest to industrial researchers. Despite the fact that the polluting components of industrial stormwater can not only include predictable contaminants such as inorganic nutrients, suspended solids and heavy metals, but can also incorporate more problematic organic contaminants like bacteria (including pathogens), pesticides, petroleum and polycyclic aromatic hydrocarbons (PAH), benzene, toluene, ethylbenzene and xylenes (Department of Water. 2007). Nevertheless, industrial stormwater is often completely overlooked as a potential source of significant contamination or of social concern.

For example, Muthukrishnan (2010) studied the role of ponds and wetlands in ameliorating the impacts of metal-contaminated stormwater runoff, Davies and Bavor (2000) studied the environmental

Corresponding author Email: lee@pranaworldgroup.com

fate of pathogens in stormwater, and Moeller (2005) showed that 40 % of U.S. water bodies are being polluted with stormwater runoff and do not meet basic water quality guidelines, however most research on stormwater such as these three representative examples centres on agriculture and urban inundation but infrequently considers industrial or industrial-scale impacts and risks.

This conclusion can be verified by surveying the published literature on industrial waste. While industrial solid waste treatment features in the vast corpus of environmental research, the majority of studies in industrial waste relate specifically to wastewater. Recent examples of comprehensive surveys of industrial wastewater include Barakat (2011), Judd (2011), and Bhandari and Ranade (2014), and the relationship of industrial wastewater to energy consumption and sustainable development has been considered (International Bank for Reconstruction and Development. 2012). However, the study of stormwater in the industrial context rarely rates a mention in relation to wastewater, necessitating studies such as the present investigation in the UK.

One overriding assumption which guides the management of industrial stormwater is the concept of the "first flush". This approach centres on a primary concern for the initial first flush of stormwater, which occurs immediately after a rain event and which it is believed will rinse most contaminants from an industrial site, with subsequent flushes of cleaner water being spontaneously discharged to receiving waterways, such as lagoons, swales, rivers and floodplains. At many sites, stormwater management systems are therefore designed to only intercept the first flush (and to treat any inherent contaminants on site before they are discharged to the sewer as "trade waste" (SA Water. 2013)) but pay little or no attention to subsequent stormwater inundation. Environmental regulators are mostly (or in some cases, only) concerned with the potential environmental impacts of the first flush, and industrial sites typically have stormwater management and mitigation plans designed to manage and treat the first flush, assuming most contaminants from sustained rain events are of no significant longer term impact to the site or the environment; such plans generally do not attempt to intercept or treat stormwater which occurs after the first flush. This approach seems justified when considering research by Wen et al. (2012) has shown that heavy metal concentrations in industrial stormwater peak shortly after the first flush but decline rapidly thereafter.

These observations in the industrial context raise questions of whether stormwater is in fact wastewater at all. For example, Reese (2012) has argued that stormwater is not wastewater. He has identified the ten primary reasons why he believes that wastewater is different to stormwater, including the observation that stormwater, unlike wastewater, has what he calls an "unlimited peak flow" and "no ultimate end-of-pipe treatment". However, his arguments are based on rain events in urban or rural circumstances (such as in urban residential and agriculture settings) but are unpersuasive in the industrial context. In fact, industrial stormwater does not have an unlimited peak flow at industrial sites (for reasons given above in relation to the first flush) and is often subject to end-of-pipe treatment, as will be discussed in the context of this research study.

Reese also maintains that stormwater (but not wastewater) is characterised by episodic discharges of non-point source pollution. But again this may be true in agriculture but is untrue at most industrial sites, including the one chosen for this study, because the point-sources of pollution are usually well defined and demarcated, often by regulatory restrictions; it would rarely be considered best practice for environmental managers at a responsible industrial site to be unclear about where contamination comes from. For the purposes of the present study, industrial stormwater at this site in London is therefore defined as a type of on-site wastewater (not simply "stormwater" in the context of Reese's framework); this conclusion becomes particularly true when both the first flush of stormwater and all in-line wastewater generated at the site are comingled and treated in a wastewater treatment plant (WTP), and are together discharged as a single point source of treated wastewater.

Of the different approaches to treating inorganic species in industrial wastewater, including heavy metals such as cadmium, copper, chromium, lead, sulphate, phosphate, and metalloids like arsenic, the direct addition of chemical agents such as calcium hydroxide ($Ca[OH]_2$), sodium hydroxide (NaOH), and magnesium oxide (MgO) features most prominently [e.g., Semerjian & Ayoub, 2003]. Where organic species are of concern, studies have focused on the direct addition of oxidising agents such as Fenton reagents,

including hydrogen peroxide (H_2O_2), potassium permanganate ($KMnO_4$), sodium persulfate ($Na_2O_8S_2$) (Neyens & Baeyens. 2003). However, filtration is rarely cited in relation to industrial wastewater and minimal research on the relationship of filtration methods and heavy metals in industrial stormwater has been recorded in the scientific literature (with S.E. Clark et al. 2004 a rare example); very few approaches even consider filtration as a viable treatment modality, although filtration generally, including biofiltration, is somewhat more common with general stormwater treatment approaches (Hatt et al. 2011).

In addition to standard chemical reagents for treating industrial wastewater, there is also growing interest and data on the use of chemical reagents derived from alumina refinery residue, both in its powdered form for use in direct addition and in a pelletised form for use in filtration. For example, this author has conducted research on a variety of industrial wastes which have been treated by chemical reagents derived from this form of residue (Fergusson. 2014a, 2014b, 2014c) and has addressed the broader sustainability issues surrounding its widespread application (Fergusson. 2014d). Similarly, Huang, et al. [2008], have researched its use in reducing phosphate in wastewater, and Burkov et al. (2012) have studied the role of alumina refinery residue (ARR) as a coagulant and absorbent in the treatment of galvanic wastewater contaminated with heavy metals.

Alumina refinery residue (ARR) is the primary byproduct of alumina refining (i.e., extracting alumina [Al_2O_3] from bauxite). ARR has several unique physical and chemical properties of relevance to this study. For example, unlike the simple ionic binding and precipitation reactions which occur through changes in pH associated with immobilizing chemicals such as $Ca(OH)_2$ and NaOH, the metals sequestered in ARR become more tightly bound as time passes. This phenomena is due to the fact that ARR is composed of a cocktail of positively and negatively charged metals and minerals, including hematite (Fe_2O_3), beohmite (ɣ-AlOOH), gibbsite ($Al[OH]_3$) and sodalite ($Na_4Al_3Si_3O_{12}Cl$), anatase (TiO_2), aragonite ($CaCo_3$), brucite ($Mg[OH]_2$), diaspore (ß-Al_2O_3.H_2O), ferrihydrite ($Fe_5O_7[OH]$.$4H_2O$), gypsum ($CaSO_4$.$2H_2O$), hydrocalumite ($Ca_2Al[OH]_7$.$3H_2O$), hydrotalcite ($Mg_6Al_2CO_3[OH]_{16}$.$4H_2O$), and p-aluminohydrocalcite ($CaAl_2[CO_3]_2[OH]_4$.$3H_2O$), which together cause long-term isomorphic substitution reactions, meaning the positively charged iron-, aluminium-, magnesium- and titanium-based molecules and negatively charged hydroxides in ARR not only initially adsorb metals but also lead to the long-term "sequestration" phenomena observed with inorganic species (Fergusson 2009).

Of significance also in these formulae is the presence of hydroxides and oxyhydroxides which contribute to the acid neutralizing capacity (ANC) of ARR, both of which have low solubility and hence slow reactivity with acid. At least 48 hours are required for complete "on contact" reactions, and tests involving the addition of sulfuric acid to these reagents have shown that about 40% of the ANC of ARR reagents is exhausted in five minutes, about 70 % in four hours, and about 95 % in 24 hours (McConchie et al. 2000). Thus, the ability of minerals in ARR which sequester trace metals is time dependent; moreover, while most initial acid neutralization and metal sequestration reactions are completed within the first 24 hours, research has also shown that the longer these reagents are left in-situ the more tightly sequestered metals become, indicating that long-term co-precipitation and isomorphic substitution reactions are occurring at a molecular level, making reversibility of reactions difficult. Such findings suggest longer term and more sustainable outcomes compared to reactions precipitated by adsorption alone.

These and other relevant phenomena identified with ARR-derived reagents at metaliferous mine sites around the world have been discussed elsewhere (Fergusson. 2012), and applications utilizing these and related reagents in the treatment of coal waste and radioactive elements, such as radium, and in industrial site remediation and flue gas scrubbing to sequester mercury, for example, have also been examined (Clark et al. 2004, 2011; Fergusson. 2013; Hutson and Attwood. 2008) other core technical issues associated with ARR have been the subject of specialist scientific research (Taylor et al. 2011).

In order to better understand the role of filtration and ARR in the treatment of contaminated industrial stormwater, the present study asked the following research questions: 1) does the direct addition of chemical reagents derived from alumina refinery residue remove heavy metals from contaminated industrial stormwater; and 2) does a filter system, incorporating pelletised reagents derived from alumina refinery residue, remove heavy metals from contaminated industrial stormwater?

2. Materials and methods

This three-part research was conducted in two phases. The first two experiments were laboratory experiments conducted on industrial stormwater samples at a private laboratory in Newcastle-upon-Tyne; the third was a field trial conducted at a site in the east of London. The contaminated stormwater used in all three experiments was generated as runoff from the smelter site, which is a producer and trader of base and minor metals, particularly lead and zinc, and a manufacturer of precious and semi-precious metal products which are exported abroad to markets in Asia. Due to manufacturing processes at the site, fine particulate lead, antimony and other metals are present as airborne dust, particularly when rotary kilns are in operation. For this reason, during first flush rain events, stormwater which collects on the hard-stand parking lots, loading bays and others impervious areas around the site becomes contaminated with these metals, albeit in relatively low concentrations. However, these concentrations of metals are enough to disqualify the collected stormwater from being automatically discharged to the local river without intervention, necessitating treatment of the first flush.

Moreover, these relatively low concentrations of airborne metals which form part of the stormwater profile are complemented with contamination of other metals derived from exposed solid waste at the site. These metals leach or dissolve in water during significant rain events, and co-mingle with stormwater runoff. Fig. 1 (left) provides photographic evidence of first flush stormwater at the site, as well as examples of the solid industrial waste which has accumulated at the site and is exposed to leaching or dissolution during storm events (right).

Fig. 1. First flush stormwater collecting on hard-stand areas at the smelter site (left) and solid lead waste on the ground at the smelter site (right).

Together, these heavy metal point (i.e., from solid on-site waste) and non-point (i.e., airborne metals) sources of contamination at the site are responsible for the presence of heavy metals in the industrial stormwater at the site, and together these two sources of metals result in concentrations which are greater than the allowable discharge limits for the site, as identified and imposed by the UK's Environment Agency.

As shown in Figure 2, a first flush at the site is channelled into two concrete-bunded collection and settling bays. In this figure accumulated particulate matter can be seen in the left-hand photograph in front of the bay, with both bays representing collection points for stormwater runoff. The accumulated first flush stormwater in each collection and settling bay is then pumped to its respective

olding lagoon, with two lagoons on the site, each with a holding capacity of about 500 kL of wastewater. The collection and settling bay in Fig. 2 (left) is pumped into lagoon #1 and the bay in Fig. 2 (right) is pumped into lagoon #2. Fig. 3 shows lagoon #1 (left) and lagoon #2 (right).

Fig. 2. Stormwater collection and settling bays, for lagoon #1 (left) and for lagoon #2 (right).

As they fill up, collected stormwater from the lagoons is pumped to an on-site centralized WTP, where it is treated using standard direct addition precipitation and coagulation agents, with solids separated using a filter press; treated water is discharged to an estuary in the nearby Thames River (which is out of site in Fig. 3, but adjacent to lagoon #2), with filtered solids (i.e., filter cake) disposed to nearby regulated landfill.

Fig. 3. Lagoon #1 (left) and lagoon #2 (right).

Test 1: Two x 20 L samples of storm water were collected from lagoons #1 and #2 and transferred to the laboratory in Newcastle-upon-Tyne. On inspection, both samples were a clear liquid with dark deposits but no odour, and both had a pH of 8.3. Pre-treated samples were analysed for pH, and for cadmium (Cd), copper (Cu), lead (Pb) and zinc (Zn) using anodic stripping voltammetry, an analytical technique that involves (i) pre-concentration of the metal phase onto a solid electrode surface (or into liquid mercury) at negative potentials, and (ii) selective oxidation of each metal phase species during an anodic potential sweep (Franke & De Zeeuw. 1976). Pre-treatment analysis showed the lagoon #1 sample contained concentrations of 0.08 mg/L Cd, 0.7 mg/L Cu, 0.23 mg/L Pb, and 0.54 mg/L Zn; lagoon #2 sample contained concentrations of 0.7 mg/L Cd, 0.21 mg/L Cu, 1.4 mg/L Pb, and 0.17 mg/L Zn.

As shown in Table 1, after 8.0 ml/L hydrochloric acid (HCl) was added to each sample to adjust pH, both samples had a pH of 6.5 immediately prior to treatment; 3.0 g/L of a finely powdered reagent called ElectroBind was added directly to each sample and mixed vigorously for ten minutes.

Table 1. Results of direct addition laboratory experiment on contaminated industrial stormwater.

Parameter	Lagoon #1 Before Direct Addition	Lagoon #1 After Direct Addition	Percent Reduction (%)	Lagoon #2 Before Direct Addition	Lagoon #2 After Direct Addition	Percent Reduction (%)	Allowable Discharge Limit
pH	6.5	8.3	—	6.5	8.5	—	6.0-9.0
Cd (mg/L)	0.08	0.0008	99	0.07	0	100	0.03
Cu (mg/L)	0.07	0	100	0.21	0.01	95	0.1
Pb (mg/L)	0.23	0.004	98	1.4	0	100	1.0
Zn (mg/L)	0.54	0.003	99	0.17	0.001	99	0.5
Total average change	—	—	99	—	—	98	—

ElectroBind is a chemical reagent derived mostly (about 90 %) from ARR and has the same or similar physical and chemical properties to those contained in the description above for ARR, including a pH of approximately 9.5. After ten minutes, filtered samples were analysed for the above analytes.

Test 2: Sample quantities and methods of collection for Test 2 were the same as Test 1, however in this Test 2 lagoon #1 pre-treated samples were analysed for pH, Cd, Cu, Pb, Zn as well as arsenic (As) and antimony (Sb) using anodic stripping voltammetry, while those of lagoon #2 were only analysed for pH, Cd, Cu, Pb and Zn. Pre-treatment analyses in the Test 2 showed that lagoon #1 and lagoon #2 samples contained metal concentrations consistent with Test 1, in addition to 0.64 mg/L As and 4.6 mg/L Sb for lagoon #1.

After the same HCl adjustment described in Test 1, both samples were passed through a gravity fed, trickling filter column containing a pelletised version of ElectroBind; chemical additives to the alumina refinery residue included calcium-based strengthening agents, as well as binders and so-called "blowing agents" which designed to increased pellet porosity. The volume of the column was 3.0 L or 3,000 cm^3. Each filter had a 0.25 bed volume/hour (bV/hr), i.e., a hydraulic retention time (HRT) of four hours.

Test 3: Between conducting Test 2 and Test 3, management at the smelter site determined that arsenic (As), antimony (Sb) and nickel (Ni) were also likely contaminants in on-site stormwater; these thee metals were therefore added to the suite of analytes for the field trial. Stormwater effluent from lagoon #1 was delivered to an on-site baffle weir and filter system after a first flush rain event. This pre-treated effluent was analysed for pH, As, Cd, Cu, Ni, Pb, Sb and Zn using anodic stripping voltammetry. The baffle weir consisted of a standard plastic-lined holding weir with seven baffles through which effluent flowed (see Figure 4, left); in the baffle weir, effluent was pre-treated for five minutes with 6.0 ml/L of HCL and 2.0 mg/L ferrous sulphate (pH = 3.0) to lower pH and add Fe ions to aid in the removal of As and Sb, both of which prior research had suggested were not readily amendable to cation exchange with standard chemicals or ARR. After pre-treatment in the baffle weir, effluent was pumped to the filter system via a calibrated peristaltic pump.

The filter system was composed of a 50 L (or 50,000 cm^3) column containing ElectroBind pellet media (see Figure 4, right). Effluent pH was 9.2 at the static feed-head of the filter system, which was designed to use trickling mode at 0.25 bV/hr, resulting in an HRT of approximately four hours. The filter system was run with fresh water for four hours prior to the delivery of lagoon effluent.

Due to the static feed-head, initial effluent distribution proved to be inefficient in dispersing liquid evenly across the entire filter head, and was therefore changed to a rotating arm-type filter head, as shown in Fig. 4 right. At the same time, a floating boom and gravel roughing filter were fitted at the filter head to help alleviate solids which were visibly present in the effluent. When the trial was restarted with these two modifications, pH adjustment was stopped in the weir so that inlet effluent pH to the filter column was approximately pH 9.2. The filter system during the field trial operated for one month, with approximately four significant rain events during the course of the trial.

The allowable discharge limits (or so-called "consent levels") for pH and metals to the Thames estuary from the smelter site, as determined and mandated by the UK Environment Agency, were: pH = 6.0-9.0; As ≤ 0.1 mg/L; Cd ≤ 0.03 mg/L; Cu ≤ 0.1 mg/L; Ni ≤ 0.1 mg/L; Pb ≤ 1.0 mg/L; Sb ≤ 0.5 mg/L; and Zn ≤ 0.5 mg/L.

3. Results and discussion

Test 1: Table 1 presents the results of Test 1. This table shows that Cd and Zn in lagoon #1 sample and the presence of Cd, Cu and Pb in lagoon #2 sample were above the allowable discharge limits for the site.

Table 1 also indicates that the direct addition of ElectroBind increased pH from 6.5 to 8.3 and 8.5 respectively for lagoons #1 and #2, and in all instances reduced metals significantly. In lagoon #1 sample, Cd decreased from 0.08 mg/L to 0.0008 mg/L, a 99 % reduction, Cu decreased from 0.07 mg/L to 0.0 mg/L, a 100 % reduction, Pb decreased from 0.23 mg/L to 0.004 mg/L, a 98 % reduction, and Zn decreased from 0.54 mg/L to 0.003 mg/L, a 99% reduction, for a total average reduction of 99%. In lagoon #2 sample, Cd decreased from 0.07 mg/L to 0.0 mg/L, a 100% reduction, Cu decreased from 0.21 mg/L to 0.01 mg/L, a 95% reduction, Pb

decreased from 1.4 mg/L to 0.0 mg/L, a 100% reduction, and Zn decreased from 0.17 mg/L to 0.001 mg/L, a 99% reduction, for a total average reduction of 98%. All results were within the allowable discharge limits for the site, with instrumentation detection limits of 0.005 mg/L for Zn, and 0.01 mg/L for all other metals.

Fig. 4. Baffle weir used during the field trial (left) and close up photograph of effluent inlet and roughing filter at the feed-head, rotating filter head, and pelletised filter media in the filter column (right).

The results of Test 1 indicate that pH can be adjusted and all heavy metals, including Cd and Zn which were above allowable discharge limits prior to treatment in this study, can be reduced to required levels for approved discharge to the Thames estuary when directly adding ElectroBind reagent, a product derived from alumina refinery residue, to contaminated industrial stormwater. Therefore, research question 1 is answered in the affirmative. Results indicate that heavy metals were reduced by an average of 98 % across all metals for both lagoons. This result is likely the first time that a chemical reagent of this type has been so reported for direct addition to stormwater.

Test 2: Table 2 presents the results of Test 2. As noted above for Test 1, the presence of Cd and Zn in lagoon #1 sample and the presence of As, Cd, Cu, Pb and Sb in lagoon #2 sample were above allowable discharge limits for the site.

The filtration of contaminated stormwater using a filter consisting of an ElectroBind pelletised media increased pH from 6.5 to 11.2 and 9.8 respectively for lagoons #1 and #2, both of which were outside allowable discharge limits. However, metals were significantly reduced in all cases. In lagoon #1 sample, As decreased from 0.64 mg/L to 0.1 mg/L an 85% reduction, Cd decreased from 0.08 mg/L to 0.007 mg/L, a 91 % reduction, Cu decreased from 0.07 mg/L to 0.02 mg/L, a 71 % reduction, Pb decreased from 0.23 mg/L to 0.002 mg/L, a 99 % reduction, Sb decreased from 4.6 mg/L to 0.02 mg/L, a 99 % reduction, and Zn decreased from 0.54 mg/L to 0.003 mg/L, a 99 % reduction, for a total average reduction on 91 %. In lagoon #2 sample, Cd decreased from 0.07 mg/L to 0.0009 mg/L, a 98 % reduction, Cu decreased from 0.21 mg/L to 0.01 mg/L, a 93 % reduction, Pb decreased from 1.4 mg/L to 0.01 mg/L, a 99 % reduction, and Zn decreased from 0.17 mg/L to 0.0007 mg/L, a 99 % reduction, for a total average reduction on 97 %. All results for heavy metals were at or below the allowable discharge limits for the site, with instrumentation detection limits of 0.001 mg/L for as, 0.005 mg/L for Zn, and 0.01 mg/L for all other metals.

The results of Test 2 indicate that filtering stormwater using a filter column containing ElectroBind pellets increased effluent pH to levels above allowable discharge limits for both lagoons #1 and #2. However, the results also indicate that heavy metals, including As, Cd, Cu, Pb, Sb and Zn which were above allowable discharge levels prior to treatment in this study, can be reduced to required levels for approved discharge to the Thames estuary when using ElectroBind in a filter system. Results indicate that heavy metals were reduced by an average of 94 % across all metals in both lagoons. Therefore, research question 2 is answered in the affirmative. This result is also likely the first time that a pelletised chemical reagent of this type has been so reported for treating stormwater using a trickling filter.

Test 3: Table 3 presents the average result of 30 data points obtained over one month in Test 3. Pre-treated lagoon #1 stormwater effluent had a pH of 9.2, which is marginally above the allowable discharge limit, and the presence of As, Cd, Cu, Pb, Sb and Zn in the effluent, as delivered to the filtration system, were also above the allowable discharge limits for the site.

The filtration of contaminated stormwater using a filter consisting of an ElectroBind pelletised media decreased pH from 9.2 to 7.4, which was within the allowable discharge limit, and all metals were reduced: As decreased from 0.34 mg/L to below the detection limit, a 100 % reduction, Cd decreased from 0.14 mg/L to 0.02 mg/L, an 86 % reduction, Cu decreased from 0.18 mg/L to 0.02 mg/L, an 89 % reduction, Ni decreased from 0.03 mg/L to 0.02 mg/L, a 33 % reduction, Pb decreased from 4.5 mg/L to 0.06 mg/L, a 99 % reduction, Sb decreased from 9.3 mg/L to 0.5 mg/L, a 95 % reduction, and Zn decreased from 1.5 mg/L to 0.16 mg/L, a 90 % reduction, for a total average reduction on 84 %. All results for heavy metals were at or below the allowable discharge limits for the site.

Of relevance also was the observation that for the duration of the field trial the removal efficiency of the pelletised media did not decline; in other words, the averages presented in Table 3 fairly represent removal efficiency of the filter system at the beginning and end of the field trial. This finding was surprising given that metal loadings on the media were accumulating throughout the trial, particularly when the rotary furnaces at the site were in operation during rain events.

The results of Test 3 indicate that filtering stormwater using a system containing ElectroBind pellets decreased effluent pH to levels which are within allowable discharge limits for lagoon #1. Results also indicate that heavy metals, including As, Cd, Cu, Ni, Pb, Sb and Zn, which were above allowable discharge levels prior to treatment in this study, can be reduced to required levels for approved discharge to the Thames estuary when using ElectroBind pellets in a filter system. Results indicate that heavy metals were reduced by an average of 84 % across all metals for lagoon #1. Therefore, along with Test 2, the results of Test 3 also mean research question 2 can be answered in the affirmative.

Further work needs to be conducted to determine the most effective and commercially viable way to reduce pH in contaminated industrial stormwater. The use of waste acid from the site (if available) or modifications to acid dosing levels need to be further examined. As the pelletised form of ElectroBind also contains calcium-based additives, these may need to be modified or the alumina refinery residue may need to be treated prior to use in this type of application, in order to address the issue of elevated pH in filtered water.

Table 2. Results of filtration laboratory experiment on contaminated industrial stormwater.

Parameter	Stormwater from Lagoon #1 Before Filtration	Stormwater from Lagoon #1 After Filtration	Precent Reduction (%)	Allowable Discharge Limit
pH	9.2	7.4	—	6.0-9.0
As (mg/L)	0.34	BDL	100	0.1
Cd (mg/L)	0.14	0.02	86	0.03
Cu (mg/L)	0.18	0.02	89	0.1
Ni (mg/L)	0.03	0.02	33	0.1
Pb (mg/L)	4.5	0.06	98	1.0
Sb (mg/L)	9.3	0.5	95	0.5
Zn (mg/L)	1.5	0.16	90	0.5
Total average change	—	—	84	—

Table 3. Results of filtration field trial on contaminated industrial stormwater.

Parameter	Stormwater from Lagoon #1 Before Filtration	Stormwater from Lagoon #1 After Filtration	Precent Reduction (%)	Allowable Discharge Limit
pH	9.2	7.4	—	6.0-9.0
As (mg/L)	0.34	BDL	100	0.1
Cd (mg/L)	0.14	0.02	86	0.03
Cu (mg/L)	0.18	0.02	89	0.1
Ni (mg/L)	0.03	0.02	33	0.1
Pb (mg/L)	4.5	0.06	98	1.0
Sb (mg/L)	9.3	0.5	95	0.5
Zn (mg/L)	1.5	0.16	90	0.5
Total average change	—	—	84	—

Moreover, as discussed above, the longer heavy metals are sequestered in the ElectroBind solid pellet matrix the more tightly bound they become, with the strength of ionic bonds increasing by as much as 40 % every six months after the pellets have been saturated with soluble metals (Fergusson. 2009). Thus, the "spent" ElectroBind pellets removed from the filtration system after use would most likely be classified as "inert" and non-hazardous, and could therefore be disposed without difficulty. For these reasons, the class of chemical reagents of which ElectroBind is a part have been identified as a sustainable waste treatment solution; they not only address the problem of treating heavy metal-laden contaminated wastewater, in this case industrial run-off, but do not result in the production of another kind of waste. In other words, rather than merely transferring or converting one form of waste into another form of waste—solving the problem of contaminated wastewater but creating another problem of contaminated waste solids, for example—these reagents provide a practical method of addressing both the immediate and long-term

problem of industrial waste. As the pelletised media also has the potential to be used in cementitious pavers, bricks and concrete, in future research it may be worth considering its use as a pervious concrete for stormwater treatment, as discussed by Weidner et al. (2012).

4. Conclusion

Further research on a number of topics is warranted. For example, research is needed to determine filter breakthrough (i.e., the point at which the filter system no longer neutralises acid or binds metals). While breakthrough did not occur during this one-month trial but may have occurred in the second or third months, when and under what conditions it will do so need to be examined more closely. If filter breakthrough occurs too early in the filtration cycle, it may make this method for treating contaminated industrial stormwater unviable and therefore commercially and operationally unattractive. Similarly, research on bed volumes and HRT, unit processes and life of the filter system, cost of the filter system, and other related topics will be important if this method can be justified outside laboratory and field trial conditions.

A thorough and more rigorous stormwater management and treatment system would require something more than a mere filter system to effectively counter the polluting effects of high-volume stormwater at an operating industrial site. For example, Moeller (2005) has identified the various elements of a sustainable stormwater management and mitigation program, including identifying and defining the nature and scope of the problem, identifying site characteristics and constraints, evaluating watershed constraints, evaluating water quality conditions, selecting suitable hydrologic controls and unit processes, selecting suitable chemical, biological and/or physical treatment systems, among other elements. Such an in-depth and comprehensive approach to industrial stormwater would more comprehensively approach some of the questions left unanswered by this research.

To further highlight this point, metals in urban runoff can occur as dissolved, colloidal and particulate-bound species, although most are present as dissolved ions or ions bound to particulates. Therefore, it will be important in future research to measure all forms of heavy metals in industrial stormwater, especially the particulate and filterable fractions, when determining their properties, fate and long-term effects on the environment. Moreover, principles of sustainable urban development require that industrial stormwater management plans should not be considered in isolation of other urban considerations, but should form part of an integrated water strategy which would include general water use and supply, litter management, and reuse of alternate water sources (e.g., City of Whittlesea, 2012).

However, for the purposes of this preliminary study, the focus on these three novel treatment methods, including filtration, was warranted and further research would generate a clearer understanding of the relationship between direct addition of chemical reagents and filtration, between contaminated industrial stormwater and heavy metals, and between each of these factors and the use of filtration systems using pelletised media derived from alumina refinery residue.

Acknowledgements

The author has received technical assistance in the preparation of this study from Virotec in Australia, with whom the author is affiliated, and one of Virotec's clients in the United Kingdom on whose stormwater and at whose site this research was conducted; ElectroBind reagent is a common law trade mark of Virotec.

References

Barakat M.A., New trends in removing heavy metals from industrial wastewater, Arabian Journal of Chemistry 4 (2012) 361-377.

Bhandari V.M., Ranade V.V., Advanced physico-chemical methods of treatment for industrial wastewaters. In V.V. Ranade and V.M. Bhandari (Eds.), Industrial Wastewater Treatment, Recycling and Reuse [pp. 83-140], Butterworth & Heinemann, Oxford, UK, (2014).

Burkov K.A., Karavan S.V. Pinchuk O.A., Red mud for purification of galvanic wastewater, Russian Journal of Applied Chemistry 85 (2012) 1838-1844.

City of Whittlesea, Stormwater management plan 2012-2017. City of Whittlesea, (2012).

Clark M.W., Akhurst D., Fergusson L., Removal of radium from groundwater using a modified bauxite refinery residue, Journal of Environmental Quality 40 (2011) 1835-1843.

Clark M.W., McConchie D., Berry J., Caldicott W., Davies-McConchie F. Castro, J., Bauxsol technology to treat acid and metals; Applications in the coal industry. In J. Skousen and T. Hilton (Eds.), proceedings of the Joint Conference the American Society of Mining and Reclamation and the 25th West Virginia Surface Mine Drainage Task Force, Morgantown, WV (2004) 292-313.

Clark S.E., Johnson P.D., Gill S., and Pratap M., Recent measurements of heavy metal removals using stormwater filters, Water Environment Federation, Alexandira, Virginia, (2004).

Davies C.M. Bavor, H.J., The fate of stormwater-associated bacteria in constructed wetland and water pollution control pond systems, Journal of Applied Microbiology 89 (2000) 349-360.

Department of Water, Contaminants in stormwater discharge, and associated sediments, at Perth's marine beaches: Beach health program 2004-2006. Department of Water, Government of Western Australia, Perth, (2007).

Erickson A.J., Weiss P.T. Gulliver J.S., Optimizing stormwater treatment practices: A handbook of assessment and maintenance, Springer Science+Business Media, New York, (2013).

Fergusson L., Commercialisation of Environmental Technologies Derived from Alumina Refinery Residues: A Ten-year Case History of Virotec, Commonwealth Scientific and Industrial Research Organisation (CSIRO), Project ATF-06-3 "Management of Bauxite Residues", Department of Resources, Energy and Tourism (DRET),

Commonwealth Government of Australia, representing part of the commitment of the Australian Government towards the Asia-Pacific Partnership on Clean Development and Climate, (2009).

Fergusson L., ViroMine technology: A solution to the world's mining megawaste, Prana World Publishing, Gold Coast, Australia, (2012).

Fergusson L., An industrial legacy now gone, Water Management and Environment 24 (2013) 40.

Fergusson L., A Sustainability Framework for the Beneficial Reuse of Alumina Refinery Residue, Journal of Multidisciplinary Engineering, Science and Technology 1 (2014a) 105-120.

Fergusson L., A 12-month Field Trial to Remediate an Exposed "Tailings Beach" in Tasmania, Resources and Environment 4 (2014b) 238-245.

Fergusson L., A Long-Term Study of Mine Site Contamination and Rehabilitation in Australia, Asian Journal of Water, Environment and Pollution 11 (2014c) 1-17.

Fergusson L., Beneficial Reuse: A Field Trial to Remediate and a Bench-scale Test to Revegetate Coal Seam Gas Dam Sediments from Queensland, American Journal of Environmental Protection 3 (2014d) 249-257.

Franke J.P. De Zeeuw R.A., Differential pulse anodic stripping volumetry as a rapid screening technique for heavy metal intoxications, Archives of Toxicology 37 (1976) 47.

Hatt B.E., Steinel A., Deletić A., an Fletcher T.D., Retention of heavy metals by stormwater filtration systems: Breakthrough analysis, Water Science and Technology 64 (2011) 1913-1919.

Huang W., Wang S., Zhu Z., Li L., Yao X., Rudolph V., Haghseresht F., Phosphate removal from wastewater using red mud, Journal of Hazardous Materials 158 (2008) 35-42.

Hutson N.S. Attwood B.C. (2008). Binding of vapour-phase mercury (Hg0) on chemically treated bauxite residues (red mud), Environmental Chemistry 5 (2008) 281-288.

International Bank for Reconstruction and Development, A primer on energy efficiency for municipal waste and wastewater utilities. Technical Report no. 001/12, Energy Sector Management Assistance Program, International Bank for Reconstruction and Development, Washington, D.C., (2012).

Judd S., The MBR book: Principles and application of membrane bioreactors in water and wastewater treatment, Butterworth & Heinemann, Oxford, UK, (2011).

McConchie D., Clark M.W., Hanahan C., Baun R., New treatments for the old problems of acid mine drainage and sulfidic mine tailings storage. Paper presented to the 5th International Symposium on Environmental Geochemistry, Capetown, South Africa, Abstracts Volume (2000) 101.

Moeller J., Stormwater management: Research frontiers. Paper presented to Emerging Technologies and Practices in Urban Stormwater Management, Northglen, Colorado, (2005).

Muthukrishnan S. (2010). Treatment of heavy metals in stormwater runoff using wet pond and wetland mesocosms. In Proceedings of the Annual International Conference on Soils, Sediments, Water and Energy 11 (2010) 126-145.

Neyens E. Baeyens J., A review of classic Fenton's peroxidation as an advanced oxidation technique, Journal of Hazardous Materials 98 (2003) 33-50.

Reese A., Ten reasons managing stormwater is different from wastewater, Stormwater Report 2 (2012) 30-33.

SA Water, Contaminated stormwater: Trade waste guideline. SA Water, Government of South Australia, Adelaide, (2013).

Taylor K., Mullett M., Adamson H., Wehrli J. Fergusson L., Application of nanofiltration technology to improve sea water neutralization of Bayer process residue, in Light Metals 2011, edited by Stephen J. Lindsay, John Wiley & Sons, Inc., Hoboken, New York (2011), pp. 79-87.

Washington State Department of Transport, Highway runoff manual. Engineering and Regional Operations, Development Division, Design Office, Washington State Department of Transport, (2014).

Weidner A., Kney A., Suleiman M.T., and Ridinger D., Stormwater using pervious concrete. In Proceedings of the Water Environment Federation, Stormwater 15 (2012) 175-189.

Wen L., Zhenyao S., Tian T., Ruimin L., Jiali Q., Temporal variation of heavy metal pollution in urban stormwater runoff, Frontiers of Environmental Science & Engineering 6 (2012) 692-700.

6

Preparation of aldehydic electrospun PAN mats for ammonia removal from wastewater

Golshan Moradi[1], Farzad Dabirian[2,*], Laleh Rajabi[1,3], Ali Ashraf Derakhshan[4]

[1]Polymer Research Center, Department of Chemical Engineering, College of Engineering, Razi University, Kermanshah, Iran.
[2]Department of Mechanical Engineering, Razi University, Kermanshah, Iran.
[3]Department of Chemistry, University of Victoria, 2329 West Mall, Vancouver, BC V6T 1Z4, Canada.
[4]Environmental Research Center, Faculty of Chemistry, Razi University, Kermanshah, Iran.

ARTICLE INFO

Keywords:
Polyacrylonitrile (PAN)
Electrospinning
Chemical modification
Ammonium adsorption

ABSTRACT

Novel electrospun polyacrylonitrile (PAN) nanofiber mats and PAN fabric were chemically modified by dissolved anhydrous stannous chloride diethyl ether saturated with hydrogen chloride to contain aldehyde groups on their surfaces, which are suitable for ammonium adsorption due to their high adsorption affinity for NH_4^+ ion. Scanning electron microscopy (SEM), and Fourier-transform infrared spectra (FT-IR) were employed to characterize the prepared adsorbents. FT-IR spectra of these adsorbents confirmed that aldehyde groups are successfully formed on the surface of these chemically modified adsorbents. The aldehydic electrospun PAN nanofiber mats and aldehydic PAN fabric were assessed for their chelating property with NH_4^+ ion from aqueous solution. The effects of contact time on the amounts of ammonium adsorbed into the prepared adsorbents were also studied. Results revealed that ammonium removal increased by increasing contact time which finally reached equilibrium at about 3.5 h and 4 h for aldehydic electrospun PAN nanofiber mats and aldehydic PAN fabric, respectively. The adsorption performance of these prepared adsorbents for ammonium adsorption with initial ammonium concentration of 300 ppm via isotherm studies was investigated. The maximum ammonium removal efficiency (% R) was 48.33 and 70 for aldehydic electrospun PAN nanofiber mats and PAN fabric, respectively. Results indicated that the adsorption of ammonium by both prepared adsorbents followed Langmuir isotherm.

---

1. Introduction

Nowadays due to the rapid population growth and development of agriculture and industrial activity, concerns in both waters-care regions and development countries have raised. It seems that the reuse of treated water is the most suitable solution for the future sustainable water cycle control (Radjenović et al. 2009).

Nitrogen is a fundamental nutrient for all the life forms. It forms a basic building block of animal and plant proteins. Although it is a fundamental nutrient, too much nitrogen can be toxic. The presence of extra amount of nitrogen in the environment has drastically disordered the natural nutrient cycle between living organisms and the solid, water and atmosphere. Nitrous oxide, nitric oxide, nitrate or ammonia/ammonium are the usual forms of nitrogen that are soluble in water, so it can exist in drinking and grounded water (Rožić et al. 2000). The extent of nitrogen contamination in water supply has raised due to the discharge of industrial and domestic wastes to the environment. The increase of nitrogen concentrations in domestic wastewater is becoming significant among the pollutants. The ammonia and ammonium ions are the most prevalent encountered nitrogenous compounds in wastewater, causes accelerated eutrophication of lakes and rivers, dissolved oxygen reduction, fish toxicity in receiving water, undesired odors and several diseases (Balci. 2004; Du et al. 2005). The word ammonia derived from the name of the old Egyptian God Ammon and had been used chemically since 1799 (Lewis. 1993). Up to now, a number of biological and physiochemical methods have been used for ammonia removal from wastewater, including air stripping (Liao et al. 1995) chemical precipitation (Li et al. 1999), ion exchange (Koon and Kaufman. 1975), and biological nitrification-denitrification (Kuai and Verstraete. 1998). Adsorption has been identified as a new economically feasible and environmentally friendly method as compared to the other methods of ammonia removal from wastewater that have been received significant attention (Liu et al. 2010). Polymer fibers with diameters in the range of micrometers (e.g. 10-100 μm) to submicron or nanometers (e.g. 10×10^{-3}-100×10^{-3} μm) are of great interest to be developed as efficient adsorbent because of flexibility in surface modification, very high surface area to volume ratio, and better mechanical performance such as better stiffness and tensile strength as compared to any other known forms of the material (Huang et al. 2003). These prominent properties made the polymer nanofibers to be a good candidate for many significant applications such as filtration (Tsai et al. 2002), wound dressing (Jin et al. 2002), medical prosthesis (Laurencin et al. 1999), and adsorbent (Lee et al. 2010). Several methods including drawing (Joachim. 1998), template synthesis (Feng et al. 2002), self-assembly (Liu et al. 1999), phase separation (Ma and Zhang. 1999), electrospinning (Deitzel et al. 2001), and others have been used for polymeric nanofiber production in recent years. However, the electrospinning technique seems to be the only method that could be used for mass production of one-by-one continuous nanofibers from various polymers (Fong and Reneker. 2001). Polyacrylonitrile (PAN) nanofibers have been widely modified to contain adsorbent groups for contamination removal from wastewater. Besides, PAN can be easily prepared into nano-scale fibrous material by electrospinning method (Kampalanowat and Supaphol. 2010). The purpose of the current contribution was to synthetize aldehydic electrospun PAN nanofiber mat and PAN fabric as NH_4^+ ions adsorbents from the reaction between electrospun PAN nanofibers mat or PAN fabric and stannous chloride suspended in dry ether in the saturated dry hydrogen chloride surroundings. Scanning electron microscopy (SEM) and Fourier-transform infrared spectroscopy (FT-IR) were used to characterize the aldehydic electrospun PAN nanofibers mat or PAN fabric. Adsorption of NH_4^+ ions by prepared aldehydic electrospun PAN nanofiber mat and PAN fabric was also examined by using NH_4^+ aqueous solutions. Adsorption performance of prepared adsorbents for ammonium removal was investigated by isotherm studies.

2. Materials and methods
2.1. Materials

Industrial Polyacrylonitrile (PAN) powder, used as the raw material for preparing the base PAN solution which was later fabricated into PAN fiber mats by electrospinning, and PAN fabric were received from Iran Polyacryle Co. The weight average molecular weight (M_w) and the number average molecular weight (M_n) of the purchased PAN were 100000 g/mol and 70000 g/mol, respectively. Dimethylformamide (DMF; ~99.98 % purity) was purchased from Merck Co. was used as a solvent. Tin (II) chloride (stannous chloride, $SnCl_2$ ~99.98 % purity), hydrogen chloride (HCL ~35-37 % purity), Barium chloride ($BaCl_2$ 99.98 % purity), Diethyl ether ($(C_2H_5)_2O$), ammonia solution (25 %) and Nessler's reagent were purchased from Merck Co. Also, deionized water was used for functionalization reaction and batch adsorption experiments. Throughout the studies all chemicals were used without further purification.

2.2. Electrospun PAN nanofibers preparation

A schematic description of electrospinning setup is shown in Fig. 1. The key components of the electrospinning system including the DC high voltage power supply, a syringe pump with controllable feed rate, polymer solution, grounded collector and nozzle. A syringe, mounted in a pump was employed as a storage for electrospinning source solution (Tang et al. 2009).

In electrospinning process, polymer solution or melt is loaded to a syringe connected to a nozzle. Then, under the application of an electric field between the nozzle tip and a collector, a droplet of polymer solution is formed at the nozzle tip. As the intensity of the electric field is increased, charge repulsion on the drop surface at the end of the nozzle tip increases and changes the hemispherical shape of a drop in a conical shape which is known as a Taylor cone. Eventually, if the applied voltage reaches a critical value, a jet of polymer solution

initiates from the Taylor cone towards the grounded collector. This critical value is strongly related to the surface tension of the polymer solution. In the other word, the higher applied voltage is required for jet initiation at higher surface tension (Lee et al. 2003). During the jet traveling towards the grounded collector, solvent evaporates and polymer fibers in the form of nonwoven mats are deposited on the collector. At the nozzle tip to collector distance, the jet undergoes frequently whipping, which stretches the fiber and considerably reduces fiber diameter (Shenoy et al. 2005). The morphology of the obtained fibers is influenced by several factors including solution properties, processing and ambient conditions (Ding et al. 2002). In this study, the 13 wt. % solution of PAN in DMF was prepared at room temperature under constant mixing for about 4 h. The prepared solution was loaded into a syringe and electrospun under fixed electric field of 22 kV and nozzle tip to collector distance of 9 cm, using DC high voltage power supply, onto an aluminum (AL) sheet wrapped around a rotating drum, which was used as a collector. Each of the electrospun fiber mats were placed in vacuo at room temperature to remove further DMF from them as possible.

2.3. Aldehydic modification of electrospun PAN nanofiber mats and PAN fabric

Aldehydic electrospun PAN nanofiber mats were prepared in similar way to that reported by Stephen (1925). The basis of this new method is the conversion of nitrile group into an aldehyde with the same number of carbon atoms. In this method, anhydrous stannous chloride powder was dissolved in diethyl ether saturated with hydrogen chloride as the most proper reducing agent. The reaction and structure of obtained aldehydic PAN is shown in Fig. 2. Briefly, 4.2 gr powdered anhydrous stannous and 100 ml diethyl ether were suspended thoroughly in a round-bottomed 250 ml flask which was then saturated with hydrogen chloride. After about 1 hour, the mixture was separated into two layers, the lower viscous layer consisting of stannous chloride dissolved in ethereal hydrogen chloride. Then, about 0.88 gr of electrospun PAN nanofiber mats was added to the mixture with vigorous shaking and then was refluxed for 5 h. After the reaction, the mixture containing electrospun PAN nanofiber mats was filtered and fiber mats were placed into 500 ml glass chamber containing 100 °C water for 1 h. Afterwards, the obtained fiber mats were washed with hot deionized water and then dried in vacuo at room temperature. For comparative purpose, aldehydic PAN fabric was also synthesized from PAN fabric at the same condition was used in the preparation of aldehydic electrospun PAN nanofiber mats.

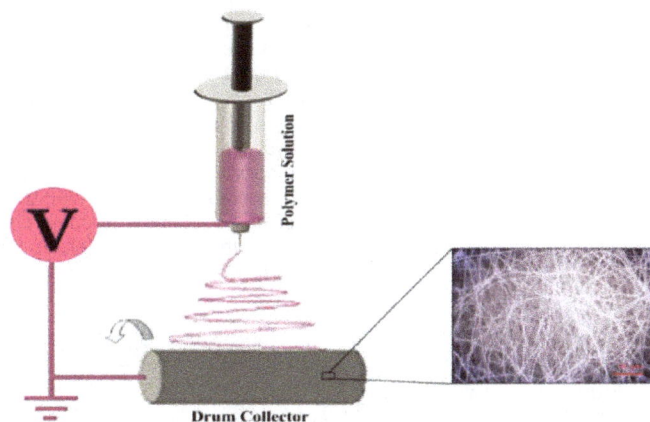

Fig. 1. A schematic diagram of electrospinning setup.

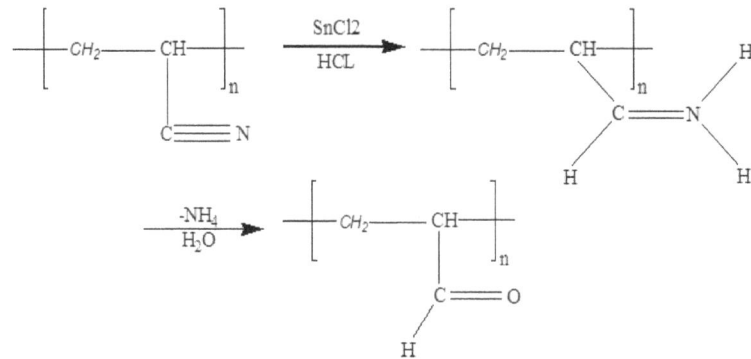

Fig. 2. The reaction and structure of obtained aldehyde PAN.

2.4. Characterization

Morphologies of both aldehydic electrospun PAN nanofiber mats and aldehydic PAN fabric were observed by Seron-AIC2300c scanning electron microscope (SEM). Each sample was coated with layer of gold to the SEM observation. The average fiber diameters of electrospun nanofiber mats were determined by measuring 100 random fibers from the SEM images using Digimizer software. The electrospun PAN nanofiber mats, aldehydic electrospun PAN nanofiber mats and aldehydic PAN fabric were also characterized using Fourier-transform infrared spectra (FT-IR) that were recorded between 400 and 4000 cm^{-1} with KBr pallets at room temperature (Bruker Alpha, Germany).

2.5. Batch adsorption experiment

Aldehydic electrospun PAN nanofiber mats and aldehydic PAN fabric were individually placed in a 100 ml conical flask containing 50 ml solution of ammonium aqueous solution. The standard solution (1000 mg/L) of ammonium was prepared through dissolving a sufficient amount of ammonium in 1000 ml deionized water. Each flask was shaken in a thermostatic shaker (Pars Azma, IN12), operating at 30 °C and 120 rpm for a certain time. Then, the adsorbents were separated by using 0.5 μm microporous membrane filters. Nessler's reagent colorimetric method was used to determine the concentration of ammonium in the aqueous solution after batch adsorption process. To study the effect of contact time, 6 batch experiments with the initial concentration of 300 ppm were done where samples were shaken from 1.5 to 4.5 h. The removal efficiency and adsorption capacity of prepared adsorbents for ammonium ion adsorption was computed from the following equations, respectively.

$$R \% = \left(\frac{C0 - Ce}{C0}\right) \times 100 \qquad (1)$$

$$q = \frac{(C0 - Ce)V}{M} \qquad (2)$$

where, R % is the removal percentage and q (mg/g) is the ammonia adsorption capacity by the adsorbent. C_0 and C_e (mg/l) are the initial and final NH_4^+ ion concentration in the testing solution, V (L) is the volume of testing solution and also M (g) is the weight of the adsorbent.

3. Results and discussion
3.1. Characterization of prepared adsorbent

Morphologies of prepared aldehydic electrospun PAN nanofiber mats and electrospun PAN nanofiber mats were observed by SEM and the results are illustrated in Fig. 3. It is clear that aldehydic electrospun PAN nanofiber mats showed identical morphology as compared to those of electrospun PAN nanofiber mats. The average fiber diameter of PAN nanofiber mats and aldehydic PAN nanofiber mats were 341.26 ± 55 nm and 345.32 ± 60 nm, respectively, indicating that the diameter of PAN nanofiber mats after aldehydic modification was essentially similar to that of PAN nanofiber mats. The FT-IR spectra of electrospun PAN and aldehydic electrospun PAN nanofiber mats and aldehydic PAN fabric are shown in Fig. 4. The FT-IR spectrum of electrospun PAN nanofiber mats showed adsorption peak of 1728.90 cm^{-1} and 1250 cm^{-1}, corresponding to the stretching vibration of carbonyl and ether groups of the methyl acrylate monomers. The frequency peak at 2244.54 cm^{-1} is related to the nitrile group (Saeed et al. 2008). The spectra of aldehydic electrospun PAN nanofiber mats showed new adsorption bands at 1732.33 cm^{-1}, 2924.56 cm^{-1} and 1633.00 cm^{-1} which are related to the C=O, C–C–H, and H–C=O groups in aldehyde, respectively. The intensity of nitrile peak in FT-IR spectra of aldehydic electrospun PAN nanofiber mats at frequency of 2244.85 cm^{-1} decreased due to the conversion of the nitrile to aldehyde group. In the case of aldehydic PAN fabric the FT-IR spectra exhibited new adsorption peak at frequencies of 2857.43 cm^{-1}, 1629.64 cm^{-1}, and 1740.03 cm^{-1} which attributes to the stretching vibration of O=C–H, C–H, and C=O in aldehyde, respectively. The adsorbing frequency peak at 1372 cm^{-1} belongs to the vibration of C–H alkane groups in the structure of aldehydic PAN fabric. The peak related to nitrile group of aldehydic PAN fabric at 2245.03 cm^{-1} also decreased in its intensity due to the conversion of nitrile to aldehyde groups. Moreover, the intensity of the adsorbing peak at 1740.03 cm^{-1} and 2857.43 cm^{-1} which are related to the aldehyde groups in prepared PAN fabric was greater than adsorbing frequency of aldehyde (1732.33 cm^{-1} and 29.24.55 cm^{-1}) in aldehydic PAN nanofiber mats.

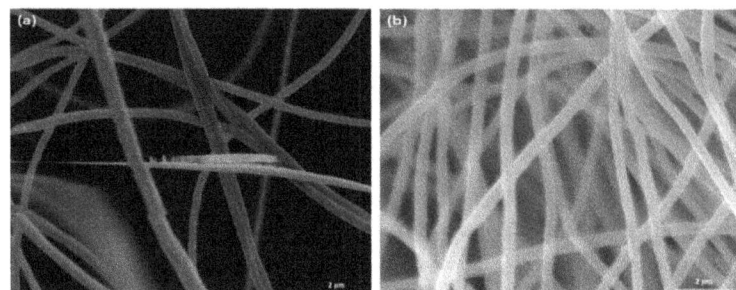

Fig. 3. SEM images of: (a) electrospun PAN nanofiber mats; (b) aldehydic electrospun PAN nanofiber mats.

3.2. Effect of contact time

Fig. 5 shows the ammonium removal efficiency (R %) and ammonium adsorption capacity (mg/g) of aldehydic electrospun PAN nanofiber mats and aldehydic PAN fabric from a 300 ppm aqueous solution of NH_4^+ as a function of time. It can be seen from the Fig. 5 that the ammonium removal efficiency (R %) of aldehydic electrospun PAN nanofibers with the maximum value of 48.33 % in all the ranges of contact time was lower than aldehydic PAN fabric. These results can be attributed to the greater amounts of unreacted nitrile group on the surface of aldehydic electrospun PAN nanofiber mats than aldehydic PAN fabric (see FT-IR spectra) that had weak interactions with ammonium ion. Therefore, considerably higher amounts of ammonium ions could be adsorbed on the surface of aldehydic PAN fabric with higher conversion of nitrile to aldehyde group. Also, it can be seen from

the Fig. 5 that at the first step of ammonium batch adsorption process, ammonium removal efficiency (R %) and ammonium adsorption capacity increased by increasing contact time, but thereafter the rate considerably leveled off and eventually reached equilibrium within 3.5 h and 4 h for aldehydic electrospun PAN nanofiber mats and aldehydic PAN fabric, respectively. This observed trend of ammonium removal from aqueous solution might be because of the fact that, at the initial step of adsorption process, due to the great number of vacant adsorbent sites and high concentration of NH_4^+ ion in aqueous solution of ammonium adsorption was swift. Thereafter, the ammonium adsorption rate decreased considerably and finally reached equilibrium because of the decrees in ammonium ion concentration in testing solution as well as adsorption sites reduction due to depletion. Our observation confirmed the past studies (Deng et al. 2003).

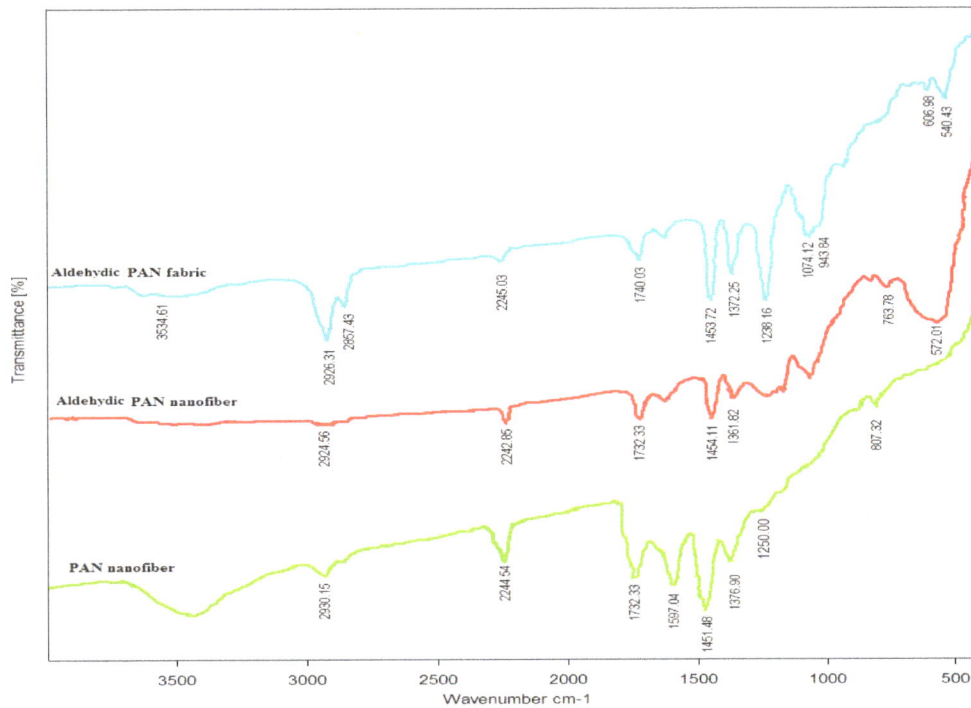

Fig. 4. FT-IR spectra of electrospun PAN nanofiber mats, aldehydic electrospun PAN nanofiber mats, and aldehydic PAN fabric.

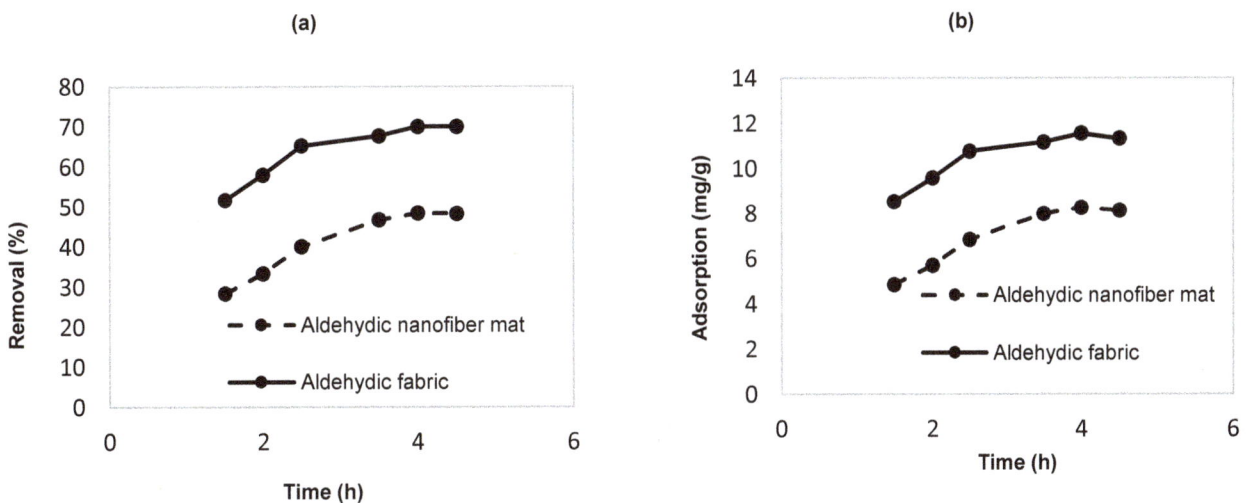

Fig. 5. (a) Ammonium removal efficiency (R %); (b) ammonium adsorption capacity (mg/g) of aldehydic electrospun PAN nanofiber mats and aldehydic PAN fabric.

3.3. Adsorption equilibrium isotherm

The experimental ammonium adsorption equilibrium data was analyzed using Langmuir isotherm model that can be written as follows:

$$\frac{Ce}{qe} = \frac{Ce}{qm} + \frac{1}{Kl\,qm} \qquad (3)$$

where, q_e is the equilibrium amount of NH_4^+ adsorbed on the aldehydic electrospun PAN nanofiber mats and aldehydic PAN fabric (mg/g), C_e is the equilibrium NH_4^+ concentration in the testing solution (mg/l). q_m and K_L are the Longmuir constants related to the maximal adsorption capacities of the NH_4^+ on the adsorbent and binding energy,

respectively. The value of q_m were calculated from the slope of the plot of C_e/q_e versus C_e (the inverse value of slope), whereas that of K_L can be calculated from the intercept of the plot [29]. The values of these parameters that were taken from the plot shown in Fig. 6, are summarized in Table 1. The linear plot confirmed that the NH_4^+ adsorption on the prepared adsorbents followed the Langmuir isotherm. The basic assumption of the Langmuir theory is that adsorption takes place on a homogeneous, flat surface of an adsorbent (Mall et al. 2006). The results showed that the adsorption of NH_4^+ on the aldehydic PAN fabric was greater than aldehydic electrospun PAN nanofiber mats that had been prepared at the same condition (see Table 1). This should be the result of the higher pore size of fabric and higher conversion of nitrile to aldehyde group in aldehydic PAN fabric.

Table 1. Langmuir constants for NH_4^+ on aldehydic PAN nanofiber mats and aldehydic PAN fabric.

Adsorbent	Langmuir model		
	q_m (mg/g)	K_L (L/mg)	r^2
Aldehyde electrospun PAN nanofiber	2.39	0.051	0.981
Aldehyde PAN fabric	5.98	0.022	0.994

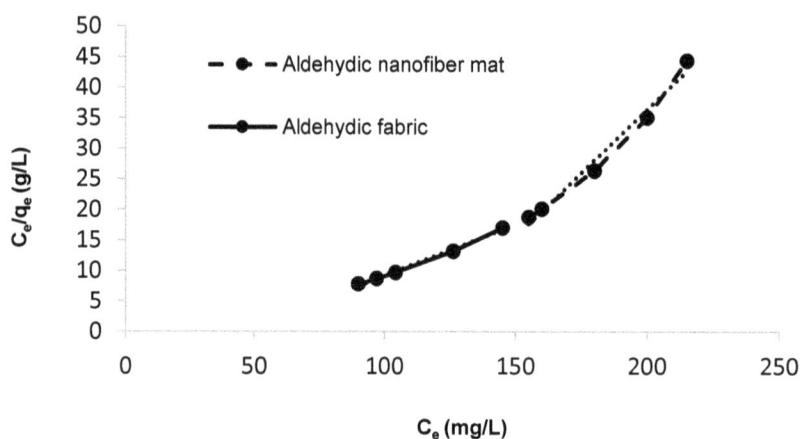

Fig. 6. Langmuir plot of aldehydic electrospun PAN nanofiber mats and aldehydic PAN fabric.

4. Conclusion

PAN nanofiber mats were prepared from 13 wt. % solution of PAN in DMF using the electrospinning technique at 20 kV and 9 cm nozzle tip to collector distance. Subsequently, the obtained electrospun PAN nanofiber mats were modified by dissolved anhydrous stannous chloride diethyl ether saturated with hydrogen chloride to produce aldehydic electrospun PAN nanofiber mats. Aldehydic PAN fabric also synthesized from PAN fabric at same condition was used in the preparation of aldehydic electrospun PAN nanofiber mats. FT-IR spectra were used to confirm the aldehydic modification of PAN nanofiber mats and fabric surface. The scanning electron microscopy (SEM) was employed to characterize the morphology of obtained fiber mats. The aldehydic electrospun PAN nanofiber mats and Aldehydic

PAN fabric were applied as adsorbents to ammonium adsorption. The initial concentration of ammonium in testing solution was 300 ppm. The contact time posed a strong effect on the ammonium removal of the prepared adsorbents. On the other word, the ammonium removal was found to increase with contact time increase and finally reached equilibrium within 3.5 and 4 h for aldehydic electrospun PAN nanofiber mats and Aldehydic PAN fabric, respectively. The ammonium removal efficiency (R %) of aldehydic electrospun PAN nanofiber mats in all the ranges of contact time was lower than aldehydic PAN fabric. The maximum ammonium removal efficiency (R %) was 48.33 and 70 % for Aldehydic electrospun PAN nanofiber mats and PAN fabric, respectively. The adsorbent of ammonium by both prepared adsorbents was fitted well with Langmuir equation.

References

Balci S., Nature of ammonium ion adsorption by sepiolite: analysis of equilibrium data with several isotherms, Water Research 38 (2004) 1129-1138.

Deitzel J., Kleinmeyer J., Harris D., Tan N.B., The effect of processing variables on the morphology of electrospun nanofibers and textiles, Polymer 42 (2001) 261-272.

Deng S., Bai R., Chen J., Behaviors and mechanisms of copper adsorption on hydrolyzed polyacrylonitrile fibers, Journal of colloid and interface science 260 (2003) 265-272.

Ding B., Kim H.Y., Lee S.C., Shao C.L., Lee D.R., Park S.J., Kwag G.B., Choi K.J., Preparation and characterization of a nanoscale poly (vinyl alcohol) fiber aggregate produced by an electrospinning method,

Journal of Polymer Science Part B: Polymer Physics 40 (2002) 1261-1268.

Du Q., Liu S., Cao Z., Wang Y., Ammonia removal from aqueous solution using natural Chinese clinoptilolite, Separation and purification technology 44 (2005) 229-234.

Feng L., Li S., Li H., Zhai J., Song Y., Jiang L., Zhu D., Super-hydrophobic surface of aligned polyacrylonitrile nanofibers, Angewandte Chemie 114 (2002) 1269-1271.

Fong H., Reneker D.H., Electrospinning and the formation of nanofibers, chapter, 2001.

Huang Z.-M., Zhang Y.-Z., Kotaki M., Ramakrishna S., A review on polymer nanofibers by electrospinning and their applications in nanocomposites, Composites science and technology 63 (2003) 2223-2253.

Jin H.-J., Fridrikh S.V., Rutledge G.C., Kaplan D.L., Electrospinning Bombyx mori silk with poly (ethylene oxide), Biomacromolecules 3 (2002) 1233-1239.

Joachim C., Drawing a single nanofibre over hundreds of microns, EPL (Europhysics Letters) 42 (1998) 215.

Kampalanonwat P., Supaphol P., Preparation and adsorption behavior of aminated electrospun polyacrylonitrile nanofiber mats for heavy metal ion removal, ACS applied materials & interfaces 2 (2010) 3619-3627.

Koon J.H., Kaufman W.J., Ammonia removal from municipal wastewaters by ion exchange, Journal (Water Pollution Control Federation) (1975) 448-465.

Kuai L., Verstraete W., Ammonium removal by the oxygen-limited autotrophic nitrification-denitrification system, Applied and environmental microbiology 64 (1998) 4500-4506.

Laurencin C.T., Ambrosio A., Borden M., Cooper Jr J., Tissue engineering: orthopedic applications, Annual review of biomedical engineering 1 (1999) 19-46.

Lee K., Kim H., Bang H., Jung Y., Lee S., The change of bead morphology formed on electrospun polystyrene fibers, Polymer 44 (2003) 4029-4034.

Lee K.J., Shiratori N., Lee G.H., Miyawaki J., Mochida I., Yoon S.-H., Jang J., Activated carbon nanofiber produced from electrospun polyacrylonitrile nanofiber as a highly efficient formaldehyde adsorbent, Carbon 48 (2010) 4248-4255.

Lewis Sr R., 1993, Hawley's Condensed Chemical Dictionary, 1139, Van Nostrand Reinhold Co, New York.

Liao P., Chen A., Lo K., Removal of nitrogen from swine manure wastewaters by ammonia stripping, Bioresource Technology 54 (1995) 17-20.

Liu G., Ding J., Qiao L., Guo A., Dymov B.P., Gleeson J.T., Hashimoto T., Saijo K., Polystyrene-block-poly (2-cinnamoylethyl methacrylate) Nanofibers—Preparation, Characterization, and Liquid Crystalline Properties, Chemistry-A European Journal 5 (1999) 2740-2749.

Liu H., Dong Y., Liu Y., Wang H., Screening of novel low-cost adsorbents from agricultural residues to remove ammonia nitrogen from aqueous solution, Journal of hazardous materials 178 (2010) 1132-1136.

Li X., Zhao Q., Hao X., Ammonium removal from landfill leachate by chemical precipitation, Waste management 19 (1999) 409-415.

Ma P.X., Zhang R., Synthetic nano-scale fibrous extracellular matrix, (1999).

Mall I.D., Srivastava V.C., Agarwal N.K., Removal of Orange-G and Methyl Violet dyes by adsorption onto bagasse fly ash—kinetic study and equilibrium isotherm analyses, Dyes and pigments 69 (2006) 210-223.

Radjenović J., Petrović M., Barceló D., Fate and distribution of pharmaceuticals in wastewater and sewage sludge of the conventional activated sludge (CAS) and advanced membrane bioreactor (MBR) treatment, Water research 43 (2009) 831-841.

Rožić M., Cerjan-Stefanović Š., Kurajica S., Vančina V., Hodžić E., Ammoniacal nitrogen removal from water by treatment with clays and zeolites, Water Research 34 (2000) 3675-3681.

Saeed K., Haider S., Oh T.-J., Park S.-Y., Preparation of amidoxime-modified polyacrylonitrile (PAN-oxime) nanofibers and their applications to metal ions adsorption, Journal of Membrane Science 322 (2008) 400-405.

Shenoy S.L., Bates W.D., Frisch H.L., Wnek G.E., Role of chain entanglements on fiber formation during electrospinning of polymer solutions: good solvent, non-specific polymer–polymer interaction limit, Polymer 46 (2005) 3372-3384.

Stephen H., CCLII.—A new synthesis of aldehydes, Journal of the Chemical Society, Transactions 127 (1925) 1874-1877.

Tang Z., Wei J., Yung L., Ji B., Ma H., Qiu C., Yoon K., Wan F., Fang D., Hsiao B.S., UV-cured poly (vinyl alcohol) ultrafiltration nanofibrous membrane based on electrospun nanofiber scaffolds, Journal of Membrane Science 328 (2009) 1-5.

Tsai P.P., Schreuder-Gibson H., Gibson P., Different electrostatic methods for making electret filters, Journal of Electrostatics 54 (2002) 333-341.

7

Hydraulic characteristics analysis of an up-flow anaerobic sludge blanket fixed film (UASB-FF) using tracer experiments

Parviz Mohammadi[1,2,*], Shaliza Ibrahim[1], Mohamad Suffian Mohamad Annuar[3]

[1]Department of Civil Engineering, Faculty of Engineering, University of Malaya, 50603 Kuala Lumpur, Malaysia.
[2]Department of Environmental Health Engineering, Public Health Faculty, Kermanshah University of Medical Science, Kermanshah, Iran.
[3]Institute of Biological Sciences, Faculty of Science, University of Malaya, 50603 Kuala Lumpur, Malaysia.

ARTICLE INFO

Keywords:
Up-flow anaerobic sludge blanket fixed film
Hydraulic characteristics
Tracer experiment

ABSTRACT

The hydraulic characteristic of an up-flow anaerobic sludge blanket fixed film (UASB-FF) were studied by changing two important hydraulic factors that can impact significantly on the hydraulic regime of the UASB-FF bioreactor: the Up-flow velocity (V_{up}) and biogas production rate (Q_g). The analysis of the reactor hydraulic performance was performed by studying hydraulic residence time distributions (RTD) obtained from tracer (Rhodamine B) experiments. The region of exploration for the process was taken as the area enclosed by V_{up} (0.5 and 3.0 m/h) and Q_g (14.87 and 7.96 l/d). Three dependent parameters viz. deviation from ideal retention time ($\Delta\tau$), dead volume percentage and Morrill dispersion index (MDI) were computed as response. The maximum $\Delta\tau$ and dead volume percentage were 33.58 min and 26 % at V_{up} of 0.5 m/h and Q_g of 14.87 l/d, respectively. While, the minimum responses (4.15 min and 19.3 %) were obtained at V_{up} of 3.0 m/h and Q_g of 7.96 l/d, respectively. The values of MDI computed at the minimum and maximum V_{up} and Q_g are identified as 11.33 and 10, respectively, showing that the hydraulic regime in UASB-FF bioreactor is a semi-complete mixing.

---

1. Introduction

The hydraulic behavior in biological reactors is of fundamental importance for the efficiency of the wastewater treatment processes. Examples of hydraulic phenomena with adverse impacts on the bioreactors performance include short circuiting streams and dead volumes. The unfavorable hydraulic situations in the bioreactors may cause lower process performance and thus higher residual concentrations in the treated wastewater. This may be particularly significant in high loaded bioreactors (Levenspiel. 2000; Fogler Scott. 2001) Comparison of actual hydraulic characteristics of a reactor measured using tracers, to the expected theoretical response can be used to assess the degree to which the ideal design has been achieved (Fogler Scott. 2001; Metcalf & Eddy. 2003).

The up-flow anaerobic sludge blanket fixed film (UASB-FF) reactors offer an alternative treatment technology to the biohydrogen production systems. Because of the widespread use of UASBs in the recent years, and on the other hand the reactor hydraulic directly affects the treatment performance, determination of hydraulic characteristic is of great importance. Poor hydraulic conditions reduce the HRT and effective volume, resulting in lower hydrogen production efficiencies of the bioreactor. Presence of a good design model for UASB reactor is a useful tool describing and predicting the UASB hydraulic regime (Yamaguchi et al. 1999; Kargi et al. 2003; Najafpour et al. 2006).

Many studies have carried out by using the tracer tests to indicate the presence of short circuiting streams and dead volume in different bioreactor comprised of three sections. The lowest section of the UASB reactor's column accommodated 67.84 % of the total working volume. The middle section of the fixed film reactor accommodated

reactors. Hydraulic characterization is performed by retention time distribution (RTD) curve (Mansouri et al. 2012; Newell et al. 1998; Williams et al. 1998; Chen et al. 2010; Martin. 2000). However, there is not any tracer study done on the hydraulic characteristics of the UASB-FF bioreactor.

Nevertheless, a few quantitative models have been proposed. In all these models, there is no common agreement on whether an UASB behaves as a plug-flow or a completely mixed reactor. Various indexes have been used to describe the mixing and hydraulic flow pattern in the different operational units.

In the present study, in order to explore the best operational conditions achieving a high hydraulic performance in an UASB-FF, the effect of two effective variables, the Up-flow velocity (V_{up}) and biogas production rate (Q_g), on the hydraulic regime of the UASB-FF were investigated. The deviation from ideal retention time ($\Delta\tau$), volume of dead space and dispersion indexes (MDI and d) as the responses were determined using the data obtained from the tracer experiments and the hydraulic characteristics.

2. Materials and methods
2.1. Bioreactor configuration

The schematic diagram of the laboratory-scale UASB-FF bioreactor (total volume of 3.5 l, working volume of 2.55 l and liquid height of 80 cm) rig set-up used in this study is shown in Fig. 1. The glass bioreactor column was fabricated with an internal diameter of 55 mm at the bottom and middle parts and 75 mm at the top part. The 14.51 % of the working volume. The top section of the bioreactor accommodated 17.65 % of the working volume consisting of a gas-solid separator and outlet zone.

Corresponding author Email: parviz8855@yahoo.com

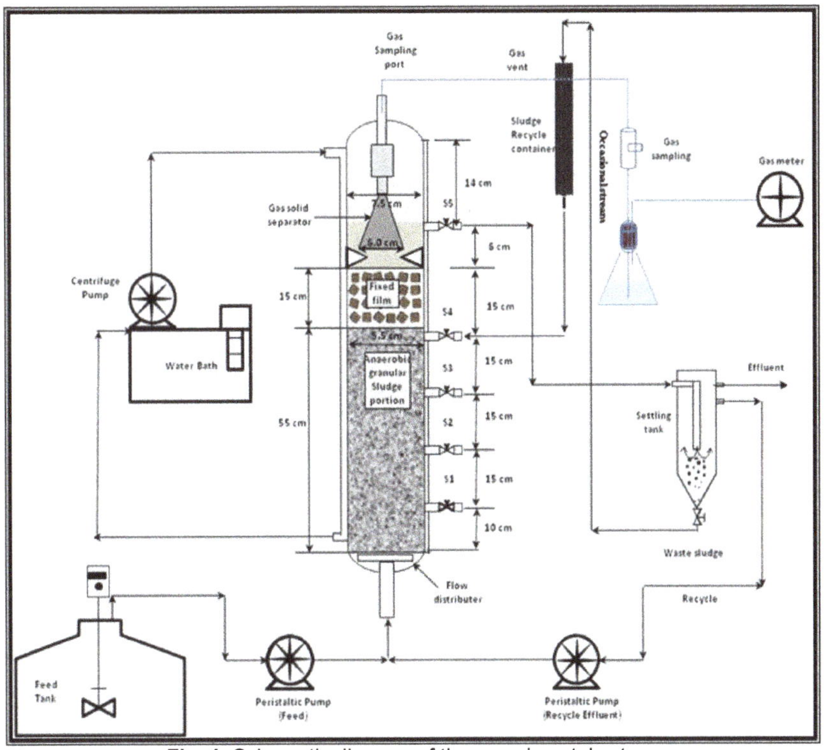

Fig. 1. Schematic diagram of the experimental set-up.

2.2. Tracer experiments procedure

In order to carry out the analysis of the hydrodynamic process within the UASB-FF, the reactor was filled with water till the top level of the outlet. Then, pulses of an inert substance was introduced into the reactor inflow and subsequently measured at the outflow. To study the flow pattern in the reactor, a technique involving analysis of tracer-response profiles was used. In this tracer study, Rhodamine B is introduced into the reactor at the influent end as a tracer and collected the tracer-output profile in the fluid leaving the reactor. Rhodamine B was used as a tracer because this substance will not be absorbed on or reacts with the exposed reactor surface, and can be detected at very low concentration using a spectrophotometric method. In each run, the influent of water into the reactor was adjusted and fed by a peristaltic pump. Subsequently, 2.5 ml of 0.01 M Rhodamine B was immediately fed into the inlet of the reactor. Effluent samples were collected immediately and then at every 5 minutes, or at even shorter time intervals if required. The collected samples were subjected to spectrophotometric reading at 554 nm. The operation was continued until no tracer was detected for at least two or three consecutive samples.

2.3. Experimental design

Up-flow velocity (V_{up}) and biogas production rate (Q_g) are two important hydraulic factors and can impact significantly on the hydraulic regime of the UASB-FF bioreactor. In this study, V_{up} and Q_g were chosen as the most critical operating factors for the hydraulic regime of the UASB-FF bioreactor. The region of exploration for the hydraulic regime was decided as the area enclosed by V_{up} (0.5 and 3.0 m/h) and Q_g (14.87 and 7.96 l/d). Selection of the values of the Qg was based on the results obtained from the earlier studies. In order to carry out a comprehensive analysis on the hydraulic regime, three dependent parameters were calculated as response. These parameters were deviated from ideal retention time ($\Delta\tau$), volume of dead space, and Morrill dispersion index (MDI).

3. Results and discussion
3.1. Hydraulic performance analysis

The hydraulic performance of the UASB-FF bioreactor is analyzed by studying water flow patterns or hydraulic residence time distributions (RTD) obtained from the tracer experiments. Fig. (2a-b) represents the modeled and experimental data curves of tracer concentration time distribution at different V_{up} (0.5 and 3.0 m/h) and Q_g (14.87 and 7.96 l/d). The ideal flow concentrations were calculated based on the equation (1):

$$C, \text{Ideal}, (\text{mmol}) = C_0 \times \text{Exp}(-\frac{t}{\tau}) \tag{1}$$

Where, C is the ideal flow concentration (mM), C_0 is the initial flow concentration of the tracer (mM), t is sampling time (min), and т is theoretical retention time (min).

The figure makes it clear that the increase in V_{up} resulted in pronounced deviation from ideal flow pattern. It can be seen that the deviation obtained at V_{up} of 0.5 m/h and Q_g of 14.87 l/d was less than the deviation obtained at V_{up} of 3.0 m/h and Q_g of 7.96 l/d. As depicted in Fig. 2b, high volume of the tracer arrives at the outlet before complete mixing with the bulk of the liquid in the reactor. It could be due to the short cut phenomenon occurred originated from the reactor geometry, inlet and outlet design and inadequate mixing (see next section). If the concentration versus time tracer response curve is defined by a series of discrete time step measurements, the theoretical mean residence time is typically approximated as:

$$\bar{t}_{\Delta c} \approx \frac{\sum t_i C_i \Delta t_i}{\sum C_i \Delta t_i} \tag{2}$$

Where, $\bar{t}_{\Delta c}$ is the mean detention time based on discrete time step measurements, (min), t_i is the time at ith measurement, (min), C_i is the concentration at ith measurement, (mmol/l), and Δt_i is the time increment about C_i, (min).

(a)

(b)

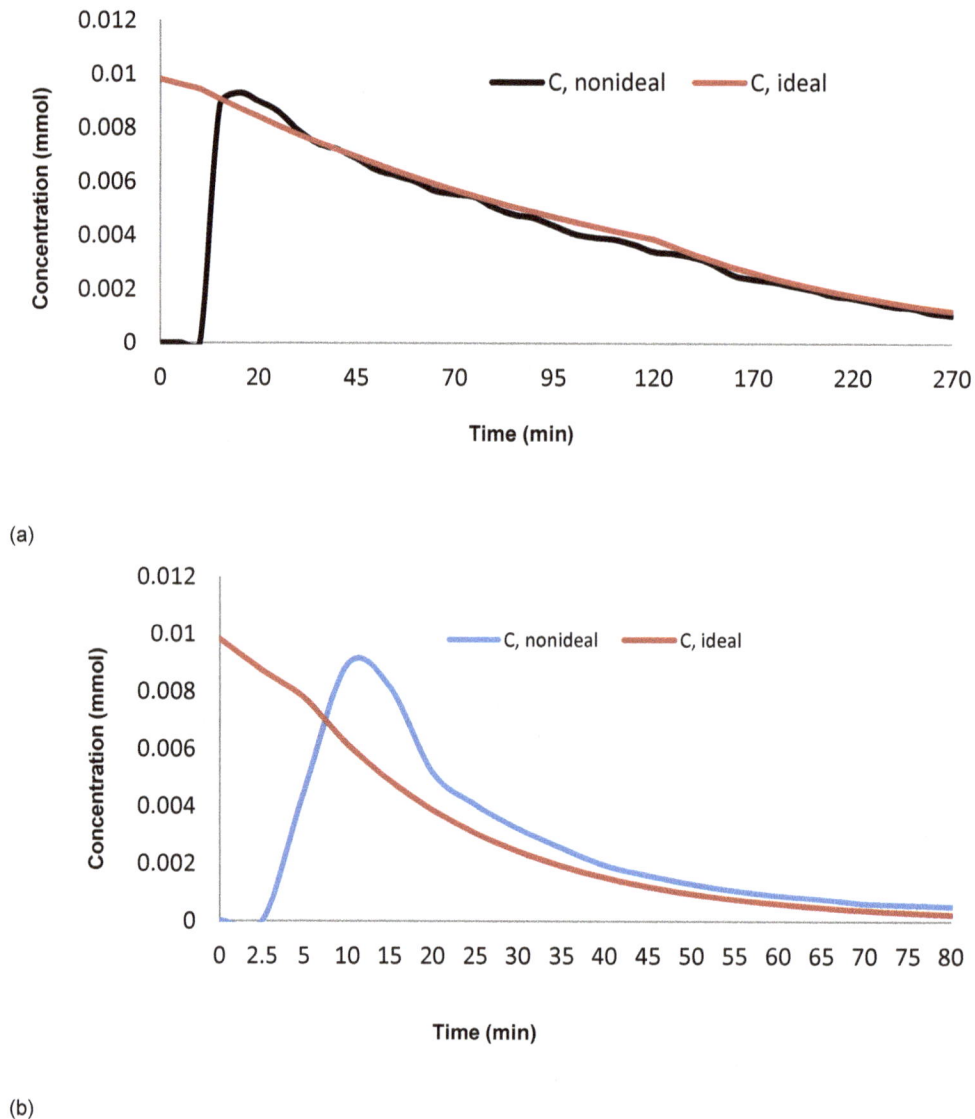

Fig. 2. Mathematical and empirical curves of tracer concentration time distribution for (a) V_{up}=0.5 m/h and Q_g=14.87l/d, (b) V_{up}=3.0 m/h and Q_g=7.96 l/d.

3.2. Statistical analysis
3.2.1. Deviation from ideal retention time (Δτ)

Deviations from the ideal retention time (Δτ) can be caused by a number of factors viz. channeling and/or recycling of fluid, short-circuiting, creation of stagnant regions in the vessel, and etc. In the short-circuiting, a portion of the flow that enters the reactor during a given time period arrives at the outlet before the bulk of the flow that entered the reactor during the same time period arrives. This non-ideal flow may be caused by density currents due to the temperature difference, wind-driven circulation pattern, inadequate mixing and poor design (Metcalf & Eddy. 2003). Ultimately, the incomplete use of the reactor volume due to above reasons can result in the decreased Δτ and reduced treatment performance. In this study the values of Δτ were obtained at 33.58 and 4.15 min for V_{up} of 0.5 m/h and Q_g of 14.87 l/d and V_{up} of 3.0 m/h and Q_g of 7.96 l/d, respectively. From the results, increasing the V_{up} (τ from 21.5 to 128.8 min) resulted in the increase in Δτ (4.15 to 33.58 min). It was attributed to the relatively low up-flow velocity of the liquid within the reactor that may create dead space and inadequate mixing. Inadequate input energy for mixing, non-mixing of some portions of the reactor contents with the incoming water and poor design of the reactor at inlet zone cause dead zones to develop within the reactor or allow short circuiting to occur (Metcalf & Eddy. 2003). To elucidate the Δτ, the volume of dead zone was also obtained as a response (Eq. 3).

$$\Delta\tau = \left(\frac{\overline{t}_{\Delta c\_\theta}}{\theta}\right) \times 1 \tag{3}$$

In general, the long tail in the RTD curve shows a discrepancy between the mean residence time and the theoretical residence time. It is possible that stagnant hydraulic zones exist near the inlet and outlet parts of the reactor or in the fixed film zone, where tracer can be trapped and slowly released. It must be noted that, in this UASB-FF reactor, the liquid volume within the fixed film part was 17 % of the total liquid volume, likely to promote a relatively high dead volume. The values observed of the response (Δτ) were 26 % and 19.3 % of the total reactor volume at V_{up} of 0.5 m/h and Q_g of 14.87 l/d and V_{up} of 3.0 m/h and Q_g of 7.96 l/d, respectively.

3.2.2. Morrill dispersion index (MDI)

One important factor that indicates the type of the reactor (complete mix or plug-flow) is Morrill Dispersion Index (MDI). MDI can be used as an indicative tool for determining features of flow patterns in reactors. These consist of the regions of stagnant fluid (dead space) and/or the possibilities of bypassing. A ratio of 90 to 10 percent values from the cumulative tracer concentration curve could be used as a measure of the dispersion index. The value for an ideal plug-flow reactor is 1.0 and about 22 for a complete-mix reactor. MDI was computed for both operational conditions studied (Metcalf & Eddy. 2003).

Morrill dispersion index,

$$MDI = \frac{P_{90}}{P_{10}} \qquad (4)$$

Where P90 is 90 percentile value and P10 is 10 percentile value from log-probability plot of the cumulative tracer concentration curve. The values of MDI computed at the minimum and maximum V_{up} and Q_g are identified as 11.33 and 10, respectively, showing that the hydraulic regime in UASB-FF bioreactor is a semi-complete mixing.

In order to further analyze the residence time distribution of the fluid in a reactor the E(t) and F(t) curves have been developed in Fig. 3. Fluid elements may require differing lengths of time to travel through the reactor. The distribution of the exit times, defined as the E(t) curve, is the residence time distribution (RTD) of the fluid. The exit concentration of a tracer species C(t) can be used to define E(t). That is:

$$E(t) = \frac{C(t)}{\int_0^\infty C(t)dt} \qquad (5)$$

Such that:

$$\int_0^\infty E(t)dt = 1 \qquad (6)$$

The F curve is defined as

$$F(t) = \int_0^t E(t)dt \qquad (7)$$

where, F(t) is the cumulative residence time distribution function. As shown in the equation (7), the F(t) curve is the integral of the E(t) curve while the E(t) curve is the derivative of F(t) curve. In the fact, F(t) represents the amount of tracer that has been in the reactor for less than the time (t). As can be observed in the Fig. 3, retention time distribution (RTD) function, E(t) and F(t) showed a higher initial values at V_{up} of 3.0 m/h compared to the values at the V_{up} of 0.5 m/h. The interaction showed that up-flow velocity and gas flow rate played an important role in the MDI of the reactor.

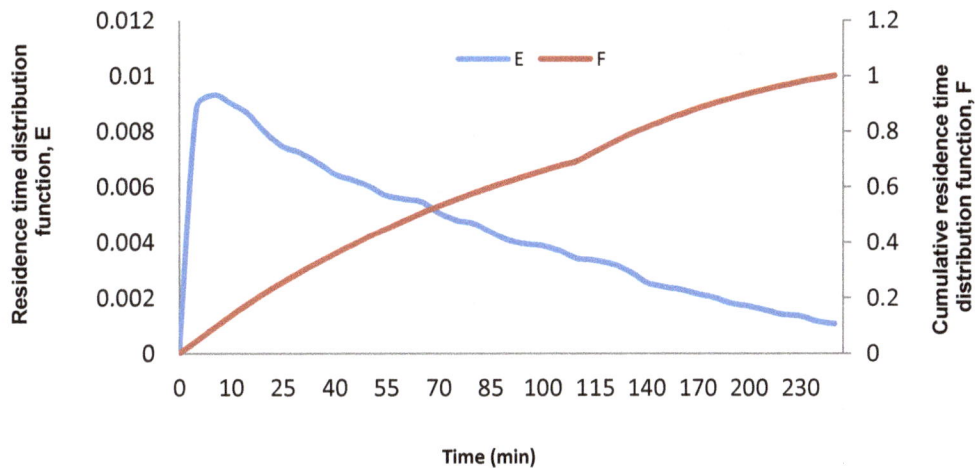

(a)

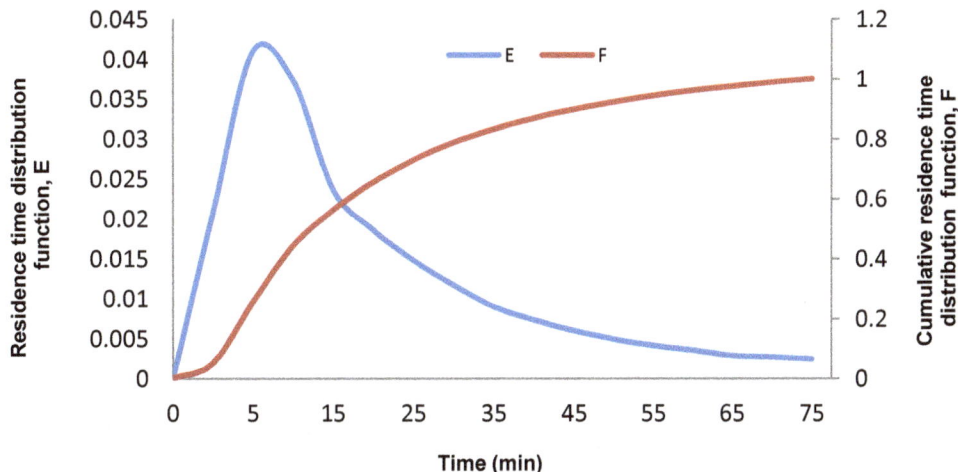

(b)

Fig. 3. E(t) and F(t) plots with respect to
(a) V_{up}=0.5 m/h and Q_g=14.87l/d, (b) V_{up}=3.0 m/h and Q_g=7.96 l/d.

4. Conclusion

The hydraulic characteristic of an up-flow anaerobic sludge blanket fixed film (UASB-FF) at various levels of the Up-flow velocity (V_{up}) and biogas production rate (Q_g) were investigated. The liquid volume within the fixed film part likely promotes a relatively high dead volume. V_{up} and Q_g were found to be influential on the deviation from ideal retention time ($\Delta \tau$). Increasing in the Q_g resulted an increase in $\Delta \tau$ while a reverse effect caused by increasing the up-flow velocity. The volume of dead space increased upon increasing the Q_g and decreasing the V_{up}. The values of MDI computed at various levels of the variables showed that the hydraulic regime in UASB-FF bioreactor is a semi-complete mixing.

References

Chen X., Zheng P., Guo Y., Mahmood Q., Tang C., Ding S., Flow patterns of super high-rate anaerobic bioreactor, Bioresource Technology 101 (2010) 7731–7735.

Fogler Scott H., Elements of chemical reaction engineering, Third Ed., Prentice Hall PTR, (2001).

Kargi F., Eker S., Performance of rotating perforated tubes biofilm reactor in biological wastewater treatment, Enzyme Microbial Technology 32 (2003) 464-471.

Levenspiel O., Chemical reactor engineering, Second Ed., Wiley, New York, (2000).

Mansouri Y., Zinatizadeh A.A., Mohammadi P., Irandoust M., Akhbari A., Davoodi R., Hydraulic characteristics analysis of an anaerobic rotatory biological contactor (AnRBC) using tracer experiments and response surface methodology (RSM), Korean Journal of Chemical Engineering 29 (2012) 891-902.

Martin A.D., Interpretation of residence time distribution data, Journal of Chemical Engineering Science 55 (2000) 5907-5911.

Metcalf & Eddy, Wastewater engineering, Fourth Ed., McGraw Hill, New York, (2003).

Najafpour G.D., Zinatizadeh A.A.L., Mohamed A.R., Isa M.H., Nasrollahzadeh H., High-rate anaerobic digestion of palm oil mill effluent in an upflow anaerobic sludge-fixed film bioreactor, Process Biochemistry 41 (2006) 370-379.

Newell B., Bailey J., Islam A., Hopkins L.P., Lant J., Characterizing bioreactor mixing with residence time distribution (RTD) tests, Water Science and Technology 37 (1998) 43-47.

Torkian A., Hashemian S.J., Alinejad K., Post treatment of Upflow Anaerobic Sludge Blanket-Treated Industrial Wastewater by a Rotating Biological Contactor, Water Environmental Research 75 (2003) 232-237.

Williams S.C., Beresford J., The effect of anaerobic zone mixing on the performance of a three-stage bardenpho plant, Water Science and Technology 38 (1998) 55-61.

Yamaguchi T., Ishida M., Suzuki T., Biodegradation of hydrocarbon by Prototheca zopfii in rotating biological contactors, Process Biochemistry 35 (1999) 403-409.

8

Removal, preconcentration and determination of methyl red in water samples using silica coated magnetic nanoparticles

Masoud Shariati-Rad[1,*], Mohsen Irandoust[1], Somayyeh Amri[1], Mostafa Feyzi[2], Fattaneh Ja'fari[2]

[1]Department of Analytical Chemistry, Faculty of Chemistry, Razi University, Kermanshah, Iran.
[2]Department of Physical Chemistry, Faculty of Chemistry, Razi University, Kermanshah, Iran.

ARTICLE INFO

ABSTRACT

Keywords:
Methyl red
Magnetic nanoparticles
Central composite design
Removal
Preconcentration

A method was developed for removal, preconcentration and spectrophotometric determination of trace amounts of methyl red based on SiO_2-coated Fe_3O_4 magnetic nanoparticles. The influence of pH, dosage of adsorbent and contact time on the adsorption of dye was explored by central composite design. The kinetic data were analyzed based on the Langmuir and Freundlich adsorption isotherms. The Langmuir model was fitted well to data and the maximum monolayer capacity q max of 49.50 mg/g was calculated. The results showed that desorption efficiencies of higher than 99 % can be achieved in a short contact time of 3 min and in one step elution using 2.0 mL of 0.1 mol L^{-1} NaOH. The magnetic nanoparticles were washed with deionized water and reused for two successive removal processes with removal efficiencies more than 90%. Then desorbed dye was determined spectrophotometrically. The calibration curve was linear in the range of 0.025–0.250 mg/l of dye with a correlation coefficient of 0.9922. The relative standard deviations obtained upon application of the method to the real samples were lower than 0.7 %. A preconcentration factor of the method was 50.

---

1. Introduction

Dye removal from wastes has been the object of many researches in the past few years because of the potential toxicity of dyes and visibility problems (Afkhami et al. 2010). These compounds are used in large quantity in many industries including textile, leather, cosmetics, paper, printing, plastic, pharmaceutical and food to color their products (Shariati et al. 2011). As a result, considerable amounts of colored wastewater are generated (Afkhami and Moosavi 2010). Many of the industrial dyes are toxic, carcinogenic, mutagenic and teratogenic. Their removal from wastewater is of great interest (Qadri et al. 2009).

The methods used to remove organic dyes and pigments from wastewaters are classified into three main categories: (i) physical (adsorption, filtration, and flotation) (Afkhami and Moosavi 2010; Qadri et al. 2009; Kannan and Sundaram 2001; Wang et al. 2006), (ii) chemical (oxidation, reduction, and electrochemical) (Arslan and Balcioglu 2001; Tsui and Chu 2001; Gutierrez et al. 2002) and (iii) biological (aerobic and anaerobic degradation) (Stolz 2001; Bell and Buckley 2003; Haghighi-Podeh et al. 2001; Kapdan and Ozturk 2005). Two most available technologies for dye removal are oxidation and adsorption. Oxidation methods are probably the best technologies to eliminate organic carbons completely but they are only effective for wastewaters with very low concentrations of organic compounds (Sun and Xu 1997). Adsorption has been found to be superior to the other techniques for removal of colors, odor, oils and organic pollutants from process or waste effluent treatments in terms of initial cost, simplicity of design and ease of operation (Juang et al. 2002). Because of its capability for efficient adsorbing of a broad range of compounds, the most efficient adsorbing of a broad range of compounds, the most commonly used adsorbent for color removal is activated carbon (Afkhami et al. 2010).

The main disadvantage of the activated carbon is its high production and treatment costs (Afkhami et al. 2010). Recently, numerous approaches have been studied for the development of alternative effective adsorbents. Some of the reported sorbents include clay materials, zeolites, siliceous material, agricultural wastes, industrial waste products and biosorbents such as chitosan and peat (Mirsha and Bajpai 2006; Aleboyeh and Aleboyeh 2006).

Magnetic nanoparticles (MNPs) have been recognized as efficient adsorbents with large specific surface area and small diffusion resistance. Moreover, the magnetic separation provides suitable route for online separation (Qadri et al. 2009).

Iron oxide MNPs are superparamagnetic. This means that when they have been adhered to the target compounds, they can quickly be removed along with them from a matrix using a magnetic field.

In the past decade, the synthesis of spinel magnetite and maghemite nanoparticles has been intensively developed not only for their great fundamental scientific interest but also for many technological applications. These applications include in biology such as extraction of genomic DNA (Xie et al. 2004), contrast agents in magnetic resonance imaging (MRI) (Bulte 2006), medical applications (such as targeted drug delivery) (Laurent et al. 2008), bioseparation (Bucak et al. 2003), separation and preconcentration of various anions and cations (White et al. 2009; Zhou et al. 2009; Tuutijarvi et al. 2009). Silica has been considered as one of the most ideal shell materials due to its chemical stability and versatility in surface modification via Si–OH groups (Santhi et al. 2010).

Application of experimental design in decolorization processes has been reported. It has been proven that it is a powerful tool for the optimization of degradation (Torrades and Garcia-Montano, 2014; Sahoo and Gupta 2012; Zuorro et al. 2013) or adsorption (de Sales et al. 2013; Ravikumar et al. 2006; Singh et al. 2011; Gomez and Pilar Callao 2008) of different dyes.

*Corresponding author E-mail: mshariati_rad@yahoo.com

In this study, SiO_2-coated Fe_3O_4 magnetic nanoparticles were synthesized and employed for removal, preconcentration and determination of methyl red in water samples using experimental design and spectrophotometry. It is the first report on the application of SiO_2-coated Fe_3O_4 magnetic nanoparticles and experimental design for preconcentration, determination and removal of methyl red in environmental water samples. The kinetics of adsorption of methyl red onto the SiO_2-coated Fe_3O_4 magnetic nanoparticles was investigated. Additionally, the recovery of the dye from the nanoparticles using different solvents is described.

2. Experimental
2.1. Reagents and materials

All the chemicals and reagents used in this work were of analytical grade. Iron nitrate $Fe(NO_3)_3.9H_2O$ (99%), tetra ethoxysilane (TEOS) (98%) and oxalic acid $H_2C_2O_4.2H_2O$ were purchased from Merck (Darmstadt, Germany). Structural formula of methyl red has been shown in Scheme 1. Double distilled water was used throughout the study. The stock 500 mg/l solution of methyl red was prepared in double distilled water and experimental solutions of the desired concentrations were obtained by successive dilutions of the stock solution with double distilled water. The initial pH was adjusted with 0.1 mol L^{-1} solutions of HCl or NaOH. All the adsorption experiments were carried out at room temperature.

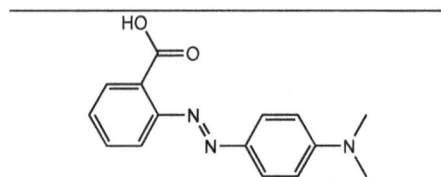

Scheme 1. Molecular structure of methyl red.

2.2. Instrumentation

An Agilent model 8453 spectrophotometer with diode array detector was used for recording spectra. A Jenway 3345 ion-meter was used for pH measurements.

2.3. Dye removal experiments

Batch-mode adsorption studies were carried out by adding 10 mg adsorbent and 10 mL dye solution of known concentration (2.5×10^{-5} mol L^{-1}) in beaker. pH of the solutions was adjusted to the desired value. The mixture solutions were shacked for appropriate adsorption time at 25 °C. After dye adsorption, SiO_2-coated Fe_3O_4 magnetic nanoparticles were quickly separated from the sample solution using a magnet. The following equation was applied to calculate the dye removal efficiency in the treatment experiments:

$$R\% = (Ci - Cr)/Ci \times 100 \qquad (1)$$

where Ci and Cr are the initial and residual concentrations of the dye in the solution, respectively.

2.4. Synthesis of SiO_2-coated Fe_3O_4 magnetic nanoparticles

The SiO_2-coated Fe_3O_4 magnetic nanoparticles were prepared using sol-gel method. Appropriate amounts of iron nitrate ($Fe(NO_3)_3.9H_2O$), tetra ethoxysilane (TEOS) and oxalic acid ($H_2C_2O_4.2H_2O$) were separately dissolved in ethanol. The three solutions were heated up to 50 °C and stirred for 20 min. The TEOS was added to the iron nitrate followed by oxalic acid addition under strong stirring at 60°C for 2 h. The precipitate composed of iron oxalate and TEOS was progressively hydrolyzed by the hydration water of iron nitrate and mainly oxalic acid, according to the following scheme:

$$Si(OC_2H_5)_4 + 4H_2O \rightarrow Si(OH)_4 + 4C_2H_5OH \qquad (2)$$

In the acidic condition (pH≈1), $Si(OH)_4$ is condensed with other materials to a homogeneous gel. Then, the monolithic gel was dried at 110 °C in vacuum for 16 h. Finally, the dried powder was calcined (450 °C for 6 h) to produce solid magnetic composite.

2.5. Sample characterization
2.5.1. X-Ray diffraction (XRD)

The XRD patterns of all the precursor and calcined samples were recorded on a Philips X'Pert (40 kV, 30 mA) X-ray diffractometer, using Kα radiation of Cu as source (λ=1.542 Å) and a nickel filter in the 2θ range of 4°-70°.

2.5.2. N_2-adsorption-desorption measurements

Using BET (Brunauer, Emmett and Teller sorption isotherm) and BJH (Barrett–Joyner–Halenda method) methods, the specific surface area, the total pore volume and the mean pore diameter were measured. For these purposes, N_2 adsorption-desorption isotherm is used at liquid nitrogen temperature (-196 ºC) using a NOVA 2200 instrument (Quantachrome, USA). Prior to the adsorption-desorption measurements, all the samples were degassed at 110 ºC in a N_2 flow for 3 h to remove the moisture and other adsorbates.

2.5.3. Scanning electron microscopy (SEM)

The morphologies of the prepared nanoparticles and their precursors were observed by means of an EM-3200 scanning electron microscope (KYKY Technology Development Ltd.).

3. Results and discussion
3.1.1. Characterization of the SiO_2-coated Fe_3O_4 magnetic nanoparticles

Characterization studies were carried out using XRD and SEM techniques. The XRD pattern of the synthesized magnetic nanoparticles is shown in Fig. 1. The actual identified phases for this sample were Fe_2SiO_4 (cubic). Furthermore, the crystallite sizes of the synthesized sample were calculated from the major diffraction peaks using the Debye–Scherrer equation (Klug and Alexander, 1974):

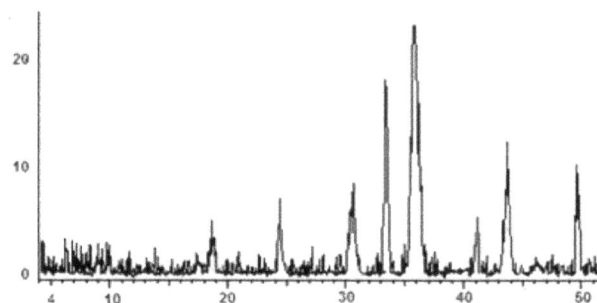

Fig. 1. XRD patterns of the SiO_2-coated Fe_3O_4 magneticnanoparticles.

$$D_C = K\lambda/\beta Cos\theta \qquad (3)$$

where β is the breadth of the observed diffraction line at its half intensity maximum, K is the so-called shape factor which usually takes a value of about 0.9 and λ is the wavelength of the X-ray source used in the XRD. The crystallite size (Dc) of the SiO_2-coated Fe_3O_4 magnetic nanoparticles was calculated to be 48 nm using the above equation.

The XRD technique may not be sufficiently sensitive to reveal the fine details of these changes. To get this, a detailed SEM study of both precursor and calcined SiO_2-coated Fe_3O_4 magnetic nanoparticles was done and the results are given in Fig. 2. SEM observations show differences in morphology of the precursor and calcined magnetic nanoparticles. The image obtained from the precursor depicts several larger agglomerations of particles (Fig. 2a) and shows that this material has a less dense and homogeneous morphology. After calcination at 450 °C for 6 h and heating rate of 3 °C min^{-1}, the morphological features became different from the precursor sample and the agglomerate size reduced greatly (Fig. 2b). It may be attributed to the covering of calcined magnetic nanoparticle surface by small crystallite of SiO_2, in agreement with XRD results.

Fig. 2. The SEM image of SiO_2-coated Fe_3O_4 magnetic nanoparticles (a) precursor and (b) calcined sample.

3.1.2. Central composite experimental design for optimization of the parameters

The experimental design technique commonly used for process analysis and modeling is central composite design (CCD) (Gunaraj and Murugan 1999; Box and Hunter 1957). Experimental design methodology involves changing all variables from one experiment to the next, simultaneously. The reason for this is that variables can influence each other and the ideal value for one of them can depend on the values of the others. In this work, we performed CCD. It is assumed that the central point for each factor is 0, and the design is symmetric around this (Brereton 2003).

CCD including the factors, their levels, and the result of each experiment are shown in Table1. Concentration of dye used in these experiments is 2.5×10^{-5} mol L^{-1}.

Table 1. Central composite design and the results of experiments.

Factor	Level		
	1	0	-1
t(min)	120	70	20
mg MNPs	10	5.5	1
pH	5	3	1

Central composite design			
Run order	t (min)	mg MNPs	pH	Removal (%)
1	120	10	1	33.6
2	70	5.5	3	51.2
3	120	10	5	89.27
4	70	10	3	64.64
5	70	5.5	5	73.17
6	20	10	5	82.4
7	20	5.5	3	48
8	70	5.5	3	58.22
9	120	1	5	67.7
10	70	5.5	3	58.4
11	70	1	3	26
12	120	5.5	3	61.6
13	20	1	5	57.6
14	70	5.5	1	9.2
15	70	5.5	3	51.36
16	120	1	1	9.2
17	70	5.5	3	59.52
18	20	1	1	6.8
19	70	5.5	3	58
20	20	10	1	27.2

Analysis of variance for the results of CCD has been given inTable2. As shown in Table 2, pH and the amount of SiO_2-coated Fe_3O_4magnetic nanoparticles (*mg* MNP) are significant factors in the removal of methyl red at 95% confidence level (calculated *p* values for these factors are smaller than 0.05). The pH of the system exerts profound influence on the adsorptive uptake of the dye presumably due to its influence on the surface properties of the adsorbent and ionization/dissociation of the dye. Very low value of the coefficient for *t* in the model indicates that time is not an important factor in the adsorption of methyl red on the SiO_2-coated Fe_3O_4 magnetic nanoparticles. This was experimentally observed.

The solutions of methyl red were rapidly decolored in contact with SiO_2-coated Fe_3O_4 magnetic nanoparticles. Among the squared and interaction terms, pH×pH is statistically important based on the *p* values. The *F* value of the regression is relatively high (with $F = 31.32$ and $p = 0$). This indicates the importance of the regression.

Table 2. Analysis of variance of the experiments in Table 1.

Term	Coefficient	t^a	p^b
Constant	-23.478	-2.39	0.038
T	-0.240	-1.07	0.309
mgMNP	5.168	2.30	0.045
pH	26.143	4.43	0.001
t ×t	0.002	1.40	0.193
mgMNP × mgMNP	-0.216	-1.20	0.258
pH× pH	-2.128	-2.33	0.042
t × mgMNP	0.000	0.04	0.965
t × pH	0.010	0.48	0.644
mgMNP × pH	0.022	0.09	0.929
Regression			
R^2 (%)	96.60		
F	31.32		

[a] Statistical t value.
[b] Probability value.

Order to gain insight about the effect of each variable, the three dimensional (3D) graphs for the responses were plotted based on the model polynomial function to analyze the variation in the response surface as shown in Fig. 3. These figures show the relationship between two variables and response (Removal %) at center level of the other variables.

It can be seen that in the solutions with weak acidity, the response is higher. This is more evident in Fig. 3c. Moreover, the removal of dye increases with the amount of SiO_2-coated Fe_3O_4 magnetic nanoparticles (mgMNP) and *t*.

In the next step, response surface optimization was used to explore the optimum conditions of the factors. Response optimization showed that the Removal% will be maximum at t = 120 min, mgMNP = 10.0 mg and pH = 5.0.

3.1.3. Mechanism of the interaction

The electrostatic interaction between methyl red and SiO_2-coated Fe_3O_4 magnetic nanoparticles is influenced by solutions pH. Partial ionization of Si-OH could start at low pH values which make the surface of the SiO_2-coated Fe_3O_4 magnetic nanoparticles negatively charged. Methyl red is positively charged at low pH values which favor the electrostatic interaction between SiO_2-coated Fe_3O_4 magnetic nanoparticles and methyl red. Increasing pH causes Si-OH to ionize which provides more electrostatic attraction sites for methyl red.

3.1.4. Study of the kinetics of adsorption

Study of the kinetics of dye adsorption onto SiO_2-coated Fe_3O_4magnetic nanoparticles is required for selection of the optimum operating conditions for the full-scale batch processes. The kinetic parameters which are helpful for the prediction of the adsorption rate give important information for designing and modeling of the adsorption processes (Afkhami and Moosavi 2010). Kinetic studies were performed in a 15 mL glass beaker where 10 mg of SiO_2-coated Fe_3O_4 magnetic nanoparticles was added to 10 mL of the dye solution with different concentrations ranging between 10 mg/l and 30 mg/l at room temperature and at pH 5.0. This was followed by shaking by a shaker at 250 rpm to ensure equilibrium is reached. At time t = 0 and equilibrium, dye concentrations were measured by UV–Vis spectrophotometry at 521 nm. The amount of adsorption at equilibrium, qt(mg/g), was calculated by:

$$qt = (Ci - Ct)V \qquad (4)$$

where C_i and C_t (in mg/l) are concentrations of dye in liquid phase at t = 0 and equilibrium after time t of incubation, respectively. V is the volume of the solution (in L) and W is the mass of dry SiO_2-coated Fe_3O_4magnetic nanoparticles used (in g).

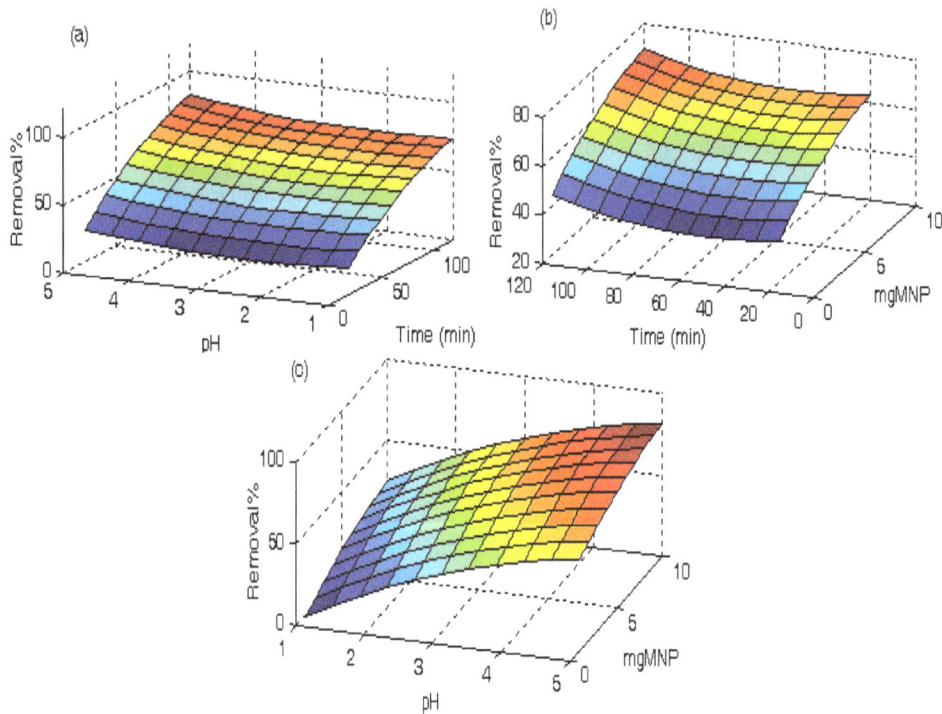

Fig. 3. Variation of response surfaces with pH and t (a), t and amount of magnetic nanoparticle (mgMNP) and (c) pH and amount of magnetic nanoparticle (mgMNP).

The removal rate was very fast during the initial stages of the adsorption process. The kinetic data for adsorption of the dye onto SiO_2-coated Fe_3O_4magnetic nanoparticles were analyzed using pseudo-second order model to find out the adsorption rate expression. The kinetics of adsorption was identified to be a pseudo-second order model. The sorption kinetics for all the initial dye concentrations was treated by Ho's pseudo-second-order rate equation (Chien and Clayton 1980):

$$\frac{t}{qt} = \frac{1}{k_2 qe^2} + \left(\frac{1}{qe}\right) \qquad (5)$$

where qt and qe are the amounts of adsorbed dye at each time and at equilibrium, respectively. k_2 is the pseudo-second order rate constant. Fitting of the pseudo-second-order kinetic model to the kinetic data is shown in Fig. 4. The pseudo-second-order rate equation constants for all of the initial concentrations used in the experiments are shown in Table 3.

Equilibrium isotherm equations are used to describe the experimental sorption data. The parameters obtained from different models provide important information on the sorption mechanisms, the surface properties and affinities of the sorbent (Shariati et al.

2011). The equilibrium adsorption isotherm model which is the number of mg adsorbed of dye per g of adsorbent (qe) versus the equilibrium concentration of adsorbate is fundamental in describing the interactive behavior between adsorbate and adsorbent (Afkhami et al. 2010). Since, more common models used to investigate the adsorption isotherm are Langmuir and Freundlich equations, these two models were fitted to the experimental data (Qadri et al. 2009).

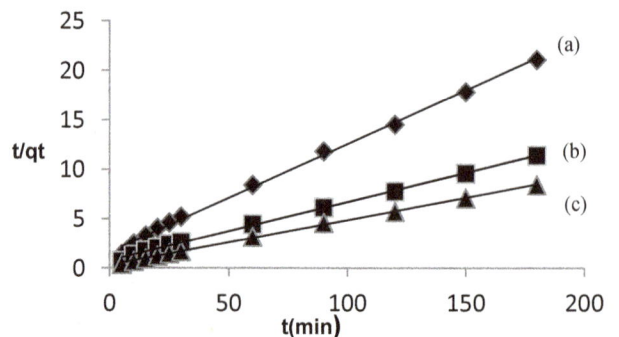

Fig. 4. Kinetics of adsorption based on the pseudo-second-order

kinetic model for initial dye concentrations of (a) 10, (b) 20 and (c) 30   mg/l.

Table 3. Values of the pseudo-second-order rate equation parameters in different initial concentration of dye.

Initial concentration(mg/l)	Equation	qe (mg/g)	k (g mg^{-1} min^{-1})	R^2
10	t/qt = 0.1088 t + 1.6663	9.19±0.16	0.0072±0.0008	0.9973
20	t/qt = 0.0591 t + 0.7963	16.92±0.24	0.0044±0.0000	0.9982
30	t/qt = 0.0454 t + 0.3364	22.03±0.23	0.0062±0.0007	0.9991

Langmuir's model does not take into account the variation in the adsorption energy, but it is the simplest description of the adsorption process. It is based on the physical hypothesis that the maximum adsorption capacity consists of a monolayer adsorption, there are no interactions between adsorbed molecules and the adsorption energy is distributed homogeneously over the entire coverage surface (Afkhami and Moosavi 2010). The equilibrium adsorption isotherm was determined using batch studies with different initial concentrations of methyl red (10–100 mg/l) at 25 °C and at pH 5.0.

The linearized form of the Langmuir isotherm, assuming monolayer adsorption on a homogeneous adsorbent surface, is expressed as (Langmuir 1918):

$$\frac{Ce}{qe} = \frac{1}{K_L q_{max}} + \left(\frac{1}{q_{max}}\right) \qquad (6)$$

where q_{max} (in mg/g) is the maximum amount of the adsorbed dye corresponding to the complete monolayer coverage and illustrates the maximum value of qe that can be attained as Ce increases. K_L(in L mg^{-1}) is the Langmuir adsorption equilibrium constant related to the energy of adsorption. Values of q_{max} and b (K_L/q_{max}) are determined from the linear regression plot of (Ce/qe) versus Ce.

The Freundlich isotherm model is an empirical equation that describes the surface heterogeneity of the sorbent. It considers multilayer adsorption with a heterogeneous energetic distribution of active sites accompanied by interactions between adsorbed molecules (Chatterjee et al. 2009). The linear form of the Freundlich isotherm is:

$$Ln(qe) = LnK_f + \frac{1}{n}Ln\ (Ce) \qquad (7)$$

where Ce is the equilibrium concentration (in mg/l), qe is the amount adsorbed at equilibrium (mg/g) and finally, K_f (in (mg/g) $(mg/l)^n$) and $1/n$ are Freundlich constants depending on the temperature and the given adsorbent–adsorbate couple. n is related to the adsorption energy distribution and K_f indicates the adsorption capacity. The values of K_f and $1/n$ can be calculated by the plotting Ln(qe) versus Ln(Ce). The intercept of the resulted line is Ln (K_f) and $1/n$ is its slope. Value of $1/n$ indicates that the adsorption intensity of dye onto the adsorbent or surface heterogeneity becomes more heterogeneous as its value gets closer to zero. A value for $1/n$ below 1 indicates a normal Langmuir isotherm while $1/n$ above 1 is indicative of the cooperative adsorption (Santhi et al. 2010).

The calculated parameters of the Langmuir and Freundlich isotherms and the correlation coefficients (r) are listed in Table 4. Table 4shows that the Langmuir isotherm equation is better fitted to experimental data (r is higher relative to r for fitting of Frendlich equation to data). It is also evident from these data that the surface of the SiO_2-coated Fe_3O_4magnetic nanoparticles is made up of homogenous adsorption patches than heterogeneous adsorption patches (Faraji et al. 2010). It is generally accepted that under a constant temperature, the n values increase with decreasing adsorption energy. This implies that the larger the n value, the stronger the adsorption intensity (Belessi et al. 2009). Values of n> 1 represent favorable adsorption conditions. In most cases, the exponent between 1<n< 10 shows beneficial adsorption (Afkhami and Moosavi 2010).

The essential feature of the Langmuir isotherm can be expressed in terms of a dimensionless constant separation factor (R_L) given by the following equation (Afkhami and Moosavi 2010):

$$R_L = \frac{1}{1 + a_L C_0} \qquad (8)$$

where a_L parameter is a coefficient related to the energy of the adsorption and increases by increasing the strength of the adsorption bond. The adsorption process can be defined as irreversible ($R_L = 0$), favorable ($0 < R_L < 1$), linear ($R_L = 1$) or unfavorable ($R_L > 1$) in terms of R_L (Afkhami et al. 2010). The calculated value of R_L for adsorption of 100 mg/l solution of methyl red is 0.806. This it is between 0 and 1, thus the adsorption of the dye onto SiO_2-coated Fe_3O_4 magnetic nanoparticles is favorable.

3.2. Analytical studies
3.2.1. Desorption and regeneration

Adsorption of methyl red onto the SiO_2-coated Fe_3O_4magnetic nanoparticles is a reversible process. Therefore, regeneration or activation of the SiO_2-coated Fe_3O_4magnetic nanoparticles to reuse is possible. Desorption of the dye from the SiO_2-coated Fe_3O_4magnetic nanoparticles was studied using different kinds of solvents. Desorption process was performed by mixing 0.01 g methyl red loaded adsorbent with a 2.0 mL volume of EtOH, pure acetic acid, HCl and NaOH solutions with concentrations of 0.1 mol L^{-1}. SiO_2-coated Fe_3O_4magnetic nanoparticles were collected magnetically from the solution. Concentration of dye in the desorbed solution was measured spectrohotometrically. Fig. 5 shows the percentage of the recovered dye. It can be concluded from Fig. 5 that a 2.0 mL volume of 0.1 mol L^{-1}NaOH solution is the most effective eluent for desorption of methyl red from SiO_2-coated Fe_3O_4magnetic nanoparticles. The results showed that desorption efficiencies higher than 99 % can be achieved in a short time of 3 min and in a one-step elution using 2 mL of 0.1 mol L^{-1} NaOH. Therefore, the dye could be desorbed from the loaded SiO_2-coated Fe_3O_4magnetic nanoparticles by changing the pH of the solution to alkaline range. The SiO_2-coated Fe_3O_4 magnetic nanoparticles were washed with deionized water and reused for two successive removal processes with removal efficiencies higher than 90 % (Fig. 6). Under higher removal cycles, removal efficiency decreases. This may be due to oxidation, losing and/or dissolving some amounts of the adsorbent during the successive steps.

Table 4. Parameters of the fitting of experimental data to the Langmuir and Freundlich isotherms equations.

Langmuir isotherm		
a_L (L/mg) K_L(L/g)		$q_{max}|=K_L/\ a_L|$ (mg/g)
$R_L r$		
0.0024±0.0006	0.12±0.02	49.5±1.76
0.806±0.200	0.9899	
Freundlich isotherm		
$K_f n r$		
8.33±1.18		2.12±0.22
0.9221		

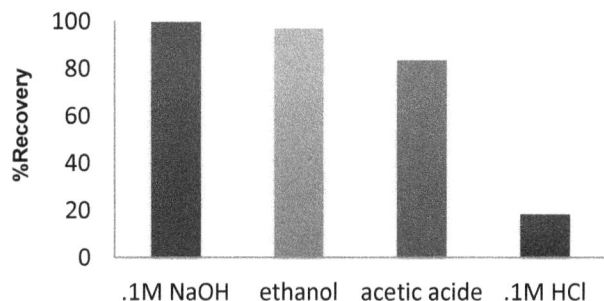

Fig. 5. Percentage of the recovered dye in desorption by different solvents.

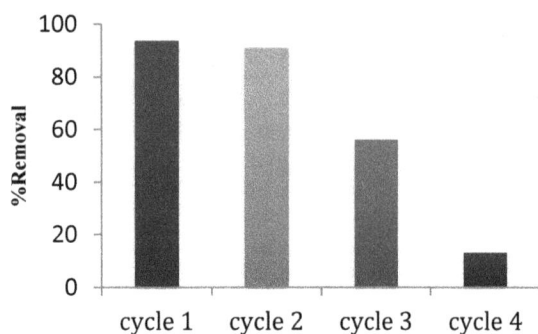

Fig. 6. Removal efficiency for reused SiO_2-coated Fe_3O_4 magnetic nanoparticles.

3.2.2. Effect of sample volume

Effect of sample volume on the adsorption of methyl red on 0.01 g of SiO_2-coated Fe_3O_4 magnetic nanoparticles was studied in the range of 10-200 mL. In order to study the effect of sample volume, 10 mL of 1 mg/l solution of methyl red was diluted to 50, 100, 150 and 200 mL with double distilled water. The results showed that the methyl red present in the volumes up to 100.0 mL was completely and quantitatively adsorbed with SiO_2-coated Fe_3O_4 magnetic nanoparticles. At higher volumes percent of recovery decreased. Therefore, a sample volume of 100.0 mL was selected for determination of trace quantities of methyl red in the samples. This volume is selected in order to increase the preconcentration factor.

3.2.3. Analytical parameters and applications

Increasing concentrations of dye were contacted with SiO_2-coated Fe_3O_4 magnetic nanoparticles in optimum adsorption conditions and then, dye was desorbed in optimum conditions. For constructing calibration curve, the spectrophotometric signal of the solution obtained by desorption process was plotted against the initial

concentration of dye. Statistical parameters of the calibration curve have been collected in Table 5. As an analytical method, the statistics `of the method in preconcentration and determination of methyl red are very good. As the amount of methyl red in 100.0 mL of the solution was concentrated to 2 mL, a preconcentration factor of 50 was achieved in this method.

The suitability of the proposed method for the analysis of natural water samples was checked by spiking samples of river water with 25, 100, 175 ng mL^{-1} of methyl red. The results have been given in Table 6. The results in Table 6 show the good accuracy (percent recoveries close to 100) and precision (RSD % below 1) of the method.

The maximum adsorption capacity (q_{max}) for the adsorption of methyl red onto SiO_2-coated Fe_3O_4 magnetic nanoparticles calculated from the Langmuir isotherm model is 49.50 mg/g. In the only reported adsorption study of methyl red based on the activated carbon as adsorbent (Santhi et al. 2010), the obtained value for q_{max} is 40.486 mg/g.

Table 5. Statistical results of the preconcentration and calibration of methyl red by the proposed method (Miller and Miller 2005).

Parameter	Characteristic
Number of samples	10
Linear range (ng mL−1)	25.0-250.0
Slope	0.0025
Standard error of slope	1.72×10^{-4}
Intercept	0.0741
Standard error of intercept	0.0266
Correlation coefficient	0.9922
Detection limit(ng mL−1)	0.174

Comparison shows that the adsorbent used in the present work have nearly 25 % higher capacity for adsorption of methyl red. Besides that, magnetic and electronic properties that cause simple magnetic separation of methyl red loaded adsorbent makes these particles as good candidate for methyl red adsorption.

Table 6. Results of the analysis of the water samples by the proposed method.

Sample	Amount added (ng mL^{-1})	Amount detected (ng mL^{-1})	RSD %	Recovery %
River water				
	0.0	n.d.[a]	-	-
	25.0	25.74±0.01[b]	0.69	97.12
	100.0	100.56±0.01[b]	0.60	99.44
	175.0	175.62±0.02[b]	0.48	99.64

[a.] Not detected.
[b.] Standard deviations were calculated based on three determinations.

4. Conclusions

Silica coated magnetic nanoparticles were synthesized and utilized for preconcentration, determination and removal of methyl red in aqueous solutions. UV–Vis absorption spectrophotometry was used to study the adsorption behavior of methyl red after treatment by the adsorbent. For adsorption of methyl red in water samples, the prepared magnetic nanoparticles can be easily dispersed and then separated by a magnet. The proposed method is novel, safe,

convenient, rapid and inexpensive for preconcentration, determination and infiltration of methyl red as a toxic compound from waste water. The maximum efficiency of the adsorbent was observed in mild conditions with pH 5.0. The pseudo-second-order kinetic model fitted well with the kinetics of the dye removal. The Langmuir model was fitted better to the dye removal data relative to the Freundlich model. The magnetic particles can be washed and recycled for two dye adsorption cycles.

References

Afkhami A., Saber-Tehrani M., Bagheri H., Modified maghemite nanoparticles as an efficient adsorbent for removing some cationic dyes from aqueous solution, Desalination 263 (2010) 240-248.

Albornoz C., Jacobo S.E., Preparation of a biocompatible magnetic film from an aqueous ferrofluid, Journal of Magnetism and Magnetic Materials 305 (2006) 12–15.

Aleboyeh A., Aleboyeh H., Effects of gap size and UV dosage on decolorization of C. I. Acid Orange 7 by UV/H_2O_2 process, Journal of Hazardous Materials 133 (2006) 167–171.

Belessi V., Romanos G., Boukos N., Lambropoulou D., Trapalis C., Removal of Reactive Red 195 from aqueous solutions by adsorption on the surface of TiO_2 nanoparticles, Journal of Hazardous Materials 170 (2009) 836–844.

Box G.E.P., Hunter J.S., Multi-factor experimental designs for exploring response surfaces, Ann. Math. Stat 28 (1957) 195-241.

Bucak S., Jones D.A., Laibinis P.E., Hatton T.A., Protein separations using colloidal magnetic nanoparticles, Biotechnology Programe 19 (2003) 477-484.

Bulte J.W.M., Intracellular endosomal magnetic labeling of cells, Methods Mol. Med 124 (2006) 419–439.

Chien S.H., Clayton W.R., Application of Elovich equation to the kinetics of phosphate release and sorption on soils, Soil Science society American Journal 44 (1980) 265–268.

De Sales P.F., Magriotis Z.M., Rossi M.A.L.S., Resende R.F., Nunes C.A., Optimization by Response Surface Methodology of the adsorption of Coomassie Blue dye on natural and acid-treated clays, Journal of Environmental Management 130 (2013) 417-428.

Faraji M., Yamini Y., Tahmasebi E., Saleh A., Nourmohammadian F., Cetyltrimethylammonium bromide-coated magnetite nanoparticles as highly efficient adsorbent for rapid removal of reactive dyes from the textile companies' wastewaters, Journal of Iranian Chemical Society 7 (2010) 130–144.

Gómez V., Callao M.P., Modeling the adsorption of dyes onto activated carbon by using experimental designs, Talanta 77 (2008) 84-89.
Juang R.S., Wu F.C., Tseng R.L., Characterization and use of activated carbons prepared from bagasse for liquid-phase adsorption, Colloids Surface: A 201 (2002) 191–199.

Klug H.P., Alexander L.E., X-ray Diffraction Procedures for Polycrystalline and Amorphous Materials, second ed., Wiley, New York, 1974.

Langmuir I., The adsorption of gases on plane surfaces of glass, mica and platinum, Journal of American Chemical Society 40 (1918) 1361–1403.

Laurent S., Forge D., Port M., Roch A., Robic C., Vander Elst L., Muller R.N., Magnetic iron oxide nanoparticles: synthesis, stabilization, vectorization physicochemical characterizations, and biological applications, Chemical Review 108 (2008) 2064–2110.

Miller J.N., Miller J.C. (Eds.), Statistics and Chemometrics for Analytical Chemistry, fifth ed., Pearson Education Limited, London, 2005, p. 114.

Mirsha A., Bajpaj M., The flocculation performances of Tamarindus Mucilage in relation to removal of vat and direct dyes, Bioresource Technology 97 (2006) 1055–1059.

Qadri S., Ganoe A., Haik Y., Removal and recovery of acridine orange from solutions by use of magnetic nanoparticle, Journal Hazardous Materials 169 (2009) 318-323.

Ravikumar K., Ramalingam S., Krishnan S., Balu K., Application of response surface methodology to optimize the process variables for Reactive Red and Acid Brown dye removal using a novel adsorbent, Dyes Pigments 70 (2006) 18-26.

Sahoo C., Gupta A.K., Optimization of photocatalytic degradation of methyl blue using silver ion doped titanium dioxide by combination of experimental design and response surface approach, Journal Hazardous Materials 215 (2012) 302-310.

Santhi T., Manonmani S., Smith T., Removal of methyl red from aqueous solution by activated carbon prepared from the anonna a squmosa seed by adsorption, Chemical Engineering Research Bulletin 14 (2010) 11-18.

Shariati S., Faraji M., Yamini Y., Rajabi A.A., Fe_3O_4 magnetic nanoparticles modified with sodium dodecyl sulfate for removal of safranin O dye from aqueous solutions, Desalination 270 (2011) 160-165.

Singh K.P., Gupta S., Singh A.K., Sinha S., Optimizing adsorption of crystal violet dye from water by magnetic nanocomposite using response surface modeling approach, Journal Hazardous Materials 186 (2011) 1462-1473.

Sun G., Xu X., Sunflower stalks as adsorbents for color removal from textile wastewater, Industrial Engineering Chemical Research 36 (1997) 808–881.

Torrades F., García-Montaño J., Using central composite experimental design to optimize the degradation of real dye wastewater by Fenton and photo-Fenton reactions, Dyes Pigments 100 (2014) 184-189.

Tuutijarvi T., Lu J., Sillanp M., Chen G., As(V) adsorption on maghemite nanoparticles, Journal Hazardous Materials 166 (2009) 1415–1420.

White B.R., Stackhouse B.T., Holcombe J.A., Magnetic-Fe_2O_3 nanoparticles coated with poly-l-cysteine for chelation of As(III), Cu(II), Cd(II), Ni(II), Pb(II) and Zn(II), Journal Hazardous Materials 161 (2009) 848–850.

Xie X., Zhang X., Yu B., Gao H., Zhang H., Fei W. Rapid extraction of genomic DNA from saliva for HLA typing on microarray based on magnetic nanobeads, Journal of Magnetism and Magnetic Materials 280 (2004) 164–168.

Zhou L., Wang Y., Liu Z., Huang Q., Characteristics of equilibrium, kinetics studies for adsorption of Hg(II), Cu(II), and Ni(II) ions by thiourea modified magnetic chitosan microspheres, Journal Hazardous Materials 161 (2009) 995–1002.

Zuorro A., Fidaleo M., Lavecchia R., Response surface methodology (RSM) analysis of photodegradation of sulfonateddiazo dye Reactive Green 19 by UV/H_2O_2 process, Journal of Environmental Management 127 (2013) 28-35.

9

Synthesis of Fe_3O_4@silica core–shellparticles and their application for removal of copper ions from water

Mohammad Eisapour Chanani, Nader Bahramifar*, Habibollah Younesi

Department of Environmental Sciences, Faculty of Natural Resources and Marin Sciences, Tarbiat Modares University, P.O. Box 46414-356, Noor, Iran.

ARTICLE INFO	ABSTRACT
Keywords: Fe_3O_4, Nano particles Heavy metal Adsorption Magnetic	The main objective of this study was to synthesize an environmentally friendly nano-structural adsorbent. These nano magnetic particles can be applied to remove heavy metal ions from industrial wastewater because the surface of the particles is covered with SiO_2, and the SiO_2 is inactive and can adsorb heavy metal ions. Tests were then conducted to study the adsorption of Cu(II) ions onto Fe_3O_4@SiO_2 from an aqueous solution for the effect of contact time, adsorbent dose, solution pH and concentration of metal ions in batch systems. The equilibrium data were analyzed using the Langmuir and Freundlich isotherm by nonlinear regression analysis and found that the adsorption isotherm data will better fit by Langmuir model. The maximum adsorption capacities of Cu (II) were 47 mg/g. Fe_3O_4@SiO_2 was regenerated and found to be suitable for reuse in successive adsorption-desorption cycles 5 times without significant loss of adsorption capacity.

1. Introduction

Nowadays, the excessive and uncontrolled discharge of heavy metal ions becomes a major problem. By means of bioconcentration, bioaccumulation and biomagnification through biologic chain and drinking water, human health will be threatened seriously by heavy metal ions. Copper is widely used in many industries, such as electroplating, paint, metal finishing, electrical, fertilizer, wood manufacturing and pigment industries. Rapid development of these industries has led to accumulation of Cu(II) ions in the environment. Unlike some organic pollutants, copper and some other toxic heavy metals are non-biodegradable and can exist for a long time in natural environment. If the level of Cu(II) ion is beyond the tolerance limit, it will cause serious environmental and public health problems. It is necessary to remove Cu (II) ion from industrial effluents prior to their discharge. Over the past decades, various methods such as precipitation, ion-exchange, and sorption have been employed to remove Cu(II) from large volumes of aqueous solution. Sorption technology is widely regarded as one of the most effective choices for the removal of heavy metal ions from aqueous solution because it is simple and cost-effective. However, most of the common sorbents, such as clay minerals, metallic, oxides and carbon materials usually suffer from either low sorption capacities or sorption efficiencies in the removal of heavy metal ions from aqueous solution. Therefore, the design and development of special sorbents with high sorption capacities for contaminants is critical for pollution management and related applications (Song et al. 2013). Recently, many research groups have explored several nanoparticles for removal heavy metals, because of the ease of modifying their surface functionality and their high surface area to-volume ratio for increased adsorption capacity and efficiency. In the last decade, magnetic nanoparticle (MNP) adsorption has attracted much interest and is an effective and widely used process because of its simplicity and easy operation (Shin et al. 2011). There have been intense interests recently in the fabrication of core-shell particles. Silica has been considered as one of the most ideal materials for protecting Fe_3O_4 MNPs due to its

reliable chemical stability, biocompatibility and versatility in surface modification. It is anticipated that incorporating silica coating on a magnetic core could attain the advantage of silica and without sacrificing the unique magnetization characteristics of Fe_3O_4. A thin and dense silica layer with a desired thickness was deposited on the surface of magnetic particles in order to protect the iron oxide core from leaching into the mother system under acidic conditions. Compared with nonmagnetic nanoparticles, the silica-magnetite nanoparticles can meet the need of rapid extraction of large volume samples by employing a strong external magnetic field (Liu et al. 2009).

In this study, our aim is to prepare these Fe_3O_4@silica core–shell and check their adsorption capability in removing Cu(II) from aqueous solution. The effects of Cu(II) ions concentration, contact time, and solution pH were studied in order to analyze the adsorption kinetics and determine the equilibrium time. Langmuir and Freundlich isotherms were applied to the experimental equilibrium data in order to explain the adsorption mechanism.

2. Materials and methods
2.1. Synthesis of Fe_3O_4@SiO₂

Magnetic Fe_3O_4@SiO_2 particles were synthesized according to the method reported by Hu et al. The whole synthesis procedure as described following: sodium silicate (1.3 g) was dissolved in deionized water (100 ml) to form a clear solution. Then the prepared Fe_3O_4 nanoparticles (0.3 g) were put into the solution. The pH value of the mixture was adjusted to 6.0 by addition of 1 M HCl and then the mixture was stirred by a nonmagnetic stirrer for 3 h. During the whole process, temperature was maintained at 80 °C. Finally, the formed Fe_3O_4@SiO_2 nanoparticles were thoroughly gathered by an external magnet and washed with deionized water. The final Fe_3O_4@SiO_2 nanoparticles were dried under vacuum at 60 °C for 8h (Hu et al. 2010).

Corresponding author Email: n.bahramifar@modares.ac.ir

2.2. Characterizations

The size and morphology of the products were characterized by a scanning electron microscope (SEM). Fourier transform infrared spectra (FTIR, 4,000–500 cm^{-1}) were obtained on a Shimadzu FT-IR 8400S. The samples were dried, mixed with KBr and pressed into a thin disc for the FTIR measurements. The concentration of Cu(II) was determined by flame atomic absorption spectrometry using a PHILIPS model PU9400.

2.3. Adsorption equilibrium experiments

The adsorption of Cu(II) ion by the magnetite nanoparticles was investigated in aqueous solution for 1h at room temperature. In general, placing 25 mg Fe$_3$O$_4$@SiO$_2$ in 50.0 mL of aqueous solution containing Cu(II) ions (50.0 mg/l), the mixture was adjusted to certain pH (pH=5) with NaOH and stirred at 200 rpm for 1h. When the adsorption process reached equilibrium, the adsorbent was separated using a magnet and the supernatant was collected and the residual concentrations of metal ions in the aliquot were determined by atomic absorption spectroscopy (AAS). The equilibrium adsorbed concentration, q$_e$ was calculated according to the equation:

$$q_e = \frac{(C_0 - C_e)V}{M} \tag{1}$$

where, q$_e$ is the equilibrium adsorption capacity of adsorbent (mg/g), C$_0$ and C$_e$ are the initial and equilibrium concentrations of the adsorbents (mg/l), respectively, M is the mass of adsorbent (g), and V is the volume of the metal ions solution (l).
The removal efficiency of the metal ions was calculated by the following equation:

$$R = \frac{(C_0 - C_t)}{C_0} \times 100 \tag{2}$$

where, R is the removal efficiency of the metal ions, C$_0$ the initial concentration and C$_t$ the concentration of the metal ions in mg/l at t time.

2.4. Effect of the adsorbent dose

In the batch sorption studies, the effects of the dose of Fe$_3$O$_4$@SiO$_2$ (10, 15, 20, 25 and 30 mg), adsorbents on adsorption of Cu(II) ion at 50 mg/L were studied. The pH of the working solution was adjusted to 5 by adding 1M HCl. A fresh dilution was carried out for each experiment.

2.5. Effect of the solution pH

A sample of Fe$_3$O$_4$@SiO$_2$ (25mg) was added to 50.0 mL of 50 mg/L Cu(II) ion solution at different pH value varied from 3 to 6 using 1 M HCl. These samples were stirred for 1h, and then adsorbents were removed by a magnet. The supernatant was also tested by AAS.

2.6. Effect of the temperature

The effects of temperature (15, 25, 35, 45°C) on metal ions adsorption were conducted with 25 mg of Fe$_3$O$_4$@SiO$_2$ adsorbent dose in 50.0 ml of 50 mg/l Cu(II) ion solution at pH 5.

2.7. Adsorption Isotherm models

The relationship between the amount of a substance adsorbed per unit mass of adsorbent at constant temperature and its concentration in the equilibrium solution is called the adsorption isotherm. Adsorption isotherm is important to describe how solutes interact with the sorbent. Developing an appropriate isotherm model for adsorption is essential to the design and optimization of adsorption processes. Several isotherm models have been developed for evaluating the equilibrium adsorption of compounds from solutions, such as Langmuir, Freundlich, Redlich–Peterson, Dubinin–Radushkevich, Sips, and Temkin (Dąbrowski, 2001). In these study the isotherm data were correlated with the Freundlich and Langmuir models. A study to determine the relationship between the Cu(II) ion adsorbed on Fe$_3$O$_4$@SiO$_2$ and those remaining in the aqueous phase was conducted. All experiments were carried out with the various initial Cu(II) concentrations (10, 20, 30, 40 and 50 mg/l) in conditions of: fixed amount of absorbent 25 mg per 50 ml solution, constant pH 5, room temperature (25 °C) and 200 rpm agitation speed. Contact time was 70 min for all equilibrium conditions. The Langmuir isotherm is based on a monolayer sorption of metal ion on the surface of the adsorbent and is described by the following equation (Sağ and Aktay. 2001):

$$q_e = \frac{q_m b C_e}{1 + b C_e} \tag{3}$$

where, q$_e$ is the adsorption capacity of the adsorbent in mg/g and Ce the concentration of metal ion in mg/l at equilibrium. The q$_m$ is the maximum adsorption capacity of the metal monolayer in mg/g, and b the constant that refers to the bonding energy of adsorption in l/mg. The Freundlich isotherm model is considered to be appropriate for describing both multilayer sorption and sorption on heterogeneous surfaces. The Freundlich model can be expressed by the following equation(Hadavifar et al. 2014):

$$q_e = K_f C_e^{1/n} \tag{4}$$

where, q$_e$ is the equilibrium adsorption capacity of the adsorbent in mg/g, Ce the liquid phase concentration in mg/l at equilibrium, K$_f$ the constant related to the adsorption capacity of the adsorbent in (mg/g) (l/mg) 1/n and n the empirical constant depicting the intensity of adsorption which varies with the heterogeneity of the adsorbent. The greater is the value of n the better its adsorption capacity. The nonlinear regression analysis was carried out with Sigma Plot software (Sigma Plot 12.0. USA) in order to predict both the K$_f$ and the n parameters.

2.8. Adsorption kinetics

The kinetic studies were carried out using 25 mg of Fe$_3$O$_4$@SiO$_2$ in 50 ml of different concentrations (10, 20, 30, 40 and 50 mg/l) of Cu(II) metal ion solutions at pH 5. In order to describe the kinetic process between aqueous and solid phase, the pseudo first-order rate was used for surface adsorption of Cu(II) ion on Fe$_3$O$_4$@SiO$_2$. This model is presented by Laguerre as follows (Heidari et al. 2009):

$$\text{Log}(q_e - q_t) = \text{Log}q_e - \frac{k_1}{2.303}t \tag{5}$$

The pseudo-second-order rate equation presented by Ho to describe the kinetic adsorption of divalent metal ion onto an absorbent is expressed as follows(Qiu et al. 2009):

$$\frac{t}{q_t} = \frac{1}{k_2 q_e^2} + \frac{t}{q_e} \tag{6}$$

where. q$_e$ and q$_t$ are the amount of adsorbed metal ion in mg/g on the adsorbent, at equilibrium and time t, respectively, while k$_1$ in min^{-1} and k$_2$ in g/mg/min are the rate constants of first and second-order adsorption, respectively.

2.9. Adsorption thermodynamics

The adsorption experiment was carried out at different temperatures (15, 25, 35 and 45 °C) to evaluate thermodynamic criteria by calculating the Gibbs free energy (ΔG) by the following equation(Hao et al. 2010):

$$\ln k_d = -\frac{\Delta G^0}{RT} = -\frac{\Delta H^0}{RT} + \frac{\Delta S^0}{R} \tag{7}$$

where, the values of ΔH° (change in enthalpy in J/mol) and ΔS° (change in entropy in J/mol/K) are obtained from the slope and intercept of ln k$_d$ vs. 1/T plots. T is the temperature in K and R is the universal gas constant (8.314 J/ mol/ K). The distribution coefficient (k$_d$) is calculated from the initial and the equilibrium concentrations (C$_0$ and C$_e$) of the metal ions, where V is the working volume in ml and W is the adsorbent mass in g (Hadavifar et al. 2014):

$$K_d = \frac{C_0 - C_e}{C_e} \times \frac{V}{W} \qquad (8)$$

The ΔG° is the change in Gibbs free energy in J/mol, calculated according to the following equation (Hao et al. 2010):

$$\Delta G^0 = -RT \ln K \qquad (9)$$

2.10. Batch desorption study

Reusability of $Fe_3O_4@SiO_2$ was determined in five adsorption–desorption cycles. To optimize the concentration of the acid, the experiments were carried out with different concentrations (0.1, 0.2 and 0.3 M) of H_2SO_4. The mixture was shaken for 1 h to reach desorption equilibrium. Then concentration of acid with more efficiency in desorption of loaded Cu(II) ion on the $Fe_3O_4@SiO_2$ was selected. After each cycle the regenerated adsorbent was washed thrice by deionized water to remove the remainder ion, from the adsorbent and $Fe_3O_4@SiO_2$ nanoparticles were separated magnetically and the supernatant was subjected to Cu(II)

measurements. The metal recovery was calculated by the following equation (Shahbazi et al. 2011):

$$\text{metal recovery} = \frac{\text{Amount of metal ions desorbed}}{\text{Amount of metal ions adsorbed}} \times 100$$

3.Results and discussions
3.1. Characterization of adsorbents

FTIR measurements were performed for Fe_3O_4 nanoparticles and $Fe_3O_4@SiO_2$ samples as shown in Fig. 1. Both spectra present absorption peak at 578 cm^{-1}, corresponding to the Fe–O vibration from the magnetite phase(Yamaura et al. 2004). Spectrum of $Fe_3O_4@SiO_2$ present the typical Si-O-Si bands of the inorganic symmetric vibration modes around 786 cm^{-1}, asymmetric stretching vibration around 1033–1100 cm^{-1} and the band at 964 cm^{-1} is assigned to the Si-O stretch that indicates the silica layer around the Fe_3O_4 (Innocenzi. 2003; Pillay et al. 2013). The morphology of the Fe_3O_4 was analyzed by a scanning electron microscope. A typical image is presented in Fig. 2, which shows many ropelike domains with relatively uniform size of approximately 50 nm.

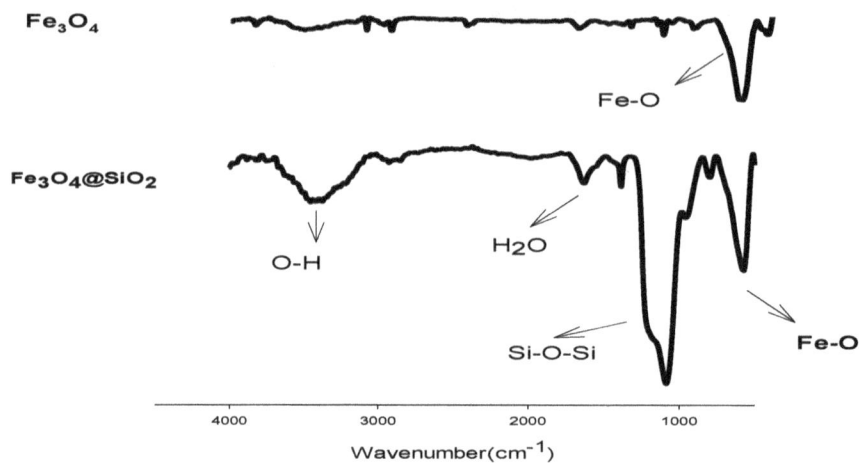

Fig. 1. IR spectrum of $Fe_3O_4@SiO_2$ nanoparticles.

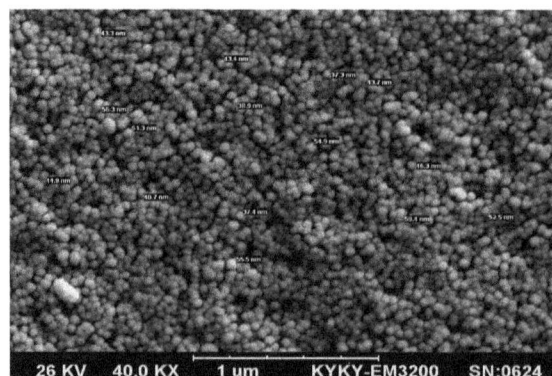

Fig. 2. Scanning electron microscopy of Fe_3O_4.

3.2. Effect of the adsorbent dose

The effect of the dose of $Fe_3O_4@SiO_2$, adsorbent on removal of Cu(II) metal ion at 50 mg/L has been studied. These results are depicted in Fig. 3. It is clear that the removal efficiency of Cu(II) metal ions increases as the amount of the adsorbent increases owing to the enhanced total surface area of the adsorbent. The results revealed that the metal removal percentage is dependent on the optimal increase of the adsorbent dose, due to a consequent increase in interference between binding sites at the higher dose or an insufficiency of metal ions in solution with respect to available binding sites. The maximum metal ions removal was attained above 25 mg dose of $Fe_3O_4@SiO_2$ for Cu(II).

3.3. Effect of solution pH

The pH of the solution plays an important role on the adsorption of ion metals. In order to determine the effect of pH on the Cu(II) ion removal by $Fe_3O_4@SiO_2$, some experiments were carried out with different pH values while the other parameters were kept constant. Fig. 4 shows the effect of the pH on the removal efficiency of Cu(II) as the function of the solution pH in the range of 3–7, obtained with the initial metal concentration of 50 mg/L and 25 mg $Fe_3O_4@SiO_2$. The metal uptake increases while increasing the pH from 3 till it reaches a maximum of 7. In acidic solution, H^+ can cause the protonation of silanol groups, which means that part of the site occupied by metal ions mareplacedplace by H^+. Besides, in acidic solution, low pH will restrain the hydrolysis of the metal ion so weaken the adsorption ability of $Fe_3O_4@SiO_2$ nanoparticles.

3.4. Adsorption isotherms

To better elucidate the adsorption mechanism of $Fe_3O_4@SiO_2$ the equilibrium experimental adsorption data were fitted by two adsorption models named Langmuir and Freundlich. Fig. 5 display the adsorption isotherms of Cu(II) on the $Fe_3O_4@SiO_2$ under varied initial metal ion concentration. As shown in this figure, the equilibrium adsorption amount depended on the heavy metal ion concentration at equilibrium, and gradually increased with the equilibrium concentration increased. The summary of all the parameters obtained from the curve-fitting results are listed in Table 1. Based on the value of correlation coefficients (R^2), it can be seen that the Langmuir isotherm is more suitable for the experimental data of different heavy metal ions adsorption, suggesting the homogenous distribution of active sites on the surface of $Fe_3O_4@SiO_2$. The maximum adsorption capacity for Cu(II) was 47 mg/g.

Table 1. Isotherm constants for the adsorption of Cu(II) onto $Fe_3O_4@SiO_2$

Langmuir			Freundlich		
q_m	b (L/mg)	R^2	n	k_f (L/g)	R^2
53.17	0.3756	0.9995	4.7	24.75	0.9956

3.5. Adsorption kinetics

Typical kinetic experimental curves for adsorption of Cu(II) on the $Fe_3O_4@SiO_2$ in different ion concentrations showed that ion adsorption increases sharply during a short contact time and slows down gradually to reach equilibrium. In order to describe the kinetics for Cu(II) ion adsorption onto the $Fe_3O_4@SiO_2$, the pseudo-first-order and pseudo-second-order kinetics are applied. The parameters of the kinetic models and the regression correlation coefficients (R^2) are listed in Table 2. The R^2 values clearly indicate the validity of the pseudo-second-order versus pseudo-first-order kinetics which is not fitted logically. As seen in Table 2, when the initial ion concentration increases from 10 to 50mg/l, the pseudo-second-order constants (k_2) decrease from 0.004 to 0.002 g/mg/min. This indicates that the available active sites on the $Fe_3O_4@SiO_2$ are saturated rapidly by Cu(II) ion, furthermore suggesting the possibility of the formation of a monolayer coverage of Cu(II) onto the adsorbent (Ahangaran et al. 2013).

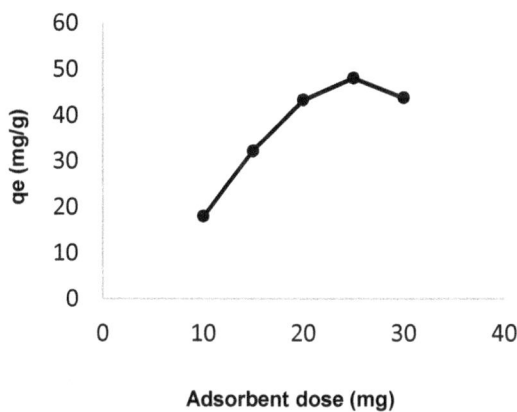

Fig. 3. The effect of adsorbent dose on the Cu(II) removal $Fe_3O_4@SiO_2$: initial Cu(II) concentration 50 mg/l, initial pH value 5, agitation time 60 min at 200 rpm and 25 °C.

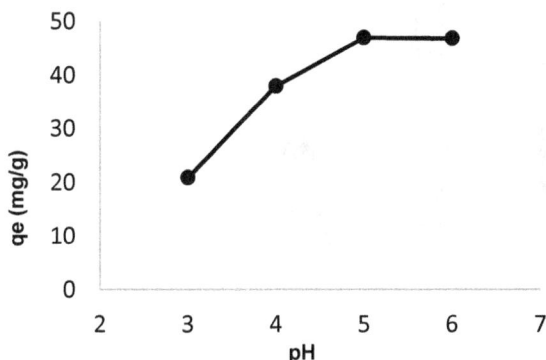

Fig. 4. Effect of pH on Cu(II) removal. Conditions: adsorbent dose 25 mg, initial concentration 50 mg/l, agitation time 60 min at 200 rpm and at 25 °C.

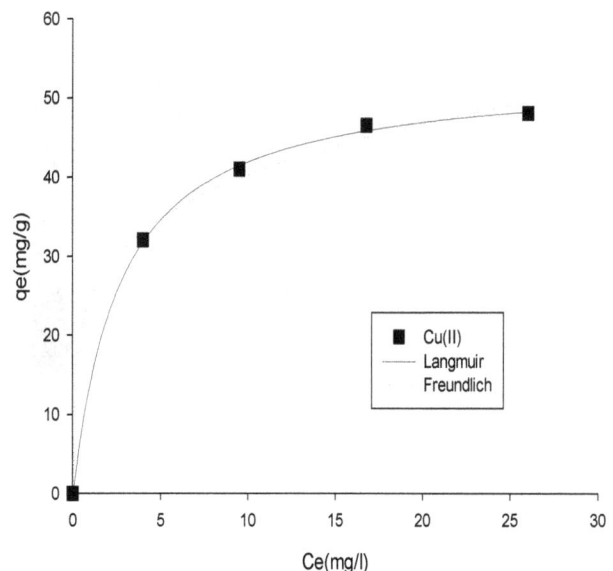

Fig. 5. Langmuir and Freundlich parameters for adsorption of Cu(II) in single solutions onto $Fe_3O_4@SiO_2$.

Table 2. Thermodynamic parameters of Cu(II) adsorption onto Fe$_3$O$_4$@SiO$_2$ at different temperatures.

T(K)	Lnk$_d$ (Lmg^1)	ΔG (kJ mol^{-1})	ΔH° (kJ mol^{-1})	ΔS° (J(mol k)$^{-1}$)
288	0.964	-23.08		
298	1.021	-25.30	3.926	21.65
308	1.073	-27.49		
318	1.118	-29.57		

3.6. Adsorption thermodynamics

The effect of temperature on the adsorption of Cu(II) ions onto Fe$_3$O$_4$@SiO$_2$ is shown by the linear plot of lnkd versus 1/T in Fig. 6 and the relative parameters and correlation coefficients calculated from Equation 7 are listed in Table 3. The negative values of ΔG confirmed that the adsorption was spontaneous, and the decreasing of ΔG as temperature rises indicated that the adsorption was more favorable at high temperatures. The positive value of ΔH^0 confirmed

the endothermic nature of adsorption which was also supported by the increase in value of Cu(II) uptake with the rise in temperature.

The positive value of ΔS^0 suggested the increasing randomness at the solid/liquid interface during the adsorption of Cu(II) ions on Fe$_3$O$_4$@SiO$_2$.

3.7. Desorption

The desorption efficiency of Fe$_3$O$_4$@SiO$_2$ was evaluated by H$_2$SO$_4$ acid treatment. The effects of concentrations of sulfuric acid (0.1, 0.2 & 0.3 mol/l) on the stability of the adsorbents and the adsorption of Cu (II) ions were investigated. The 0.1 M H$_2$SO$_4$ was more effective than other concentrations of H$_2$SO$_4$. The adsorbent was reused in five successive adsorption–desorption cycles as can be seen in Fig. 7. However, the Cu(II) adsorption capacity decreased from 88.25 % in the initial cycle to 81.72 % in the final cycle that revealed a slight loss of adsorption capability (6.6 %).

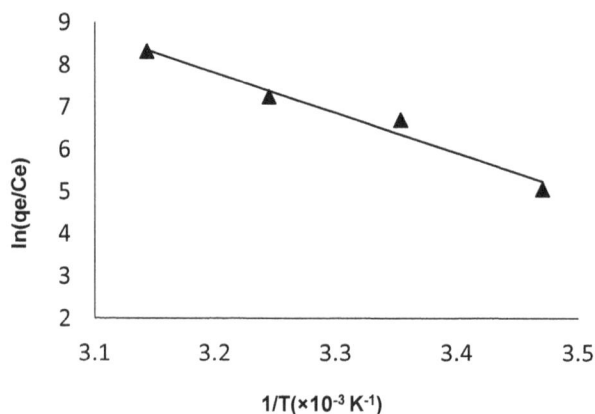

Fig. 6. Plot of ln K vs. 1/T to predict thermodynamic parameters for the adsorption of Cu(II) ion onto Fe$_3$O$_4$@SiO$_2$.

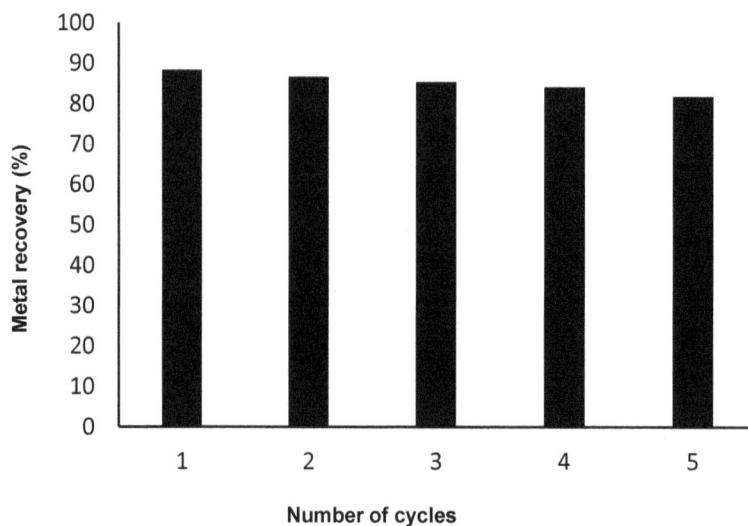

Fig. 7. The number of adsorption–desorption cycle of Cu(II) on Fe$_3$O$_4$@SiO$_2$.

Table 3. The kinetic parameters of different models (for Cu(II) ions adsorption onto $Fe_3O_4@SiO_2$.

Metal conc.(mg/l)	qe_{exp}	pseudo-first-order			pseudo-second-order		
		$k_1(min^{-1})$	$qe_1(mg/g)$	R^2	$k_2 (g/mg/min)$	qe_2	R^2
10	19.96118	0.05204	14.2790	0.9862	0.004205	22.7790	0.998
20	32.0305	0.045139	21.7971	0.9908	0.002302	37.03704	0.997
30	41.0003	0.049284	27.2144	0.9912	0.002109	46.72897	0.999
40	46.5064	0.075538	29.0268	0.6773	0.002887	51.28205	0.997
50	47.9934	0.05366	20.3704	0.8518	0.003413	52.08333	0.994

4. Conclusion

Silica core shell magnetic nanoparticles, $Fe_3O_4@SiO_2$ were successfully prepared for the removal of copper ion from synthetic wastewater. The adsorption experiments revealed that the $Fe_3O_4@SiO_2$ has a good capability for Cu(II) ion removal from aqueous solution. At desired conditions of the batch system (pH 5 and adsorbent dose of 25 mg/l) the maximum adsorption capacity of $Fe_3O_4@SiO_2$ for Cu(II) ion removal was achieved at 47.0 mg/g. This maximum capacity is higher than other reports on the magnetic adsorbents as compared in Table 4 (Chang and Chen. 2005; Hu et al. 2006; Rao et al. 2007; Shukla et al. 2009). The adsorption-desorption experiments showed a small loss in the adsorption capacity for Cu(II) after five cycles. Finally, we found out that our new synthesized $Fe_3O_4@SiO_2$ is a good adsorbent for copper ion removal from wastewater.

Acknowledgements

Authors wish to thank the Iranian Mines & Mining Industries Development & Renovation and Tarbiat Modares University that have financed this research work. The authors wish also to thank the Iranian Nano Technology Initiative Council and Mrs. Haghdoust (Technical assistant of the Laboratory) for her assistance in Laboratory works.

Table 4. Comparison of adsorption capacity of various adsorbents for Cu(II).

Sorbent	Adsorption capacity(mg/g)	Refs.
Maghemite nanoparticle	27.7	Hu et al. 2006
Chitosan-bound Fe_3O_4 magnetic nanoparticles	21.5	Chang and Chen. 2005
Carbon nanotubes	24.49	Rao et al. 2007
Oxidized coir	6.99	Shukla et al. 2009
$Fe_3O_4@SiO_2$	47	this work

References

Ahangaran F., Hassanzadeh A., Nouri S., Surface modification of $Fe_3O_4@$ SiO_2 microsphere by silane coupling agent, International Nano Letters 3 (2013)1-5.

Chang Y.C., Chen D.H., Preparation and adsorption properties of monodisperse chitosan-bound Fe_3O_4 magnetic nanoparticles for removal of Cu (II) ions, Journal of Colloid and Interface Science 283 (2005) 446-451.

Dąbrowski A., Adsorption—from theory to practice, Advances in Colloid and Interface Science 93 (2001) 135-224.

Hadavifar M., Bahramifar N., Younesi H., Li Q., Adsorption of mercury ions from synthetic and real wastewater aqueous solution by functionalized multi-walled carbon nanotube with both amino and thiolated groups, Chemical Engineering Journal 237 (2014) 217-228.

Hao Y.M., Man C., Hu Z.B., Effective removal of Cu (II) ions from aqueous solution by amino-functionalized magnetic nanoparticles, Journal of Hazardous materials 184 (2010) 392-399.

Heidari A., Younesi H., Mehraban Z., Removal of Ni (II), Cd (II), and Pb (II) from a ternary aqueous solution by amino functionalized mesoporous and nano mesoporous silica, Chemical Engineering Journal 153 (2009) 70-79.

Hu H., Wang Z., Pan L., Synthesis of monodisperse $Fe_3O_4@$ silica core–shell microspheres and their application for removal of heavy metal ions from water, Journal of Alloys and Compounds 492 (2010) 656-661.

Hu J., Chen G., Lo I.M., Selective removal of heavy metals from industrial wastewater using maghemite nanoparticle: performance and mechanisms, Journal of Environment Engineering 132 (2006) 709-715.

Innocenzi P., Infrared spectroscopy of sol–gel derived silica-based films: a spectra-microstructure overview, Journal of Non-Crystalline Solids 316 (2003) 309-319.

Liu J., Sun Z., Deng Y., Zou Y., Li C., Guo X., Zhao D., Highly Water-Dispersible Biocompatible Magnetite Particles with Low Cytotoxicity Stabilized by Citrate Groups, Angewandte Chemie 121 (2009) 5989-5993.

Pillay K., Cukrowska E.M., Coville N. J., Improved uptake of mercury by sulphur-containing carbon nanotubes, Microchemical Journal 108 (2013) 124-130.

Qiu H., Lv L., Pan B.C., Zhang Q.J., Zhang W.M., Zhang Q.X., Critical review in adsorption kinetic models, Journal of Zhejiang University Science A 10 (2009) 716-724.

Sağ Y., Aktay Y., Application of equilibrium and mass transfer models to dynamic removal of Cr (VI) ions by chitin in packed column reactor, Process Biochemistry 36 (2001) 1187-1197.

Rao G.P., Lu C., Su F., Sorption of divalent metal ions from aqueous solution by carbon nanotubes: a review, Separation and Purification Technology 58 (2007) 224-231.

Shahbazi A., Younesi H., Badiei A., Functionalized SBA-15 mesoporous silica by melamine-based dendrimer amines for adsorptive characteristics of Pb (II), Cu (II) and Cd (II) heavy metal ions in batch and fixed bed column, Chemical Engineering Journal 168 (2011) 505-518.

Shin K.Y., Hong J.Y., Jang J., Heavy metal ion adsorption behavior in nitrogen-doped magnetic carbon nanoparticles: Isotherms and kinetic study, Journal of Hazardous materials 190 (2011) 36-44.

Shukla S.R., Gaikar V.G., Pai R.S., Suryavanshi U.S., "Batch and column adsorption of Cu (II) on unmodified and oxidized coir, Separation Science and Technology 44 (2009) 40-62.

Song W., Hu J., Zhao Y., Shao D., Li J., Efficient removal of cobalt from aqueous solution using β-cyclodextrin modified graphene oxide, RSC Advances 3 (2013) 9514-9524.

Yamaura M., Camilo R.L., Sampaio L.C., Macedo M.A., Nakamura M., Toma H.E., Preparation and characterization of (3-aminopropyl) triethoxysilane-coated magnetite nanoparticles, Journal of Magnetism and Magnetic Materials 279 (2004) 210-217.

Yuan Q., Li N., Chi Y., Geng W., Yan W., Zhao Y., Dong B., Effect of large pore size of multifunctional mesoporous microsphere on removal of heavy metal ions, Journal of Hazardous materials 254 (2013)157-165.

Xiaoli Z., Shi Y., Wang T., Cai Y., Jiang G., "Preparation of silica-magnetite nanoparticle mixed hemimicelle sorbents for extraction of several typical phenolic compounds from environmental water samples, Journal of Chromatography 1188 (2008) 140-147.

Simultaneous saccharification and fermentation (SSF) of rice cooker wastewater by using Aspergillus niger and Saccharomyces cerevisiae for ethanol production

Masoud Hatami, Habibollah Younesi[*], Nader Bahramifar

Department of Environmental Science, Faculty of Natural Resources, Tarbiat Modares University, P.O. Box 46414-356, Tehran, Iran.

ARTICLE INFO	ABSTRACT
Keywords: Ethanol production Rice wastewater SSF A. niger S. cerevisiae	This work examined the simultaneous saccharification and fermentation (SSF) process for the biological conversion rice wastewater into ethanol using co-culture of Aspergillus niger (A. niger) and Saccharomyces cerevisiae (S. cerevisiae) in batch condition. In this study, The A. niger and S. cerevisiae were used for hydrolysis and production of ethanol from rice wastewater, respectively. The Effects of fermentation parameters such as pH (4, 4.5, 5 and 5.5), temperature (25, 30, 35 and 40 °C), incubation period (12 to 72 h), incubation time (12 to 72 h) and nitrogen source on SSF were evaluated. The results showed that among the optimal parameters of pH 5, temperature 35 °C, incubation period 36 h, incubation time 36 h and nitrogen source of $(NH_4)_2SO_4$ were obtained in ethanol production by SSF process. Under these optimized conditions, maximum ethanol production and product yield were 16.97 g/l and 0.36 g/g, respectively.

1. Introduction

One of the most important issues in the 21 century in global communities is the growing demand for energy and providing low-cost and suitable raw materials for production it to be used in different industries, specifically in transportation sectors. In recent years this issue has become evident with an increase in crude oil price. Although, fossil fuel has more than 80 percent of energy consumption in the world's. But in recent decades they have caused many problems for society, including: uneven distribution of the world, Environmental pollution, such as emissions of CO_2, SO_2 and NO_X and Implied Global warming caused by increase greenhouse gases. In today everyone knows that use of fossil fuels led to increasing the global warming. Therefore the most feasible way to meet this growing demand and reduce global warming is by utilizing alternative energies (Najafpour et al. 2004).

Amongst the alternative energies, one of the most important energy sources in near future is biomass. Biofuel is a renewable energy source produced from biomass, which can be used as a substitute for petroleum fuels (Ghorbani et al. 2011).The benefits of biofuels over other fuels such as fossil fuel, include greater energy security, reduced environmental impact, reducing greenhouse gas emissions and Provide Kyoto Protocol, foreign exchange savings and socioeconomic issues. Among the different biofuels, ethanol, due to Inexpensive and appropriate resources, the ability to produce of various resources of sugars, starch and cellulose has attracted the most attention. Ethanol can be used as a biofuel environmentally friendly due to high octane number and existence oxygen in chemical structure. It could be used alone as a fuel or alternative to MTBE in gasoline and to replace MTBE in gasoline. And also it can be used as a carrier of oxygen in gasoline to increase the oxygen content. Ultimately this practice lead to better fuel oxidation and thus reduce the exhaust gas from (Cardona & Sanchez. 2006).

Accordingly, many researchers have investigated the production of ethanol from various sources, such as sugar, starch, cellulose and or amylolytic enzymes. These enzymes could be economically produced by microorganisms (Nigam & Singh. 1995). For example, Ghorbani and coworkers (2011) from cane molasses and Kadar et al. (2004) of industrial wastes. Starchy materials and effluent generated from starch processing units are the abundant, available and inexpensive substrates that can be used as suitable raw materials for ethanol production (Verma et al. 2000). These materials can be easily hydrolysed to fermentable sugars by acid ethanol production from starchy materials such as rice cooker wastewater is a two-step process. The first stage is the saccharification process, which during it the starch converted into simple sugars such as glucose by amylolytic enzymes or acid.

The second stage is the fermentation of sugar derived from hydrolysis in which the sugars derived from saccharification converted to ethanol by microorganisms.

Rice cooker wastewater is one of the most important and-common-effluent urban wastewater. That is produced amount daily and without any use or treatment are disposal into municipal wastewater systems and household. Arriving the effluent to aquatic ecosystems due to high nutrient loads and various actions and anions, causing environmental problems, such as increase COD and BOD ecosystems and creates irreversible damage. The effluent due to high organic matter can be used as an excellent op tion for ethanol production, which is in fact used to be a sustainable biofuel. Since some of the microorganisms used in the fermentation of ethanol, especially, S. cerevisiae and Z. mobilis, are lack of amylolytic enzymes and unable to directly convert the starch into ethanol (Ang et al. 2001; Gupta et al. 2003). Therefore, when using starchy materials as substrates for ethanol production in first must complex structure them broken down into fermentable sugars. There are several technologies available for the conversion of starchy materials to ethanol. The main difference between these technologies are the catalyst and methods used for the brake-down of starch in the fermentable sugars (Kádár et al. 2004). Amongst the different catalyst and methods such as Acid hydrolysis and Separate Fermentation, SSF and Separate hydrolysis and Fermentation, Simultaneous saccharification and fermentation with mixed microorganisms such as an amylolytic microorganism and an ethanol-fermenting microorganism is an effective method for the direct fermentation of starch. The advantages of the simultaneous saccharification and fermentation are that a two-stage process for the conversion of starch into ethanol are realized in one reactor and the glucose produced is rapidly converted into ethanol, and the reduced end-product inhibition of the enzymatic hydrolysis, and the reduced investment costs (Beschkov et al. 1984). However, in this system the ethanol yield decreased because much starch was consumed by the growth of amylolytic microorganisms (Nakamura et al. 2000).

The purpose of the present study was to investigate simultaneous saccharification and fermentation (SSF) process of ethanol production from rice cooker wastewater by using Aspergillus niger (A. niger) and Saccharomyces cerevisiae (S. cerevisiae). Also the effect type of nitrogen source on the production of glucose and ethanol in this process by S. cerevisiae and A. niger were evaluated. The glucose production, volumetric ethanol productivity and yield process parameters were examined to describe the consumption of rice cooker wastewater and the production of ethanol.

2. Materials and methods
2.1. Microorganisms

Saccharification and fermentation was performed by A. niger and S. cerevisiae Respectively (Persian Type Culture Collection, PTCC 5010) supplied from the Research and Technology of Ministry of Sciences (Iran) in the form of freeze-dried culture in the form of freeze-dried culture.

2.2. Media

The medium for the A. niger cultivations were as follows: It was cultured in a sterilized liquid medium, propagated in nutrient agar and then stored in a refrigerator at 48 °C. The composition of the growth medium was (in g/l): sucrose, 50; NH_4NO_3, 2; KH_2PO_4, 0.15; $MgSO_4$, 0.15; $FeSO_4.7H_2O$, 0.005; $MnSO_4.H_2O$, 0.016; $CoCl_2.6H_2O$, 2.9; $ZnSO_4.7H_2O$, 0.0014. The medium was sterilized by autoclaving at a pressure of 1atm and a temperature of 1218 °C for 20 min. The temperature and the pH of the growth medium were at ambient temperature 30 °C and pH 5.5, without shaking. The fungal cells were grown for 5 days (end of the exponential phase) and then filtered (0.451 m pore size). The medium for S. cerevisiae cultivations were as follows: The culture was maintained on a sterilized solid Potato Dextrose Agar (PDA) medium in a 20 ml-test tubes and transferred to fresh medium every six months. The culture was incubated at 30°C for 1 - 3 days and stored at 4 °C until use. Before starting the experiment, the microorganism was inoculated under sterile conditions into glass test tubes containing the same solid culture medium. These tubes were then kept in an incubator at 30 °C for 16 h in order to obtain cells at the same growth stage for every experiment. The composition of the media was (in g/l): glucose, 15; $(NH_4)_2SO_4$, 9; $MgSO_4.7H_2O$, 2.5; yeast extract, 1; KH_2PO_4, 10; K_2HPO_4, 5. The media were sterilized in an autoclave (Reyhan Teb, F2000, Iran) at 121 °C and 1 atm for 20 min. The pH was maintained at 4.5 by the addition of either 1 N NaOH or 1 N HCl when necessary.

2.3. Characteristic of the rice cooker wastewater

Rice cooker waste water is one of the most abundant urban and household wastewater effluents that is produced from different places such as restaurants, hotels and houses. The supply of rice cooker wastewater as municipal effluent, used in this study and provided by restaurant's university of Natural Resources and Marine Sciences, Tarbiat Modares University (Noor, Iran).

2.4. Simultaneous Saccharification and Fermentation (SSF)

Saccharification and fermentation of rice cooker wastewater were performance simultaneously in the batch culture at temperature 30 °C and pH 5 by using the A. niger and S. cerevisiae for hydrolysis fermentation. For this experiment prior to SSF, in first rice cooker wastewater was concentrated to the glucose that it arrived to the 50 g/l. The SSF experiments were performed in 500 ml E-flasks. Each flask contained 200 ml of culture medium in which the concentrations of nutrients were (in g/l): $(NH_4)_2SO_4$, 9; $MgSO_4.7H_2O$, 2.5; yeast extract, 2; KH_2PO_4, 10; K_2HPO_4, 5; $FeSO_4.7H_2O$, 0.005; $MnSO_4.H_2O$, 0.016; $CoCl_2.6H_2O$, 2.9; $ZnSO_4.7H_2O$, 0.0014. The A. niger fungal was

inoculated on fermentation medium. The sampling time for reducing sugar concentration analysis was 12, 24, 48 and 72 h and then the medium was incubated with three present S. cerevisiae in different times (12, 24, 36, 48 and 60 h) to the fermentation medium and the sampling time for ethanol production analysis was every 12 h, respectively. The flasks were incubated in a rotary shaker at 30 °C for 72 h samples were withdrawn regularly every 12 h, centrifuged in a laboratory desktop centrifuged for 15 min at 6000 × g, and the supernatants were analyzed for determination glucose and ethanol. All experiments were performed in triplicate and the average values are presented.

2.4.1. Effect of pH and temperature on ethanol production

Effect of pH and temperature on ethanol fermentation by the process of simultaneous saccharification and fermentation using mixed cultures of S. cerevisiae and A. niger was carried out by varying the pH, (4, 4.5, 5 and 5.5) and temperatures (25, 30, 32 and 35 °C).

2.5. Effect of incubation time on glucose and ethanol production

For determining the effect of incubation time on glucose and ethanol production, 200 ml of rice cooker wastewater substrate was dispersed to 500 ml E-flasks. In first step the A. niger was incubated in the medium and every 12 h samples were taken to determine the amount of glucose produced by A. niger. And in second step the S. cerevisiae was incubated in different times (12, 24, 36, 48, 60 and 72 h) every 12 h samples were taken to determine the amount of glucose consumption and ethanol production.

2.6. Effect of nitrogen source on glucose and ethanol production

The effects of nitrogen source ($(NH_4)_2SO_4$, NH_4NO_3 and NH_4Cl) were investigated on the hydrolysis of rice cooker wastewater and ethanol production by A. niger and S. cerevisiae at a pH and temperatures were adjusted to 5 and 30 °C, respectively. The flasks were incubated in an orbital shaker with a speed of 120 rpm. The reaction time was set 72 h. Samples were periodically withdrawn up to 12 h to monitor the extent of glucose and ethanol produced by A. niger and S. cerevisiae, respectively. Under this condition, the S. cerevisiae was incubated into medium after 36 h as A. niger.

2.7. Analytical methods

In this study, samples for analysis of glucose and ethanol contents were first centrifuged for 15 min at 6000 × g and the supernatants were analyzed for determination glucose and ethanol. Reducing sugar concentration was estimated by DNS method (Miller. 1959). Ethanol concentration was determined by gas chromatography (Philips, PU440, US) using flame ionization detector and with software (Clarity 4.2, Data Apex Czech Republic) used to analyze the liquid samples. The column used was PEG 20 M (glass column) 1.5 m and 1/8 mm (Philips, USA). Temperature programming was employed for the liquid analysis in GC. During the analysis, the column temperature was initially maintained at 120 °C and after 2 min the oven temperature was increased at a rate of 10 °C/min until it reached to 150 °C. The injector and detector temperatures were maintained at 220 °C. Nitrogen was used as the carrier gas at 30 ml/min. Acetone (1 %, v/v) was used as an internal standard with concentration of 20 ml/ml per sample. The injection sample volume was 2 μl. Each set of the experiment and the data points were repeated three times. The reported value was the average.

3. Results and discussion

Starchy materials and effluent generated from the starch generating unit such as rice mil are the cheap and abundant substrates that could be used as potential raw materials for ethanol fermentation. But these materials require a reaction of starch with water (hydrolysis) to break down the starch into fermentable sugars (saccharification). In the present study the ethanol production of rice cooker wastewater, as a Starchy material, was investigated in SSF experiments. By using the A. niger and S. cerevisiae for saccharification and fermentation respectively at a pH and temperatures were adjusted to 5 and 30 °C, respectively. The flasks were incubated in an orbital shaker with a speed of 120 rpm. The reaction time was set 72 h.

3.1. Effect of pH and temperature on ethanol production

Fig.1. shows the result of the effect of pH and temperature on ethanol concentration. Effects of temperature to determine the optimum temperature, experiments were conducted at different temperatures (25 to 40 °C). It was observed that the ethanol production increased (11 g/l) by increase in temperature up to 30 °C (Fig. 1) for rice cooker wastewater, but above this the productivity decreased though there was an increase in SSF of the rice cooker wastewater. However, temperatures beyond 30 °C showed a fall in ethanol production which is in line with the findings of Sharma et al. (2007), who reported optimum temperature for simultaneous saccharification and fermentation of kinnow waste and banana peels was found to be 30 °C with maximum ethanol yield of 0.376 g/g and fermentation efficiency of 74.11 % (Sharma et al. 2007). Verma et al. (2000) also reported 30 °C as the optimum temperature for maximum ethanol production using starch employing co-culture of amylolytic yeast and S. cerevisiae (Verma et al. 2000). Among the physical parameters, the pH of growth medium has played an important role by inducing morphological change in the organism and in enzyme secretion. This result shows the optimum pH range of 5 gave the optimum yield of glucose. This corroborates the results of Aderemi et al. (2008) at which the optimum glucose yields were obtained range between at pH 4.5 and 5. The productivity decreased by decrease or increase in pH of the medium. This may be due to the low activity of enzyme that is involved in the process.

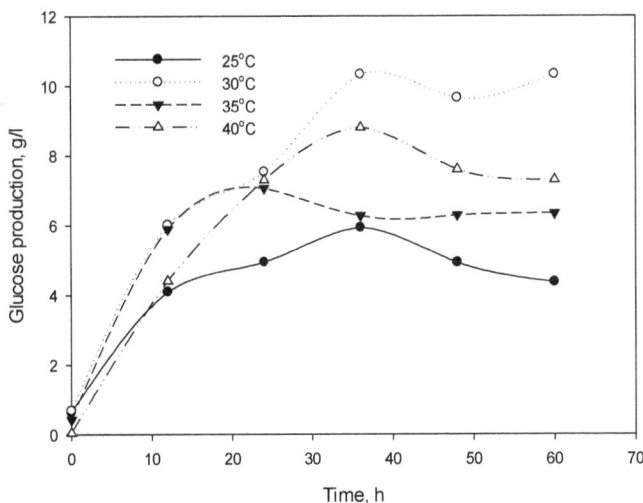

Fig.1. Effect of pH on co-culture A. niger and S. cerevisiae cell activity on rice cooker wastewater for ethanol production.

The results of the effect of different pH on the ability of the fungus in hydrolysis of effluent showed that pH 5 have greatest effect on glucose production by A. niger in temperature 35 °C. The results of this experiments according to Sohail et al. (2009) that they studied the effect of pH and different temperatures on growth of A. niger of the cellulose producing. Although most fungi are active in the range of pH 4 to 6, but it is necessary to increase the efficiency of their operations, the appropriate pH was optimized. Because changing in pH causes changes in protein composition of the cell wall. For example, reducing pH causes the membrane fatty acids to become more saturated

forms. Thus, this is getting stronger and impenetrable thereby reducing microbial growth and activities brings. Pedersen et al. (2000) have studies the effect of pH 2.5 to 6 on the growth and production of starch degrading enzymes. Their results showed that changes pH does not have much impact on growth and ability to produce enzymes glucoamylase. However, their results showed that the maximum amount of enzymes produced was at pH 4 to 5.5, And that the highest enzyme production was observed at pH 5.

3.2. Effect of incubation time on glucose and ethanol production

In SSF process, the incubation time is the limiting factor for fungus and yeast growth or ethanol production. Therefore, the effect of incubation time on Saccharification and fermentation using co-culture A. niger and S. cerevisiae was examined at 0 to 72 h, at a pH and temperatures to 5 and 30 °C, respectively. Fig. 2 shows the reducing sugar and ethanol concentration produced from enzymatic saccharification and fermentation by A. niger and S. cerevisiae. It was found that the reducing sugar concentrations increased with increase in time up to 48 h but above this the productivity decreased. The results show that the maximum reducing sugar concentrations were 27.2 g/l in during 48 h after incubation of A. niger though there was an increase in saccharification of the rice cooker wastewater. This can be caused by breaking starch and glucose production piece, and then consumption piece is by yeast. Results revealed that A. niger can be utilized 53 percent of 50 grams per liter of total sugar. In fact, this show has the ability to hydrolysis of rice cooker wastewater by A. niger. And glucose produces for ethanol production by S. cerevisiae. The result of incubation time S. cerevisiae for ethanol production showed when the incubation time were varied (12 to 72 h) these results were different (Table 1). The effect incubation time of S. cerevisiae on ethanol production shows that the maximum ethanol produces in 36 h after incubation of A. niger 11.26 g/l. (Fig. 2).There might be an increase in saccharification over the period making glucose available to S. cerevisiae for fermentation. Olofsson et al. (2008) reported that enzymatic hydrolysis of the solid fraction has a large control over the total rate of ethanol production in SSF. In a similar study carried out on effect of incubation period on ethanol productivity, Sharma et al. (2007), has reported maximum ethanol yield and fermentation efficiency of 0.397 g/g and 77.84 percent, respectively after 36 h of incubation at 30 °C using mixed culture of S. cerevisiae and P. tannophilus. Reported the maximum ethanol production (0.398 g/g) of at 48 h incubation of employing process of Co-culture of S. cerevisiae G and P. tannophilus MTCC 1077 along with enzymes exhibited. Some authors have reported maximum glucose yield after 48 h of incubation from starchy materials.

3.3. Effect nitrogen source on glucose and ethanol production

Nitrogen is one of the main elements found in many macromolecules of living organisms, playing a central role in structure and function (Magananik. 2005; Najafpour et al. 2004). Type and concentration of nitrogen source effect on fungal growth because it is not only important for metabolic rates in the cells but it is also the basic part of cell protein. In this study the effect of nitrogen source on glucose and ethanol production by A and S were investigated. $(NH_4)_2SO_4$ give the best results, producing 24.63 g/l glucose directly from rice cooker wastewater after 48 h. Also results show NH_4Cl has lowest effect on glucose production.

Table 1. Effect of incubation period and incubation time on ethanol production in SSF process.

Time, h	Incubation time	Incubation period
	Ethanol concentration, g/l	Ethanol concentration, g/l
12	3.56	3.86
24	6.7	5.9
36	12.34	13.6
48	11.26	12.2
60	10.87	11.98
72	10.16	12.34

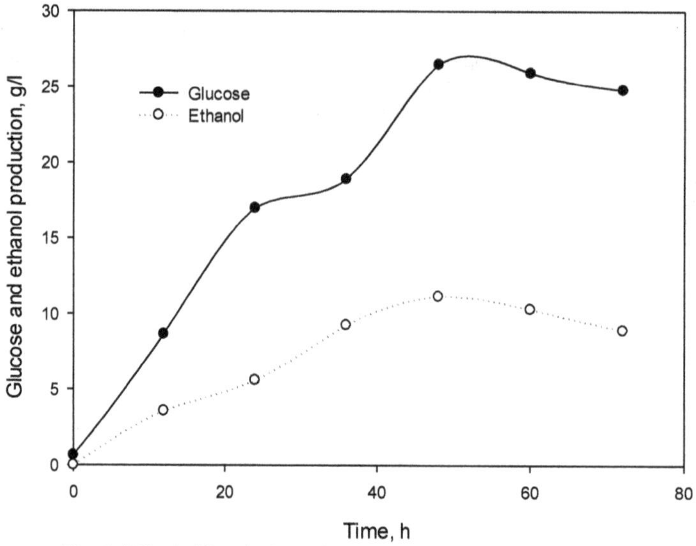

Fig. 2. Effect of incubations time on glucose and ethanol production.

According to Acourene of Ammouche (2010), the nature and relative concentration of nitrogen source are important in formation of α-amylase. Among the organic sources, the yeast extract is the best nitrogen source that increased in the amylase activity. They express that among inorganic nitrogen source ammonium nitrate also enhanced α-amylase activity relatively and but ammonium chloride, repressed the enzymes production.

Nitrogen and complexity of the nitrogen source, strongly affect the glucose fermentation. Effect of nitrogen source was investigated on ethanol production in (SSF) co-culture A. niger and S. cerevisiae the result show that NH₄NO₃ have the highest effect on ethanol production

in co-culture A, and S. with 16.97 g/l. Also ammonium sulfate Compared to NH₄NO₃ has lower effect on ethanol production. According Júnior et al. (2008) ammonium sulfate always induced poorer fermentation performance, with lower biomass and ethanol production, and loss of yeast viability. Their result is similar to that of this study. Among the nitrogen source NH₄Cl have the lower effect on ethanol production. In general, during the SSF experiment with both A. niger and S. cerevisiae, the rate of hydrolysis was lower than the rate of glucose consumption by the yeast cell, which resulted in glucose complete consumption in the fermentation broth.

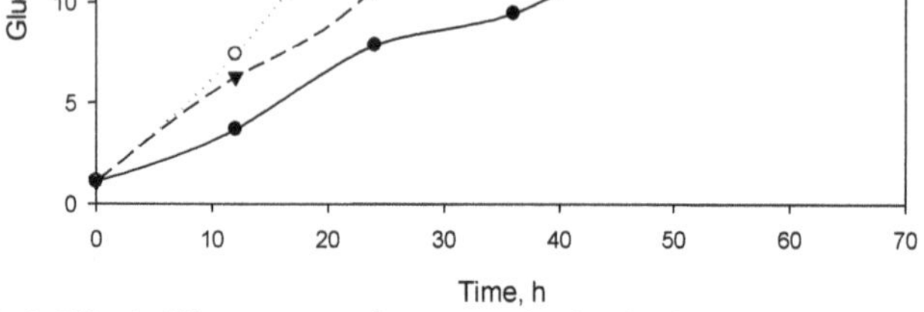

Fig.3. Effect of N course on glucose production by A. niger.

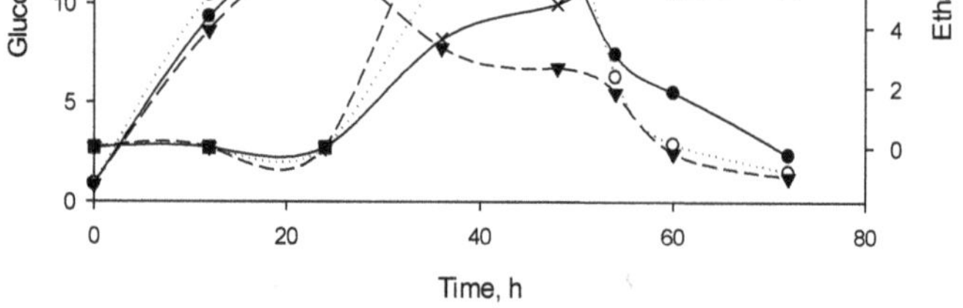

Fig. 4. Effect of nitrogen source on glucose and ethanol production by A. niger and S. cerevisiae in SSF process.

The result show that the between type nitrogen source and maximum glucose and ethanol production sole A. niger and co-culture A. niger and S. cerevisiae have different Fig.4. This may be due to affected structural complexity of the nitrogen source on fungus and yeast metabolism (Ghorbani & Younesi. 2013; Ghorbani et al. 2011). According Messias et al. (2008) and Cruz et al. (2002) have shown that the structural complexity of the nitrogen source strongly affected yeast metabolism.

4. Conclusion

This work thus demonstrates the simultaneous saccharification and fermentation (SSF) process for the biological conversion rice wastewater into ethanol using co-culture of A. niger and S. cerevisiae in batch condition. The process fairly described glucose liberation from SSF and biomass growth, total sugar consumption, ethanol formation, accumulation of ethanol inhibition from fermentation in a batch experiments. The results of this study indicated that the

maximum substrate consumption rate was inhibited by formation of ethanol in batch condition. The results showed that highest glucose production observed using (NH₄)₂SO₄ as nitrogen source in SSF process. From an engineering point of view, these alternatives exhibited comparable biological activity in comparison to the ethanol obtained using the more costly feedstocks and has the potential to become an environmentally and economically acceptable technology for biofuel production.

Acknowledgement

The present research was made possible through a university grant, sponsored by Ministry of Science, Iran, Tarbiat Modares University (TMU). The authors wish to thank Mrs. Haghdoust (Technical assistant of Environmental Laboratory) for her assistance, Tarbiat Modares University and Ministry of Science for their financial support

References

Ang D., Aziz S., Yusof H., Karim M., Ariff A., Uchiyama K., Shioya S., Partial purification and characterization of amylolytic enzymes obtained from direct fermentation of sago starch to ethanol by recombinant yeast, Pakistan Journal of Biological Science 3 (2001) 266-270.

Cardona C.A., Sanchez O.J., Energy consumption analysis of integated flowsheets for production of fuel ethanol from lignocellulosic biomass, Energy 31 (2006) 2111-2123.

Cruz S.H., Cilli E.M., Ernandes J.R., Structural com-plexity of thenitrogen source and influence on yeast growth and fermentation, Journal of the Institute of Brewing 108 (2002) 54-61.

Ghorbani F., Younesi H., The Kinetics of Ethanol Production from Cane Molasses by Saccharomyces cerevisiae in a Batch Bioreactor, Energy Sources, Part A: Recovery, Utilization, and Environmental Effects 35 (2013) 1073-1083.

Ghorbani F., Younesi H., Esmaeili Sari A., Najafpour G., Cane molasses fermentation for continuous ethanol production in an immobilized cells reactor by Saccharomyces cerevisiae, Renewable Energy 36 (2011) 503-509.

Gupta R., Gigras P., Mohapatra H., Goswami V.K., Chauhan B., Microbial α-amylases: a biotechnological perspective, Process Biochemistry 38 (2003) 1599-1616.

Júnior M.M., Batistote M., Ernandes J.R., Glucose and Fructose Fermentation by Wine Yeasts in Media Containing Structurally Complex Nitrogen Sources, Journal of the Institute Brewing 114 (2008) 199-204.

Kádár Z., Szengyel Z., Réczey K., Simultaneous saccharification and fermentation (SSF) of industrialwastes for the production of ethanol, Industrial Crops and Products 20 (2004) 103-110.

Magananik B., The transduction of the nitrogen regulationsignal in Saccharomyces cerevisiae, Proceedings of the National Academy of Sciences USA 102 (2005) 16537-16538.

Miller G.L., Use of dinitr osalicylic acid reagen t for determination of reducing sugar, Analytical Chemistry 31(1959) 426- 428.

Najafpour G., Younesi H., Ku Ismail K.S., Ethanol fermentation in an immobilized cell reactor using Saccharomyces cerevisiae, Bioresource Technology 92 (2004) 251-260.

Nakamura Y., Kobayashi F., Ohnaga M., Sawada T., Alcohol fermentation of starch by a genetic recombinant yeast having glucoamylase activity, Biotechnology and bioengineering 53 (2000) 21-25.

Nigam P., Singh D., Enzyme and microbial systems involved in starch processing, Enzyme and Microbial Technology 17 (1995) 770-778.

Olofsson K., Bertilsson M., Liden G., A short review on SSF-an interesting process option for ethanol production from lignocellulosic feedstocks, Biotechnol. for Biofuels 1 (2008) 1-14.

Peterson H.N.J., The influence of nitrogen sources on the a-amylase productivity of Aspergillus oryzae in continuous cultures, Applied Microbial Biotechnology 1 (2000) 278-281.

Sharma N., Kalra K.L., Oberoi H.S., Bansal S., Optimization of fermentation parameters for production of ethanol from kinnow waste and banana peels by simultaneous saccharifi cation and fermentation, Indian Journal of Microbiology 47 (2007) 310-316.

Sohail R.S.M., Ahmad A., Khan S.A., Cellulase production from Aspergillus niger MS82: effect of temperature and pH, New Biotechnology (2009) 25 437-441.

Verma G., Nigam P., Singh D., Chaudhary K., Bioconversion of starch to ethanol in a single-step process by coculture of amylolytic yeasts and Saccharomyces cerevisiae, Bioresource Technology 72 (2000) 261-266.

11

Interaction between diazinon and nitrate pollutant through membrane technology

Peyman Mahmoodi, Mehrdad Farhadian*, Ali Reza Solaimany Nazar, Amin Noroozi

Department of Chemical Engineering, University of Isfahan, Isfahan, Iran.

ARTICLE INFO

Keywords:
Contaminated water
Diazinon
Nitrate
Agricultural wastewater
Environment
Nanofiltration

ABSTRACT

The efficiency of diazinon (as insecticides) and nitrate (related to nitrogen fertilizer) removal from contaminated water is investigated through NF membrane technique. The effects of nitrate concentration (40-160 mg/l), diazinon concentration (10-1000 µg/l) and pH (5-9) on the efficiency of a commercial polyamide nanofilter membrane at a constant pressure of (800 KPa) are investigated. The response surface method (Box-Behenken design) is applied in design of experiment. As the diazinon concentration and pH are enhanced, the contaminant removal efficiency increases from 85 % to 90 %; while nitrate concentration has an opposite effect (removal efficiency reduces about 10 %). The regression models obtained for nitrate and diazinon rejection show good fitting to the experimental results (r-squared equal to 94 % and 98 %, respectively). The models are able to predict the evolution of diazinon and nitrate as a function of concentration and pH at a constant pressure.

1. Introduction

Agriculture is an important industry in many parts of the world while the water pollutants wastewater can cause environmental contamination. This type of wastewater contains different pollutants such as: insecticides, pesticides, organic pollutants, chemical and animal fertilizers (Ongley 1996).

Nitrate as a carcinogenic compound is a key pollutant found in water runoff. According to World Health Organization standard, the maximum allowable concentration of nitrate in drinking water is 50 mg/L (World Health Organization 2003). Diazinon is one of the insecticides and pesticides used in agriculture. This compound is toxic and according to Canadian Standards, the maximum allowable concentration in drinking water is 20 µg /L (Moreno et al. 2005).

In recent years, great attention is paid to the use of nanofiltration (NF) process for simultaneous removal of organic and inorganic contaminants, water hardness, insecticides, heavy metals, nitrate, and micro-pollutants in one step (Kaya et al. 2010). NF is a membrane process occurring between ultrafiltration (UF) and reverse osmosis (RO). In this method, the separation mechanisms are based on the size of molecules, differences in diffusivity and solubility of feed components, and the electrical interaction between the surface of the membrane and ions present in the feed (Bellona and Drewes 2005).

In case of ionic mixtures, the electrostatic interactions with other anions may cause reduced nitrate rejection, particularly in presence of less permeable anions in solution according to Donnan exclusion. The nitrate removal efficiency can be changed in present of chloride and sulphate (Santafé-Moros and Gozálvez-Zafrilla 2010; Paugam et al. 2004). Tepus et al. (2009) investigated the comparison of nitrate and atrazine and dimethyleatrazine (as pesticides) removal from contaminated water using NF membrane. They used a commercial nanofilter membrane (DK- GE Osmonics Desal), and the results indicate that the present of pesticides reduced the nitrate removal efficiency. A limited number of studies are performed to determine the effect of diazinon on the nitrate ion rejection by NF membranes. Due to the possibility of simultaneous presence of nitrate and diazinon in contaminated water within agricultural wastewater, the focuses of this study is to investigate the effect of diazinon and nitrate concentrations

and pH of solution on the simultaneous removal of the pollutants by applying a commercial NF. For this purpose, the response surface methodology is applied to optimize the response of the amounts of nitrate and diazinon removal.

2. Materials and methods
2.1. Chemicals

Potassium nitrate is supplied by Merck (Germany). Hydrochloric acid and sodium hydroxide are used to adjust the pH. Commercial emulsion of 60 % diazinon pesticide is supplied by Giah Sam Company (Iran). The nitrate and diazinon solutions are prepared using distilled water.

2.2. Experimental set-up

All experiments are performed in a NF pilot plant (Fig. 1), equipped with a spiral wound polyamide membrane (developed by Noshirvani University of Technology, Iran) operated in a continuous flow mode. The characteristics of the membrane are presented in Table 1. Two diaphragm pumps with a capacity of 1.6 liters per minute at a maximum outlet pressure of 8.5 bars are used in the setup.

2.3. Methods

The factors and the selected levels are based on the actual levels in water resources (see Table 2). For all experiments, feed temperature and optimum pressure are set at 20±1 °C, and 800 kPa, respectively. The recovery rate is regulated at 75±2 percent. All the experimental results are obtained when the steady state is achieved. All measurements are performed according to standard methods relevant to water and wastewater (Arnold et al. 2003). The nitrate concentrations are measured by Jasco V-570 spectrophotometer according to standard method (4500B) (Arnold et al. 2003). The diazinon concentrations in contaminated water are measured by high performance liquid chromatography (HPLC- KNAUER model-Germany). The HPLC-column used is a C18 column, 15cm in length and 0.46 cm in internal diameter. The mobile phase was acetonitrile:water (70:30). The UV detector is operated at a wavelength of 220 nm. The diazinon and nitrate removal efficiency by nanofilter are determined as follow:

$$R,\% = [1-(\frac{C_p}{C_0})]\times100 \quad (1)$$

where, R represents the removal percentage of diazinon or nitrate and C_p and C_0 are the concentrations of the pollutant in the permeate and the feed water, respectively.

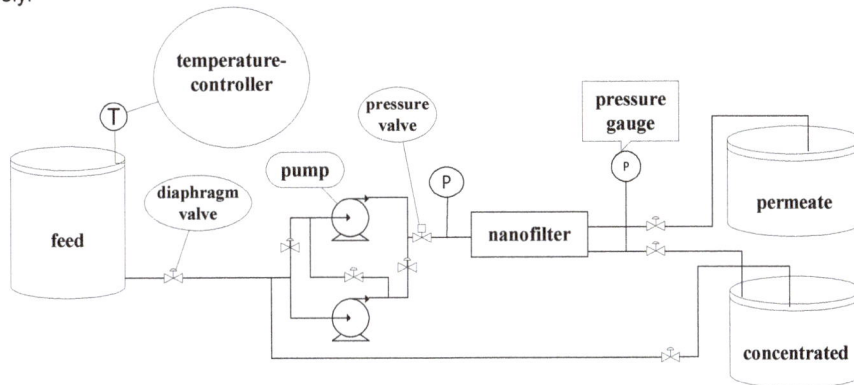

Fig. 1. Schematic view of the experimental setup.

Table 1. The specifications of the commercial polyamide nanofilter membrane.

Specification	Allowed range
Maximum operating pressure (bar)	20
Maximum operating temperature (ºC)	50
pH range	3-12
Active surface (m²)	0.35
Isoelectric point	4.6
Surface charge	Negative

Table 2. Factors and selected levels.

Factor	Level 1	Level 2	Level 3
Diazinon concentration (µg/L)	10±1	100±5	1000±10
Nitrate concentration (mg/L)	40±2	80±3	160±5
pH	5±0.1	7±0.1	9±0.1

2.4. Response surface methodology

Response surface methodology is an effective method for optimizing the responses (Myers and Montgomery 2002). In this method, the Box-Behnken design is used to optimize responses. This design includes three level factors, and a three-time implementation of the experiments in the central surface, in order to obtain the experimental error. A second order polynomial is presented by the design approach to fit the experimental data as (Myers and Montgomery 2002).

$$Y = b_0 + b_1X_1 + b_2X_2 + b_3X_3 + b_{11}X_1^2 + b_{22}X_2^2$$
$$+ b_{33}X_3^2 + b_{12}X_1X_2 + b_{13}X_1X_3 + b_{23}X_2X_3 \quad (2)$$

where, X1, X2, X3 represent the coded levels of the independent variables and b0, bi, bij (i,j=1,2,3) the coefficient estimates, and b0 is the interception, bi the linear terms, biithe quadric terms and bij is the interaction terms.

The statistical analyses of the results are obtained through Design-Expert released 8.0.1 software. The confidence level was selected at 95%. In this study, the objective is to maximize the pollutants removal efficiency.

3. Results and discussion

The levels of independent variables according to the Box-Behnken method and the nitrate rejection percent (Y_1) and the diazinon rejection percent (Y_2) responses for all experiments are presented in Table 3. In order to avoid systematic bias, the experiments are carried out on a random basis.

Table 3. Experimental design (conditions and responses) for nitrate and diazinon rejection.

Run	Nitrate concentration in mg/L	Diazinon concentration in µg/L	pH	Nitrate Rejection Y1 (%)	Diazinon Rejection Y2 (%)
1	80±3 (0)	100±5 (0)	7±0.1 (0)	88.2	91.5
2	40±2 (-1)	100±5 (0)	5±0.1 (-1)	81.4	87.1
3	40±2 (-1)	100±5 (0)	9±0.1 (1)	93.5	94.7
4	40±2 (-1)	1000±10 (1)	7±0.1 (0)	90.2	93.5
5	40±2 (-1)	10±1 (-1)	7±0.1 (0)	81.1	81.4
6	160±5 (1)	100±5 (0)	5±0.1 (-1)	77.3	88.2
7	160±5 (1)	100±5 (0)	9±0.1 (1)	90.8	95.6
8	80±3 (0)	10±1 (-1)	9±0.1 (1)	85.6	91.2
9	80±3 (0)	100±5 (0)	7±0.1 (0)	85.9	92
10	80±3 (0)	1000±10 (1)	9±0.1 (1)	96.5	97
11	80±3 (0)	1000±10 (1)	5±0.1 (-1)	79.8	86.3
12	80±3 (0)	100±5 (0)	7±0.1 (0)	89.3	89.8
13	80±3 (0)	10±1 (-1)	5±0.1 (-1)	72.3	77.4
14	160±5 (1)	10±1 (-1)	7±0.1 (0)	78.6	82.3
15	160±5 (1)	1000±10 (1)	7±0.1 (0)	87.6	94.1

3.1. Analysis of experimental data

The analysis of variance is shown in Table 4. A factor is significant when its effect on response is inevitable and cannot be neglected. The effect of any factor is significant when its P-value is less than 0.05, which means that there is only 5% probability of error if a non-

significant factor is considered as a significant one. There is no evidence of 'lack-of-fit' since for diazinon and nitrate removal production P-values are 0.158 and 0.822 (≥0.05), respectively.

The greater F-value shows a greater effect of the factor on the response. For removal of diazinon from contaminated water, pH and diazinon concentration effects are significant while nitrate concentration is not significant. The pH of Solution (X_1) followed by diazinon concentration of (X_2), and nitrate concentration of (X_3) have the greatest effects on the nitrate removal from contaminated water. Besides, there are no interacting effects among the mentioned factors.

Table 4. Analysis of variance for nitrate (A) and diazinon (B) rejection

Source	d.f.	Seq SS	Adj MS	F	P
(A) Nitrate rejection					
Model	9	609.14	67.68	38.4	0.0004
X3-Nitrate concentration	1	17.7	17.7	10.04	0.0248
X2-Diazinon concentration	1	166.53	166.53	94.49	0.0002
X1-pH	1	386.42	386.42	219.25	<0.0001
X3X2	1	0.0025	0.0025	0.00142	0.9714
X3X1	1	0.49	0.49	0.28	0.6206
X2X1	1	2.89	2.89	1.64	0.2565
X3X3	1	1.39	1.39	0.79	0.4159
X2X2	1	29.21	29.21	16.57	0.0096
X1X1	1	7.63	7.63	4.33	0.0920
Residual error	5	8.81	1.76		
Lack of Fit	3	2.79	0.93	0.31	0.8216
Pure Error	2	6.02	3.01		
Total	14	617.95	1.39		
(B) Diazinon rejection					
Model	9	427.88	47.54	9.68	0.0111
X3-Nitrate concentration	1	8.53	8.53	0.31	0.6006
X2-Diazinon concentration	1	186.24	186.24	37.93	0.0016
X1-pH	1	195.03	195.03	39.72	0.0015
X3X2	1	0.022	0.022	0.00458	0.9487
X3X1	1	0.01	0.01	0.00204	0.9658
X2X1	1	2.4	2.4	0.49	0.5155
X3X3	1	0.021	0.021	0.00423	0.9507
X2X2	1	41.44	41.44	8.44	0.0336
X1X1	1	0.19	0.19	0.038	0.8530
Residual error	5	24.55	4.91		
Lack of Fit	3	21.89	7.30	5.49	0.1580
Pure Error	2	2.66	1.33		
Total	14	452.43			

The mathematical model based on actual values for diazinon and nitrate removal percentages are expressed through Eqs. (3) and (4) as follows, respectively:

$$Y_1 = 91.1 + 0.438X_3 + 4.825X_2 + 4.938X_1 - 0.075X_2X_3 \qquad (3)$$
$$- 0.050X_1X_3 - 0.775X_1X_2 + 0.075X_3^2 - 3.350X_2^2 + 0.225X_1^2$$

$$Y_2 = 87.8 - 1.488X_3 + 4.563X_2 + 6.950X_1 - 0.025X_2X_3 \qquad (4)$$
$$+ 0.350X_1X_3 + 0.850X_1X_2 - 0.613X_3^2 - 2.813X_2^2 - 1.438X_1^2$$

The regression parameter R^2 is applied to determine the agreement in comparison of the experimental responses to the ones estimated by Box-Behnken method. For diazinon and nitrate rejection, R^2 statistic parameter is 94.6 % and 98.6 %, respectively. Due to their proximity to unity, the proposed models are accurate and acceptable.

3.2. Diazinon removal

The contour plots for diazinon removal at varying pH/diazinon concentration values (a), pH/nitrate concentration values (b) and diazinon concentration/nitrate concentration values (c) are illustrated in Fig. 2, respectively. It should be noted that the third factor, in all these cases, is held constant at the center point, i.e. nitrate concentration of 80 mg/L, diazinon concentration of 100 µg/L and pH of solution 7, respectively. Furthermore, the graphs in Fig. 2 indicate that, in general, pH has a significant effect on diazinon rejection. With an increase in pH, the membrane surface swells casing a decrease in pore size; therefore, molecular transfer becomes difficult, that in turn increases the removal percentage. The results are in agreement with the other reported results (Ahmad et al. 2008).

As the diazinon concentration increases, its removal percentage increases. The reason is that the membrane pores are able to allow a limited number of pesticide molecules to cross and as the number of molecules per unit volume increases; their crossing becomes more difficult. Due to the same implementation time for every experiment in different concentrations, the number of molecules that cannot cross membrane pores increases; therefore, the diazinon removal percentage by the membrane increases. The same similar trend mechanism is presented by other researchers (Tepus et al. 2009; Ko˘sutic´ et al. 2005).

The data indicated that with an increase in nitrate concentration, the diazinon removal efficiency increased slightly (Fig. 2b and 2c). The reason here is that when nitrate concentration increases, cation concentration increases as well; hence cation adsorption on membrane surface increases (since membrane surface charge is negative) and as a result, the repulsion between the membrane surface and the nitrate ion decreases and the nitrate ions are able to pass easily through the membrane pores. The effect of nitrate concentration and reduction of the removal efficiency of pollutants by nanofilter membrane are reported by other researchers (Richards et al. 2010; Santafe-Moros et al. 2007).

3.3. Nitrate removal

The contour plots for nitrate removal at varying pH/diazinon concentration values (a), pH/nitrate concentration values (b) and diazinon concentration/nitrate concentration values (c), can be observed in Fig. 3, respectively.

The results reveal that the removal percentage decreases with an increase in nitrate concentration. Since the electrical charge of the commercial nanofilter is negative at the operating conditions (isoelectric point of the membrane is 4.6), the removal efficiency of nitrate increases with an increase in pH, as illustrated in Figs. 3-b, c. The results indicate that an increase in pH from 5 to 9, the removal percentage of nitrate increases from about 75% to over 90%. Similar results are reported by Richards et al. (2010).

Here the results indicate that the removal efficiency increases as diazinon concentration increases. The diazinon molecule radius of 0.834 nm (Ko˘sutic´ et al. 2005) is greater than the nitrate ions radius of 0.128 nm (Wang et al. 2005), thus, an increase in diazinon concentration increases the space barrier, which in turn increases the removal efficiency (Tepus et al. 2009; Plakas and karabelas 2012).

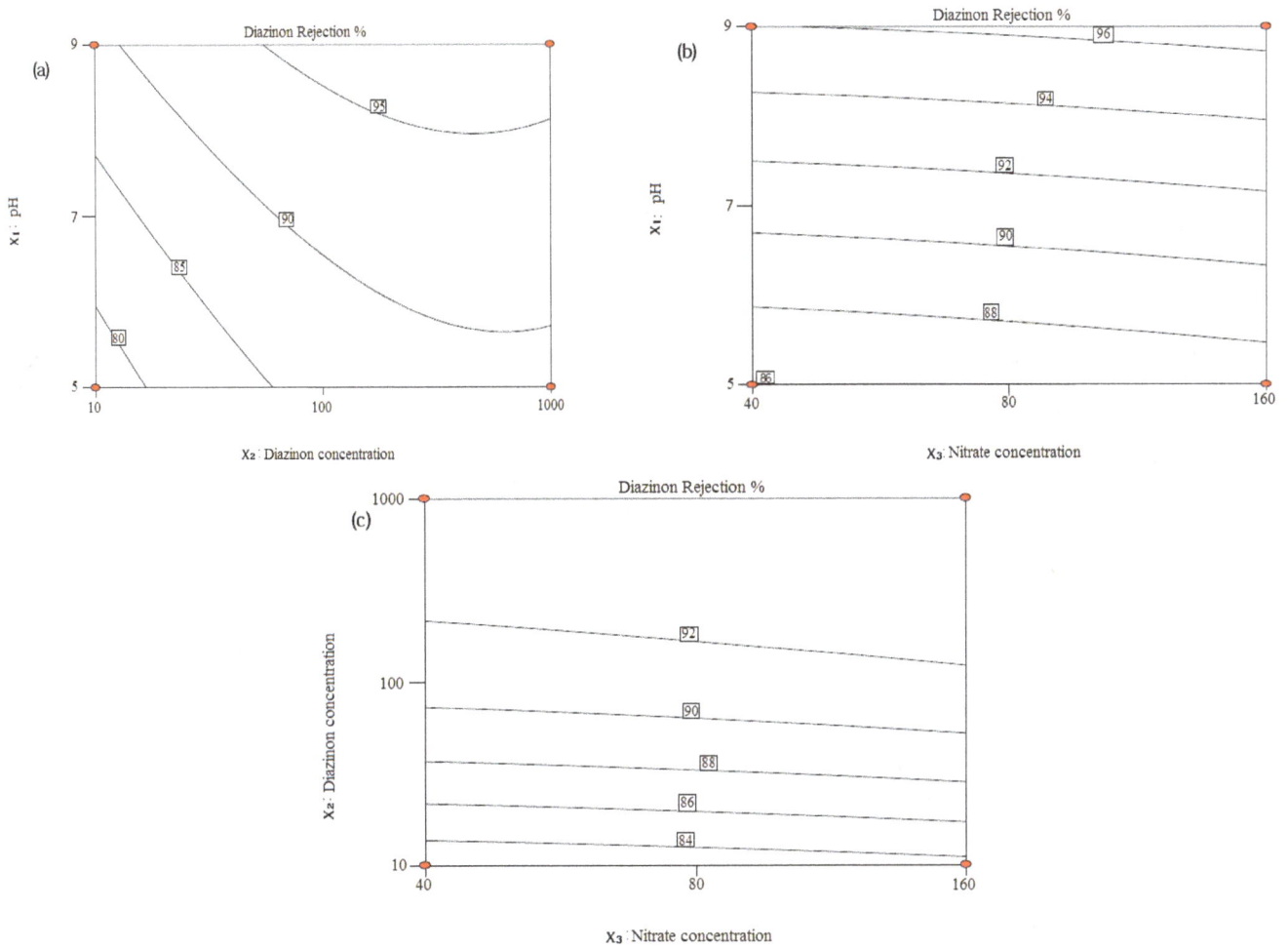

Fig. 2. Contour plots of the removal percentage of diazinon; (a): the effect of pH and diazinon concentration on the removal efficiency at constant nitrate concentration of 80 mg/L; (b): the effect of pH and nitrate concentration on the removal efficiency at constant diazinon concentration of 100 µg/L; (c): the effect of diazinon and nitrate concentrations on the removal efficiency of diazinon at pH=7.

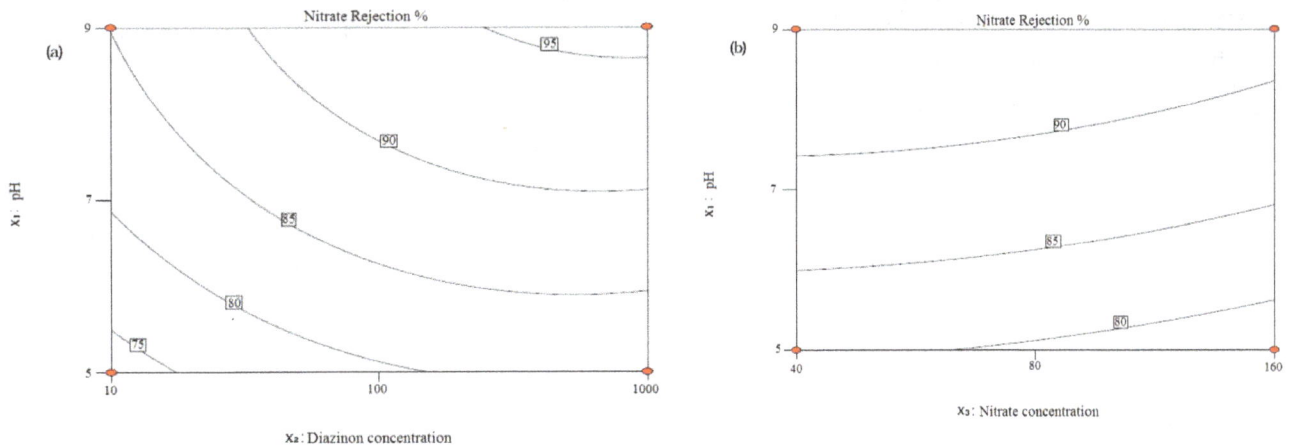

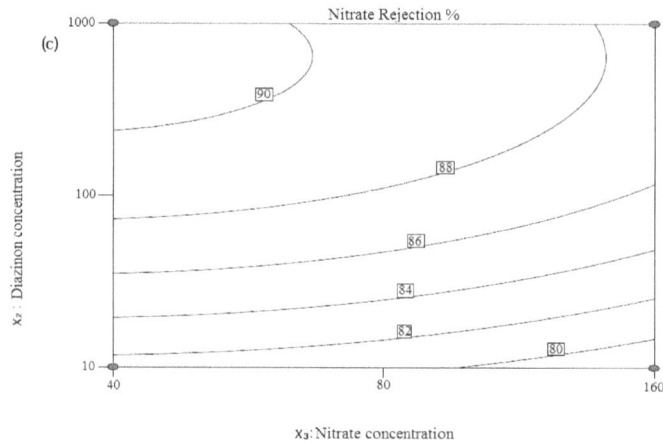

Fig. 3. Contour plots of the removal percentage of nitrate; (a): the effect of pH and diazinon concentration on the removal efficiency at constant nitrate concentration of 80 mg/L; (b): the effect of pH and nitrate concentration on the removal efficiency at constant diazinon concentration of 100μg/L; (c): the effect of diazinon and nitrate concentrations on the removal efficiency of nitrate at pH=7.

4. Conclusions

The nanofiltration process using a commercial spiral polyamide nanofilter is efficient for simultaneous removal of nitrate and diazinon from contaminated agricultural wastewater. The physical and chemical properties of water and the characteristics of membrane have great impacts on the nanofiltration system performance. Higher levels of diazinon and nitrate concentrations can be rejected as diazinon concentration increases at maximum pH of 9, with adjustment of the nitrate concentration close to 45 mg/L.

References

Ahmad L., Abd A.S., Tan L.R., Shukor S., The Role of pH in Nanofiltration of Atrazine and Dimethoate from Aqueos Solution, Journal of Hazardous Materials 154 (2008) 633-638.

Arnold E., Greenbeag A., Michael j., TARAS A., Standard Methods for the Examination of Water and Waste Water- American Public Health Association, 18 ed., American Water works Association, (2003).

Bellona C., Drewes J.E., The role of membrane surface charge and solute pysico-chemical properties in the rejection of organic acids by NF membranes, Journal of Membrane Science 249 (2005) 227-234.

Kaya Y., Gonder Z.B., Vergili I., Barlas H., The effect of transmembrane pressure and pH on treatment of paper machine process waters by using a two-step nanofiltration process: Flux decline analysis, Desalination 250 (2010) 150-157.

Ko˘sutic´ K., Fura˘c L., Sipos L., Kunst B., Removal of arsenic and pesticides from drinking water by nanofiltration membranes, Separation and Purification Technology 42 (2005) 137–144.

Moreno B., Gomez M.A., Ramos A., Gonzalez J., Lopez J., Hontoria E., Influence of inocula over start up of a denitrifying submerged filter applied to nitrate contaminated, groundwater treatment, Journal of Hazardous Materials 127 (2005) 180-186.

Myers R.H., Montgomery D.C., Response Surface Methodology, Process and Product Optimization Using Designed Experiments, 2nd ed., John Wiley and Sons Inc., (2002).

Ongley E.D., Control of water pollution from agriculture, GEMS/Water Collaborating Centre, (1996).

Paugam L., Diawara C.K., Schlumpf J.P., Jaouen P., Quéméneura F., Transfer of monovalent anions and nitrates especially through nanofiltration membranes in brackish water conditions, Separation and Purification Technology 40 (2004) 237-242.

Plakas K.V., karabelas A.J., Removal of Pesticides from Water by NF and RO Membranes, Desalination (2012) 255-265.

Richards L., Vuarchère M., Schäfer A.I., Impact of pH on the removal of fluoride, nitrate and boron by nanofiltration/reverse osmosis, Desalination 261 (2010) 331-337.

Santafe-Moros A., Gozalvez-Zafrilla J.M., Lora-Garcia J., Nitrate removal from ternary ionic solutions by tight nanofiltration membrane, Desalination 204 (2007) 63-71.

Santafé-Moros A., Gozálvez-Zafrilla J.M., Nanofiltration study of the interaction between bicarbonate and nitrate ions, Desalination 250 (2010) 773–777.

Tepus B., Siminic M., Petrinic I., comparison between nitrate and pesticides removal from ground water using adsorbents and NF and RO membranes, Journal of Hazardous Materials 170 (2009) 1210-1217.

Wang D.X., Su M., Yu Z.Y., Wang X.L., Ando M., Shintani T., Separation performance of a nanofiltration membrane influenced by species and concentration of ions, Desalination 175 (2005) 219-225.

World Health Organization (WHO), Chemical fact sheets, in, www.who.int/water_sanitation_health/dwq/GDW12rev1and2.pdf, 2003.

12

A kinetic and thermodynamic study of methylene blue removal from aqueous solution by modified montmorillonite

Adeleh Afroozan, Ali Mohammad-khah*, Farhad Shirini

Department of Chemistry, Faculty of Sciences, University of Guilan, Rasht, Iran.

ARTICLE INFO	ABSTRACT
Keywords: Na$^+$-Montmorillonite Methylene blue Removal Kinetic Thermodynamic	In current study, sulfonic acid-functionalized ordered nanoporous Na$^+$-Montmorillonite (SANM) has been utilized as the adsorbent for the removal of a cationic dye, methylene blue (MB), from aqueous solution using the batch adsorption technique under different conditions such as temperature, adsorbent dosage, initial dye concentration, contact time, and pH solution. The optimum sorption conditions were found as following: contact time 10 min, initial dye concentration 800 mg/L, adsorbent dose 0.3 g and temperature 25 °C. The results indicate that the process is pH independent. The sorption capacity was 500 mg/g for this dye. Different thermodynamic parameters i.e., changes in standard free energy, enthalpy, and entropy have also been evaluated. The ΔH_{ads} and ΔS_{ads} values are thus found to be +38240 (J/mol) and ΔS_{ads}138.43 J/K, respectively, while the ΔG_{ads} values is -3012.14 J in 298 K and it has been found that the reaction was spontaneous and endothermic in nature. On the other hand, Kinetic parameters have been investigated with pseudo first and second order. The result of experimental data indicates that pseudo second order equation fit better than the other.

1. Introduction

Dyes, usually having a synthetic origin, are generally characterized by complete aromatic molecular structures which afford physicochemical, thermal, and optical stability (Dogan et al. 2006). The discharge of wastewaters containing dyes into river and streams is easily noticed since dyes are highly visible (Mohammed et al. 2014). Not only do they damage the aesthetic nature of the environment, but also they are commonly toxic to aquatic life (Akar et al. 2006). Various treatment processes including physical separation, chemical oxidation, and biological degradation have been widely investigated to remove dyes from wastewaters. Although there are many methods for the removal of the dyes, it is difficult to treat the wastewater through using traditional methods because most of the synthetic dyes are stable to photo-degradation, bio-degradation and oxidizing agents, etc. (Mahanta et al. 2008). The most commonly used methods for color removal are biological and chemical precipitation. Although biological procedures are widely utilized in the removal of color, they are very inefficient (Preethi et al. 2006) because of the low biodegradability of dyes (Vadivelan and Vasanth Kumar. 2005). Many techniques such as ion exchange (Raghu and Basha. 2007), chemical precipitation (Zhu et al. 2007), coagulation (Raghu and Basha. 2007), ozonation (Yan et al. 2009) and adsorption (Ansari et al. 2012) have been developed to remove these dyes from aqueous solution. Among these techniques, adsorption is considered as a preferred and effective technique due to the fact that it can deal with various concentrations of dyes and does not induce the formation of hazardous materials. The most widely employed adsorbent is activated carbon (Mohammad-khah and Ansari. 2009) due to its pore structure, high efficiency, and adsorption capacity for some dyes, but the cost and difficulty in separation after adsorption hinders its large-scale application (Uddin et al. 2009a).

In recent year, there has been considerable growth of interest in using environmentally friend adsorbent to enhance removal of dye from wastewater one of which is clay. Montmorillonite (MMT) is the most often studied swelling clay mineral. The MMT layers have permanent negative charges because of the isomorphous substitution of, for instance, Mg^{2+} for Al^{3+} or, rarely, Al^{3+} for Si^{4+}.

These net negative charges are balanced by exchangeable cations such as sodium and potassium between the clay layers in the gallery spaces (Shirini et al. 2012a).

In this study, functionalized surface of clay with Sulfonic acid (SANM) has been used as a newly reported derivative of Na$^+$-Montmorillonite in which the number of negative charges is increased in enhancing the adsorption efficiency of methylene blue (MB). The aim of this work is to improve sorption characteristics of montmorillonite via modification using sulfonic acid. Montmorillonite was found a proper choice because it is available as clay. It was also found that a suitable substrate for the sorption of sulfonic acid in surface and negative charges are more and more in clays surface and convert this to best choice for dye removal because of efficiency and short contact time.

2. Materials and methods
2.1. Materials and reagents

Chemicals were purchased from Southern Clay Products, Fluka, Merck, and Aldrich chemical companies. A Metrohm pH meter (Model 827), with a combined double junction glass electrode, was used for pH measurements. The pH adjustments were carried out using diluted NaOH and HCl solutions. Methylene blue dye, termed as MB (chemical formula= $C_{16}H_{18}N_3SCl$, MW= 319.85 g/mol), was employed as a typical cationic dye for the current investigation. An UV-Vis spectrophotometer (Perkin-Elmer 35) was employed for the determination of residual MB concentration.

2.2. Preparation of SANM

A 500 mL suction flask, charged with 2.5 g Na$^+$-montmorillonite (Southern Clay Products) and 10 mL CHCl$_3$, was equipped with a

Corresponding author Email: mohammadkhah@guilan.ac.ir

constant pressure dropping funnel containing chlorosulfonic acid (0.50 g, 9 mmol) and a gas inlet tube for passing HCl gas through an adsorbing solution i.e. water. Chlorosulfonic acid was added drop wise over a period of 30 min while the reaction mixture was stirred slowly in an ice bath. After addition was completed, the mixture was stirred for additional 30 min to remove all HCl. Then, the mixture was filtered and the solid residue was washed with methanol (20 mL) and dried at room temperature to obtain SANM as a white powder (Shirini et al. 2012a).

2.3. Characterization of SANM

SANM was prepared and characterized as the reported method in the literature (Shirini et al. 2012a; Shirini et al. 2012b).

2.4. Adsorption study

Batch adsorption experiments were carried out using a shaker. The effects of contact time, dye concentration and temperature on adsorption were investigated. For each experimental run, 0.3 g adsorbent and 50 mL of MB solution of known concentration were transferred into a flask, and agitated by shaker at a constant speed of 140 rpm with a required adsorption time. At predetermined time intervals, the solutions were centrifuged at 50 rpm for 10 min. The adsorption kinetics was determined by analyzing adsorption capacity from the aqueous solution at different time intervals. For adsorption isotherms, MB solution of different concentrations in the range of 200-1800 mg/L was agitated till the equilibrium was achieved. The effect of temperature on the adsorption characteristics was investigated by determining the adsorption isotherms at 303 to 353 K. Concentrations of dyes were determined by finding out the absorbance characteristic wavelength using UV-spectrophotometer. A standard solution of the dye was taken and the absorbance was determined at different wavelengths to obtain a plot of absorbance versus wavelength. The wavelength corresponding to maximum absorbance (λ_{max}) was determined from this plot. The λ_{max} for MB were found to be 640 nm. Calibration curves were plotted between absorbance and concentration of the dye solution. The adsorbed amounts (q) of MB were calculated by the following equation:

$$q = \frac{(C_0 - C_e)}{m} \qquad (1)$$

where, C_0 and C_e are the initial and equilibrium concentrations of dye (mg/L), m is the mass of sorbent (g), and V is the volume of solution (L).

3. Results and discussion
3.1. Effect of initial MB concentration

The initial concentration of adsorbate in solution provides an important driving force in overcoming mass transfer resistance between the aqueous and the solid phases (Dogan et al. 2006).

Equilibrium adsorption studies have been performed to determine the capacity of the adsorbent; and the equilibrium is established when the concentration of adsorbate in the bulk solution is in dynamic balance with that on the surface (Bouberka et al. 2006). The effect of initial MB concentration on the adsorption onto SANM and the percentage dye removal are shown in Fig. 1. Hence, it appears that more MB was retained by the adsorbent and the adsorption mechanism also became more efficient, as the initial dye concentration increased, the percentage removal was higher at low concentration (Gupta et al. 2006). Moreover, the percentage of MB removal by the SANM sample at equilibrium indicates that the initial dye concentration has an important influence on the adsorption capacity of the SANM (Fig. 1). For example, at higher solution concentrations, 1600 and 1800 mg/L, the MB uptake shows a slight decrease. This result was obtained by 0.3 g of adsorbent in 10 min contact time.

3.2. Effect of contact time

The contact time being necessary to reach equilibrium depends on the initial dye concentration. It has been shown that the adsorption capacity increases with this concentration and the rate of adsorption on the surface should be proportional to a driving force times an area

(Almeida et al. 2009). The adsorption uptake versus contact time was investigated to find out the equilibrium time for maximum adsorption, as illustrated in Fig. 2. It is clear that the adsorption capacity increased promptly in the first time slot (0-10 min) but, after the contact time reached to 10 min, the adsorption capacity remained almost unchanged. Thereby, the equilibrium time of the studied experimental system was 10 min. Thus, a contact time of less than 10 min was invariably sufficient to reach to the equilibrium. The fast uptake of the dye molecules is due to solute transfer, as there are only adsorbate and sorbent interactions with negligible interference from solute–solute interactions. The initial rate of adsorption was therefore greater for high initial MB concentrations, the resistance to the dye uptake diminishing as the mass transfer driving force increased (Dogan et al. 2006).

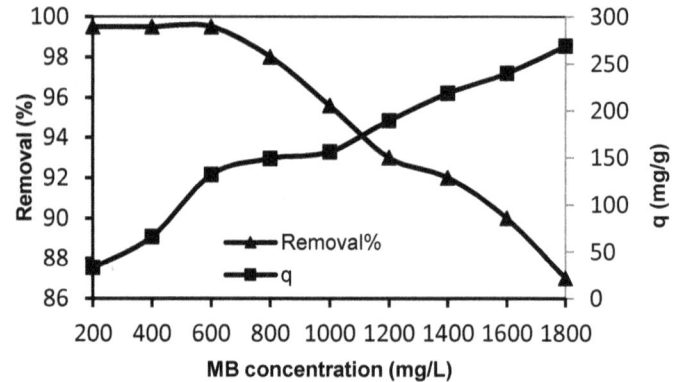

Fig. 1. Effect of initial MB concentration on the adsorption onto SANM (T=298 K, adsorbent = 0.3 g, V = 0.05 L).

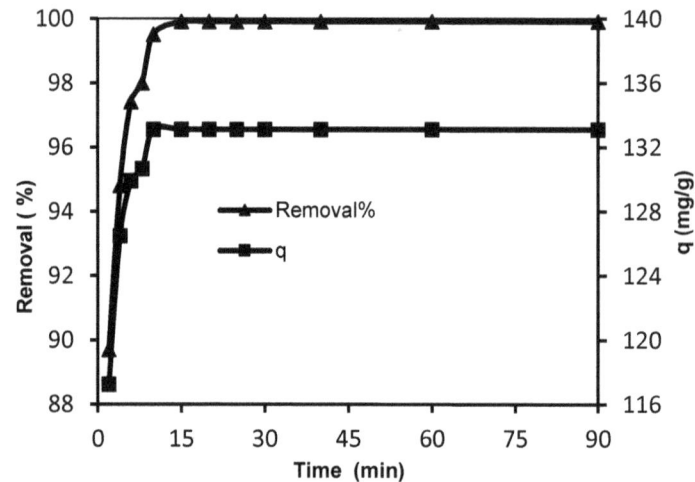

Fig. 2. Effect of contact time on the adsorption of MB onto SANM (T=298 K, C_0=800 mg/L, adsorbent = 0.3 g, V = 0.05 L).

3.3. Effect of adsorbent dosage

The effect of SANM concentration on MB adsorption at a contact time of 10 min was studied by varying the adsorbent dose from 0.1 to 0.75 g in 800 mg/L MB solution. The results are shown in Fig. 3. Increasing the adsorbent dosage enhances the percentage of removal of MB. Increased adsorbent concentration implies a greater surface area of SANM and, consequently, a greater number of possible binding sites. At adsorbent doses greater than 0.3 g, there was few changes in either the rate of attaining adsorption equilibrium of MB or the percentage removal of MB.

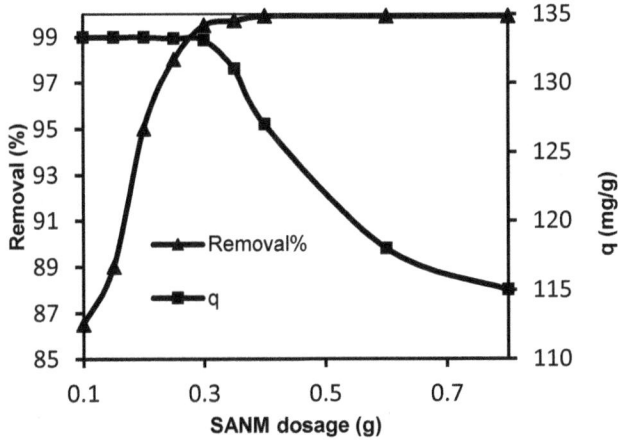

Fig. 3. Effect of SANM dosage on MB adsorption at a contact time of 10 min (T=298 K, C_0=800 mg/L, V = 0.05 L).

3.4. Effect of pH

Many studies suggest that pH is an important factor in the adsorption process (Seki and Yurdakoc. 2006). Some experiments were performed at 25 °C with 200 mg/L solutions to study the MB adsorption on SANM as a function of solution pH. The pH change by the addition of appropriate amounts of 0.1 M NaOH and HCl solutions. As it can be seen, this process is pH independent and initial pH=5 of MB is appropriate.

3.5. Adsorption isotherms

The adsorption isotherm suggests how an adsorbate distributes between the liquid phase and the solid phase when the adsorption process reaches to an equilibrium state and analysis of the equilibrium data is essential to optimize the adsorption system. Certain constants of an adsorption isotherm express the surface properties and affinity of the adsorbent. The adsorption equilibrium data was analyzed by Langmuir and Freundlich isotherms to evaluate the adsorption capacities of adsorbent. The Langmuir isotherm assumes that molecules adsorbed on an adsorbent do not react with each other and it is valid for monolayer adsorption on the surface of the adsorbent including a finite number of identical sites. A basic assumption is that sorption takes place at specific homogeneous sites within the adsorbent (Langmuir. 1918; Liu et al. 2013). A well-known linear form of Langmuir equation can be expressed as:

$$\frac{1}{q_e} = \frac{1}{q_m} + \frac{1}{bq_mC_e} \tag{2}$$

where, q_m represents the amount of adsorbate required to form a monolayer (mg/g) and q_e is adsorbate uptake at equilibrium (mg/g). Ce is the equilibrium concentration (mg/L), and b is Langmuir constant (L/mg). The values of b and q_m can be calculated from the intercept and the slope of the plot $1/q_e$ versus $1/C_e$. An important feature of the Langmuir isotherm is R_L which is a dimensionless equilibrium parameter, also known as the separation factor, defined as (Hall et al. 1966).

$$R_L = \frac{1}{1+bC_0} \tag{3}$$

where, C_0 is the initial concentration of MB (mg/g). The value of R_L in the range of 0-1 indicates favorable adsorption, while R_L>1 represents unfavorable adsorption; R_L=1 hints linear adsorption while the adsorption process is irreversible in the event of R_L=0. The results given in Table 1 show that the adsorption of methylene blue onto SANM is favorable.

As also seen in Table 1, the adsorption capacities of SANM for methylene blue is 500 mg/g. Previously, some researchers investigated several adsorbents such as montmorillonite clay (Almeida et al. 2009), Luffacylindrica fibers (Demir et al. 2008), tea waste (uddin et al. 2009b), MMT/$CoFe_2O_4$ composite (Lunhong et al. 2011), Sewage Sludge Based Granular Activated Carbon (Liu et al. 2013) and activated carbon (Legrouri et al. 2005) for the removal of methylene blue from aqueous solutions. By comparing the results obtained in this study with those in the previously reported works, Table 2 on adsorption capacities of various low-cost adsorbent and activated carbon in aqueous solution for methylene blue, it can be stated that the findings of the current study are really good.

The Freundlich isotherm is valid for multilayer adsorption on the surface of an adsorbent, indicating non-ideal adsorption on heterogeneous surfaces. The linear form of the Freundlich equation is given as follows (Hall et al. 1966).

$$\log q_e = \log K_F + \frac{1}{n} \log C_e \tag{4}$$

where, K_F and $1/n$ are Freundlich constants correlated to adsorption capacity and adsorption intensity of an adsorbent, respectively. The intercept and slope of the plot of log q_e against log C_e can be obtained K_F and $1/n$. Correlation coefficients (R^2) and other parameters computed by fitting the experimental equilibrium data to Langmuir and Freundlich isotherm equations were tabulated in Table 1. The best-fit model was determined corresponding to R^2 value. From Table 1, checking the R^2 values indicated that Langmuir model was more suitable for the experimental data due to higher R^2 values, nearly close to unit at this temperature, suggesting the monolayer coverage of MB on the surfaces of adsorbent (Langmuir. 1918). Furthermore, the experimental adsorption capacities (q_e) were close to theory maximum adsorption capacities (q_m), and the time-rate curves substantially validated the conclusion mentioned above. The values of R_L were in the range of 0-1, indicating that the adsorption process was favorable.

Table 1. Isotherm constants for MB on adsorbents.

Langmuir model				Freundlich model		
q_m (mg/g)	b(L/mg)	R^2	R_L	k_f (mg/g)(L/mg)$^{1/n}$	N	R^2
500	0.01	0.984	0.11	28.5	2.35	0.946

3.6. Thermodynamic studies

The thermodynamic parameters are important for a better understanding of the effect of temperature on adsorption (Seki and Yurdakoc, 2006). Hence, K_C can be used to estimate the enthalpy change accompanying adsorption, ΔH_{ads}, i.e., the standard enthalpy change of adsorption at a fixed surface coverage (Rytwo et al. 2006). The thermodynamic parameters ΔH_{ads}, ΔS_{ads}, and ΔG_{ads} associated with the adsorption process can be determined using the following equations. The standard Gibbs' free energy change of adsorption, ΔG_{ads} can be related to the equilibrium, K_C, by:

$$\Delta G_{ads} = -RTLn\,K_C \tag{5}$$

where, R is the gas constant (R = 8.314 J/mol K). A convenient form of the van't Hoff equation (Seki and Yurdakoc. 2006) then relates K_C to the standard enthalpy and entropy changes of adsorption, respectively.

$$Ln\,K_C = \frac{\Delta S°}{R} - \frac{\Delta H°}{RT} \tag{6}$$

On the basis of a plot of ln K_C versus $1/T$ (Eq. 6), ΔH_{ads} can be estimated from the slope and ΔS_{ads} from the intercept of what should be a straight line passing through the points. Fig. 4 shows just such a plot with a correlation coefficient of 0.987. The ΔH_{ads} and ΔS_{ads}values are thus found to be +38240 (J/mol) and ΔS_{ads}138.43 J/K, respectively, while the ΔG_{ads} value is -3012.14 J in 298 K. The positive value of ΔH_{ads}confirms the endothermic nature of the adsorption process, as has been found in most cases.

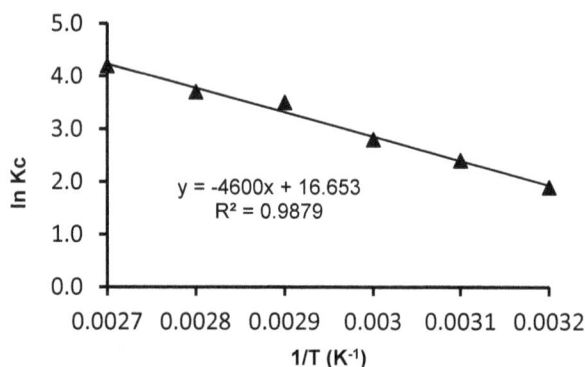

Fig. 4. Plot of lnK_c versus 1/T of adsorption of MB onto SANM (T=298 K, C_0=800 mg/L, adsorbent = 0.3 g, V = 0.05 L, contact time=10 min).

This feature may be an indication of the occurrence of monolayer adsorption (Rytwo et al. 2006). The positive value of ΔS_{ads} corresponds to an increased degree of freedom in the system as a result of adsorption of the MB molecules. There are various possible explanations. Another issue is that structural changes take place as a result of interactions of MB molecules with active groups in the SANM surface. Elsewhere comes the suggestion of increased randomness at the solid-solution interface reflecting principally the extra translational entropy gained by the solvent molecules previously adsorbed on the clay but displaced by theadsorbate species (Tahir et al. 2006).

3.7. Adsorption kinetics

In order to investigate the mechanism of sorption and potential rate controlling steps such as mass transport and chemical reaction processes, kinetic models have been used to test experimental data. These kinetic models included the pseudo-first order equation, the pseudo-second order equation. The pseudo-first order equation expressed as follows:

$$\log (q_e-q_t)=\log q_e-\frac{k_1 \, t}{2.303} \qquad (7)$$

The parameter k_1 (1/min) is the rate constant of the pseudo-first-order model. Due to equation (7) the experimental data, the equilibrium sorption capacity, q_e, must be known. In many cases, q_e is unknown and, as chemisorptions tend to become immeasurably slow, the amount sorbed is still significantly smaller than the equilibrium amount. If the rate of sorption is a second order mechanism, the pseudo-second order chemisorptions kinetic rate equation is expressed as:

$$\frac{t}{q_t}=\frac{1}{k_2 \, q_e^2}+(\frac{1}{q_e}) \, t \qquad (8)$$

k_2 (g /mg min) is the rate constant of the pseudo-second-order models, q_e and $k_2$can be obtained from the plot t/q_t versus t.

From the pseudo-second-order rate constants, k_2, at a different temperature and using the Arrhenius equation (Eq. 9), it is possible to gain some insight into the type of adsorption.

$$Ln \, k_2 = Ln \, A - \frac{E_a}{RT} \qquad (9)$$

Here E_a is the activation energy (J/ mol), k_2 the pseudo second-order rate constant for adsorption (g /mol s), A the temperature-independent Arrhenius factor (g/mol s), R the gas constant (8.314 J/Kmol), and T the solution temperature (K).The slope of the plot of ln k_2 vs. 1/T can then be used to evaluate E_a. Low activation energies (5-40 kJ/ mol) are characteristic of physical adsorption, while higher ones (40–800 kJ/mol) suggest chemisorptions (Almeida et al. 2009).The present results give Ea = +53.98 kJ/mol for the adsorption of MB onto SANM (Fig. 5), indicating that the adsorption is chemisorptions.

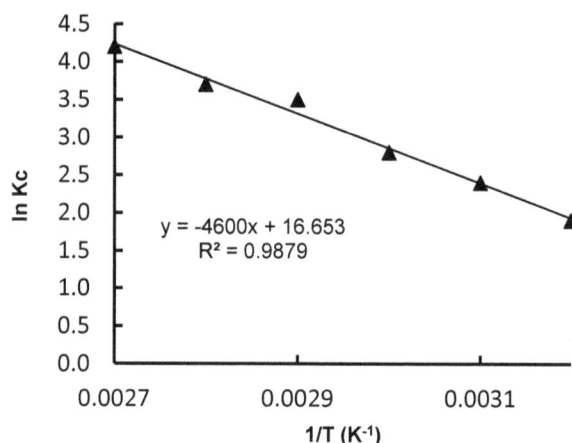

Fig. 5. Plot of lnK_c versus 1/T of adsorption of MB onto SANM (T=298 K, C_0=800 mg/L, adsorbent = 0.3 g, V = 0.05 L, contact time=10 min).

Based on the parameters in Table 3, the R^2 values of pseudo-second-order model were found to be higher than the other models, which indicated that the experimental data was fitted to pseudo-second-order kinetic model for MB onto SANM. In addition, the calculated q_e values from pseudo-second-order model were close to those of experimental q_{exp}, validating the results mentioned above.

Table 2. Previously reported adsorption capacities of variousadsorbents for methylene blue.

Adsorbent	Adsorption capacity (mg/g)	Initial concentration of MB (mg/L)	Contact time (min)	references
SANM	500	800	10	This work
Activated carbon	435	200	1200	(Legrouri et al. 2005)
Luffacylindrica fibers	156.6	47-52	1500	(Demir et al. 2008)
Montmorillonite clay	86.32-348.87	200-1000	30	(Almeida et al. 2009)
Tea waste	85.16	20-50	300	(Uddin et al. 2009b)
MMT/CoFe$_2$O$_4$ composite	97.75	100	40	(Lunhong et al. 2011)
SSGAC*	131.8	400	1440	(Liu et al. 2013)

* Sewage Sludge Based Granular Activated Carbon

Table 3. Kinetic parameters for the removal of MB onto SANM.

Pseudo-first order			Pseudo-second order		
R^2	K_1(1/min)	q_{exp} (mg/g)	R^2	K_2 (g/mg.min)	q_{exp} (mg/g)
0.973	0.32	129.91	0.999	0.025	142.85

4. Conclusion

In this article, the authors have reported a new, efficient, and environmentally friendly adsorbent (SANM) in water treatment which is independent to pH and reaches to equilibrium in a contact time of less than 10 min, low cost of the sorbent, and simple procedure. This process obeyed pseudo-second-order kinetics with an activation energy of +53.98 kJ/mol, consistent with the description of the process as involving chemisorptions.

Acknowledgement

Financial support for this work by the research affair of the University of Guilan, Rasht, Iran, is gratefully acknowledged.

References

Akar T., demir T., kiran A., Ozcan I., Ozcan A., tunail A.S., Biosorption potential of Neurosporacrassa cells for decolorization of Acid Red 57 (AR57) dye, Journal of Chemical Technology and Biotechnology 81 (2006) 1100-1106.

Almeida C.A.P., Debacher N.A., Downsc A.J., Cotteta L., Mello C.A.D., Removal of methylene blue from colored effluents by adsorption on montmorillonite clay, Journal of Colloid and Interface Science 332 (2009) 46-53.

Ansari R., Seyghali B., Mohammadkhah A., Zanjanchi M.A., Highly Efficient Adsorption of Anionic Dyes from Aqueous Solutions Using Sawdust Modified by Cationic Surfactant ofCetyltrimethylammonium Bromide, Journal of Surfactants and Detergent 15 (2012) 557-565.

Bouberka Z., Khenifi A., Benderdouche N., Derriche Z., Removal of Supranol Yellow 4GL by adsorption onto Cr-intercalated montmorillonite, Journal of Hazardous Material 133 (2006) 154-161.

Demir H., Top A., Balkose D., Ulku S., Dye adsorption behavior of Luffacylindrica fibers,Journal of Hazardous Material 153 (2008) 389-394.

Dogan M., Alkan M., Demirbas O., Ozdemir Y., Ozmetin C., Adsorption kinetics of maxilon blue GRL onto sepiolite from aqueous solutions,Journal of chemical engineering 124 (2006) 89-101.

Gupta V.K., Mohan D., Saini V.K., Studies on the interaction of some azo dyes (naphthol red-J and direct orange) with nontronite mineral, Journal of Colloid and Interface Science 298 (2006) 79-86.

Hall K.R., Eagleton L.C., Acrivos A., Vermeulen T., Pore- and Solid-Diffusion Kinetics in Fixed-Bed Adsorption under Constant-Pattern Conditions, Industrial & Engineering Chemistry fundamentals 5 (1966) 212-223.

Langmuir I., The adsorption of gases on plane surfaces of glass, mica and platinum, Journal of American Chemical Society 40 (1918) 1361-1403.

Legrouri K., Khouyab E., Ezzinea M., Hannachea H., Denoyelc R., Pallierd R., Naslaind R., Production of activated carbon from a new precursormolasses by activation with sulphuric acid, Journal of Hazardous Material 118 (2005) 259-263.

Liu L., Lin Y., Liu Y., Zhu H., He Q., Removal of Methylene Blue from Aqueous Solutions by Sewage Sludge Based Granular Activated Carbon: Adsorption Equilibrium, Kinetics, and Thermodynamics,Journal of chemical engineering 58 (2013) 2248-2253.

Lunhong A., Zhou Y., Jiang J., Removal of methylene blue from aqueous solution by montmorillonite/CoFe2O4 composite with magnetic separation performance, Desalination 266 (2011) 72-77.

Mahanta D., Madras G., Radhakrishnan S., Patil S., Adsorption of Sulfonated Dyes by Polyaniline Emeraldine Salt and Its Kinetics,Journal of Physical Chemistry B112 (2008) 10153-10157.

Mohammad-khah A., Ansari R., Activated Charcoal: Preparation, characterization and Applications: A review article, International Journal of ChemTech Research 1 (2009) 859-864.

Mohammed M.A., Shitu A., Ibrahim A., Removal of Methylene Blue Using Low Cost Adsorbent: A Review, Research Journal of Chemical Science 4 (2014) 91-102.

Preethi S., Sivasamy A., Sivanesan S., Ramamurthi V., Swaminathan G., Removal of Safranin Basic Dye from Aqueous Solutions by Adsorption onto Corncob Activated Carbon, Industrial and Engineering Chemistry Research 45 (2006) 7627-7632.

Raghu S., Basha A., Chemical or electrochemical techniques followed by ion exchange for recycle of textile dye wastewater, Journal of Hazardous Material 149 (2007) 324-330.

Rytwo G., Huterer-Harari R., Dultz S., Gonen, Y., Adsorption of fast green and erythrosin-B to montmorillonite modified with crystal violet,Journal of Thermal Analysis andCalorimetry 84 (2006) 225-231.

Seki Y., Yurdakoc K., Adsorption of Promethazine hydrochloride with KSF Montmorillonite, Adsorption 12 (2006) 89-100.

Shirini F., Mamaghani M., Atghia S.V., A mild and efficient method for the chemoselectivetrimethylsilylation of alcohols and phenols and deprotection of silyl ethers using sulfonic acid-functionalized orderednanoporous Na+ montmorillonite, Applied Clay Science 58 (2012a) 67-72.

Shirini F., Mamaghani M., Atghia S.V., Sulfonic acid-functionalized ordered nanoporous Na+-montmorillonite (SANM): Anovel, efficient and recyclable catalyst for the chemoselective N-Boc protection ofamines in solventless media,Catalysis Communication 12 (2012b) 1088-1094.

Tahir S.S., Rauf N., Removal of a cationic dye from aqueous solutions by adsorption onto bentonite clay, Chemosphere 63 (2006) 1842-1848.

Uddin Md.T., Islam Md.A., Mahmud S., Rukanuzzaman Md., Adsorptive removal of methylene blue by tea waste, Journal of Hazardous Material 164 (2009b) 53-60.

Uddin Md.T.,Rukanuzzaman Md., Rahman Khan Md. M., Islam Md.A., Adsorption of methylene blue from aqueous solution by jackfruit (Artocarpusheteropyllus) leaf powder: a fixed-bed column study, Journal of Enviromental Management 90 (2009a) 3443-3450.

Vadivelan V., Vasanth Kumar K., Equilibrium, kinetics, mechanism, and process design for the sorption of methylene blue onto rice husk, Journal of Colloid and Interface Science 286 (2005) 90-100.

Yan L., Shuai Q., Gong X., Gu Q., Yu H., Synthesis of microporous cationic hydrogel of hydroxypropyl cellulose (HPC) and its application on anionic dye removal, Journal of Clean - Soil, Air, Water 37 (2009) 392-398.

Zhu M.X., Lee L., Wang H.H., Wang Z., Removal of an anionic dye by adsorption/precipitation processes using alkaline white mud, Journal of Hazardous Material 149 (2007) 735-741.

13

The potential utilization of grey water for irrigation: A case study on the Kwame Nkrumah University of Science and Technology, Kumasi Campus

Godfred Owusu-Boateng*, Victoria Adjei

Faculty of Renewable and Natural Resources, Kwame Nkrumah University of Science and Technology, Kumasi, Ghana.

ARTICLE INFO	ABSTRACT
Keywords: Grey water KNUST Water Quality Irrigation	The problems of shortages and quality deterioration of water, have led to an increased interest in the reuse of treated grey water in many parts of the world. This study examined the suitability of locally available materials (beach sand, oyster shells, and charcoal) to treat grey water samples collected weekly from three halls of residence (Unity Hall, Africa Hall, and Independence Hall) on Kwame Nkrumah University of Science and Technology (KNUST) campus for irrigation. Beach sand, oyster shells, and charcoal were employed in the construction of three vertical flow-through filter systems, each consisting of PVC pipes of height 100 cm and internal diameter 5.08 cm. The grey water samples were filtered and the levels of physicochemical parameters (pH, conductivity, TDS and salinity), nutrient and microbial counts determined over a three-week period. Results indicate that the measured physico-chemical parameters treated grey water were within the permissible limits for irrigation water. Also filtration process is effective in reducing phosphate, the total and faecal coliform levels in grey water from the halls of residence. These observations suggest that treated grey water from KNUST campus would support production when used as irrigation water.

1. Introduction

Nations are endowed with resources for survival, growth, and development. Among these resources is water, one of nature's most important gifts to mankind. It is a finite and an invaluable resource. The importance of water to all living things underpins the global concern for its security. However, with the advent of population explosion, urbanization, and the increased competition among the different uses (domestic, industrial, and agricultural) of water, water is becoming a rare resource in the world without the possibility for an increase in its supply. It is estimated that in the year 2000, the urban areas of Ghana generated about 763,698 m³ of wastewater each day, resulting in approximately 280 million m³ of the wastewater over the entire year. Regional capitals accounted for another 180 million m³ of generated wastewater. Only a small fraction of the generated urban wastewater is collected, and an even smaller proportion is being treated. Moreover, less than 25% of the 46 industrial and municipal treatment plants in Ghana were functional. Treatment plants for municipal wastewater are operated by local governments, and most of them are stabilization ponds. A biological treatment plant which was built in the late 1990s at Accra's Korle Lagoon only handles about 8% of Accra's wastewater (Awuah and Abrokwa 2008).

In recent years, researchers have begun showing greater interest in treatment and reuse of wastewater. Wastewater management in Ghana has undergone a paradigm shift in terms of how

Wastewater is valued and managed. Waste water is being used in some countries for agriculture. Agricultural water needs represent the greatest percentage of global water use, and wastewater reuse is an attractive alternative with good potential to supplement freshwater supplies Finley et al (2009). According to the WHO (2006), such use is promoted by some factors mainly:

- Increasing water scarcity and stress, and degradation of freshwater resources due to from improper disposal of wastewater;

- Increased demand for food as a result of population increase;
- Environmental sustainability and eliminating poverty and hunger as prescribed by the Millennium Development Goals.
- Recognition of the value and importance of wastewater such as the nutrients content

Improper wastewater disposal which pollutes surface water and groundwater resources and degenerate water supply problems have created a serious conundrum for governments and the nation at large. These conditions among others such as improper wastewater disposal which pollutes surface water and groundwater resources and degenerate water supply problems have created a serious conundrum for governments and the nation at large. There is therefore the need to search for sustainable water management schemes by encouraging wastewater reuse while re-evaluating the uses to which potable water should be put.

Generally, grey water is used to refer to all wastewater generated from households with the exception of human solid or liquid wastes (referred to as sewage or black water). It is termed as such because of its cloudy appearance and from its status as being between fresh, potable water and sewage water (black water). Jefferson et al. (2004) found that, although similar in organic content to full domestic wastewater, grey water contains fewer solids and is less turbid than domestic wastewater, suggesting that more of its contaminants are dissolved. The use of grey water for agricultural and irrigation is occurring more frequently because of water scarcity and population growth (Keraita et al. 2003). According to Friedler (2004), grey water makes up about 60-70% of domestic wastewater volume in developed countries. Grey water reuse is one such strategy which is already in use by millions of farmers worldwide, and is estimated that 10% of the world's population consumes foods irrigated with grey water (WHO, 2006).

Grey water reuse is a strategy that simultaneously alleviates the environmental and economic concerns of water usage and helps

*Corresponding author E-mail: boateng.irnr@knust.edu.gh

reduce the alarming global water menace. Generally, the reuse of grey water is particularly important to;

- Ease the mounting demand on limited freshwater reserves by substituting the use of freshwater for non-potable uses such as irrigation, dishwashing, laundry, car washing and toilet flushing since grey water is an inevitable wastewater that is generated daily irrespective of water availability in a locality.
- Reduce conflicts over water and reduce the demand for new water supply projects by substituting the demand for potable water.
- Reduce strain on septic tanks or treatment plants. Grey water reuse greatly extends the useful life and capacity of septic systems. For municipal treatment systems, decreased wastewater flow means higher treatment effectiveness and lower costs.
- Reclaim otherwise wasted nutrients. Loss of nutrients through wastewater disposal in rivers and streams is a highly significant form of erosion. Reclaiming nutrients in grey water helps to maintain the fertility of the land.
- Support plant growth because it contains nutrients, mainly nitrogen and phosphorus and has potential for use on agricultural crops, hence subsidizing commercial fertilizer use.
- Improve food production capacity all year round since grey water is available at all times, be it the dry or wet season.
- Economically provide significant savings in freshwater use and management. Thus, grey water reuse will go a long way to reduce the cost of obtaining freshwater and also help reduce payment of water bills.

The public is quite skeptical about the health safety or implications of grey water when reused (Brown and Davies 2007). Perceived health risk, perceived cost, operation regime, and environmental awareness religious and cultural values Laban (2010) are some of the factors that guide acceptability of grey water by the public.

The indispensable utilization of water for domestic hygiene including washing, bathing, etc comes with the production of grey water. When handled and managed well, it becomes an asset otherwise it is a menace. Potential uses may be available. Kwame Nkrumah University of Science and technology, being a large community of learning has a large population and therefore generates large quantities of grey water the quality of which and hence the possible uses including irrigation farms in the university community including the university's demonstration farms are not known. Determination of physicochemical and microbiological quality of grey water generated from the students' halls of residence and construction of a filter system for treating grey water samples collected are likely to bring to light the potentials of grey water.

2. Description of the study area

The study was executed on the Kwame Nkrumah University of Science and Technology (KNUST) which has laboratory for running various physicochemical and microbiological tests and availability of grey water and irrigation farm. Three out of the seven halls of residence were selected for the study. These halls were the Africa Hall (an all-female hall), Unity Hall (an all-male hall), and Independence Hall (a male-female hall). The choice of these halls introduced a balance in the composition of the halls of study.

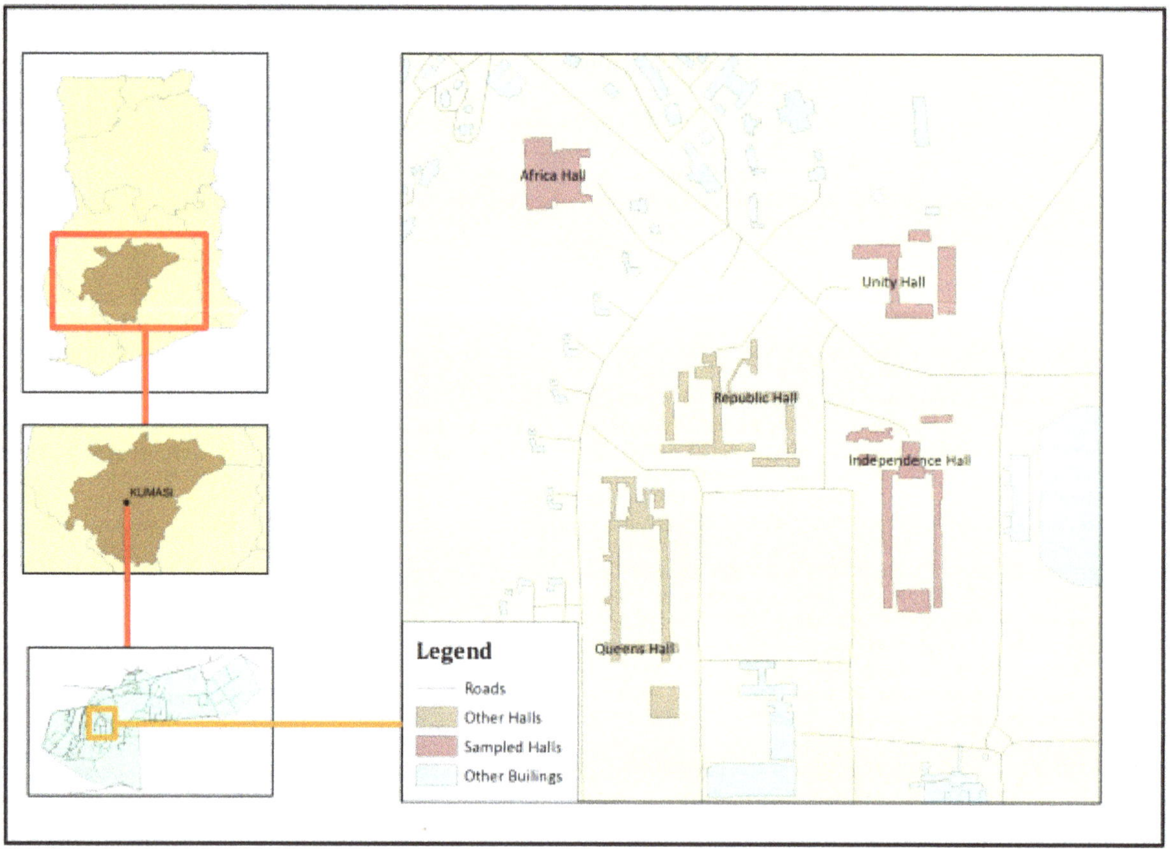

Fig. 1. Map of KNUST with grey water some sampling Sites (halls).

3. Methodology

The presence of contaminants of water for various designated uses necessitate primary treatment including filtration (Morel and Diener 2006). Three vertical flow-through sand filter systems were constructed using PVC pipes of internal diameter of 5.08 cm (2 inch) and a height of 100 cm. One end of each PVC pipe was closed with a 1 mm mesh size net and the other end left opened for the water feed. They were then clamped firmly into position and correctly labelled. Each sampling hall was designated a single filter unit.

3.1. Selection and pretreatment of filter media

According to Khalaphallah (2012), properly constructed sand filtration system consists of a tank, a bed of fine sand, and a layer of coarse aggregate (e. g. gravels) to support the sand. Therefore, white beach sand, crushed oyster shells and charcoal were use in the construction of filtration system. The beach sand and oyster shells were thoroughly backwashed to ensure that they were free from any residual salts, organic matter or dirt. This was followed by sun drying to allow the solar radiation to inactivate any possible attached pathogen.

3.2. Packing of filter media

Each treatment unit was first and foremost underlain with a 15cm thick layer of crushed oyster shells of average size 7.49 mm. Aside supporting the sand, the oyster shells were to effectively withhold large particles of unbroken organic matter. Above the layer of oyster shells was laid a 15 cm thick middle layer consisting of charcoal of average size 11.92 mm. This layer of charcoal was introduced to trap carbon-based impurities (organic compounds) as well as inorganic and odorous compounds. This layer was then overlaid with a 50cm thick layer of beach sand.

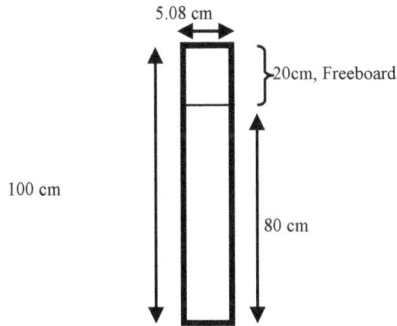

Fig. 2. Schematic diagram of filter column.

3.3. Collection of grey water samples

Grey water samples were collected over a three-week period from three halls of residence on campus- Independence Hall, Unity Hall and Africa Hall. Grey water samples were collected weekly from each hall. The levels of some physicochemical parameters (pH, conductivity, total dissolved solids (TDS), and salinity) of the untreated grey water were measured in situ with the multi-parameter probe prior to collection. The samples were collected into 1.5 litre plastic sampling bottles by placing them directly at the mouth of the receiving drains of each hall. The samples were transported on ice to the Faculty of Renewable Natural Resources for treatment.

3.4. Treatment of samples

Grey water samples from each of the halls were each fed into their respective filter units. The filtrate (treated grey water) was collected into well labeled receiving containers. Again, the multi-parameter probe was used to measure the levels of the physicochemical properties (pH, conductivity, total dissolved solids (TDS), and salinity) of the treated grey water. The entire sampling and treatment process was repeated weekly.

3.5. Laboratory analysis

Both the treated and untreated grey water samples of each sampled hall were analyzed for nutrient (nitrate and phosphate) and microbial counts (total and faecal coliform) loads.

3.6. Statistical analysis

All results were expressed as the mean ± standard deviation and displayed in bar charts. The raw untreated and grey water samples were subjected to the Student's t-test to determine whether their means were statistically different from each other, using the GraphPad Prism 5 software.

4. Results and discussion

Water quality issues that can create real or perceived problems in agriculture include nutrient and sodium concentrations, heavy metals, and the presence of contaminants such as human and animal pathogens (Toze 2006). The ranges of the levels of the tested physicochemical parameters of the untreated and treated grey water over the entire sampling period are tabulated (Table 1). Adsorption of trace organic and inorganic compounds by the micro pores of the various filter media in the treatment system is thought to be the primary physicochemical mechanism for the removal of refractory compounds. Generally, these physico-chemical interactions are a function of surface area: the larger the surface area of the filter material, the greater the interaction.

4.1. Physicochemical parameters
4.1.1. pH

pH is an index of the concentration of hydrogen ions (H^+) in the water. It is defined as -log [H^+]. According to Bauderet al.,(2011), the lower the concentration of hydrogen ions in the water, the higher the pH value is or the more basic the water is and vice versa. Results of pH consistently increased for all treatments over the entire study period with Unity Hall recording the highest pH values at 9.16 and 9.09 for weeks one and two respectively. Generally, the normal pH range for irrigation water is from 6.5-8.5 (FAO, 1985); (USEPA, 200). Values for pH of untreated grey water for week one ranged from 6.77 to 7.88 with Independence Hall recording the highest pH value while Africa Hall recorded the least pH value. pH values of untreated grey water recorded for weeks two and three were in the ranges 7.57-8.43 and 7.19-7.43 respectively with Unity Hall recording the highest pH value in both weeks. Most of the activities involved washing. This made the grey water rich in surfactants (Gross et al., 2007). The observed pH ranges may be accounted for by the presence of soapy and soapless detergents which are often alkaline and therefore driving the pH to between 7 and 8 (Jefferson et al., 2004).

Table 1. Levels of the physicochemical parameters of untreated and treated grey water samples.

	Parameters	Africa Hall	Independence Hall	Unity Hall
Untreated Grey water	pH	6.77-7.57	7.19-8.19	7.04-8.43
	Conductivity(μS/cm)	246-319	242-652	195-402
	TDS (mg/L)	141-206	135-331	98-164
	Salinity(PSU)	0.07-0.15	0.12-0.31	0.09-0.16
Treated Grey water	pH	8.13-8.41	8.29-8.38	8.45-9.16
	Conductivity(μS/cm)	374-724	363-1355	424-1117
	TDS (mg/L)	234-368	260-429	218-295
	Salinity(PSU)	0.23-0.35	0.33-0.60	0.26-0.53

For the treated grey water samples, pH values recorded for the individual sampling weeks were: Week one (8.29-9.16), Week two (8.19-9.09) and Week three (8.13-8.45) respectively (Figure 3). The highest pH value for the treated grey water was recorded by Unity hall in week one, whiles Africa hall recorded the least pH value as 8.13 in week three. However, the pH values recorded for the treated grey water relatively increased over the entire sampling period. High pH's above 8.5 are often caused by high bicarbonate (HCO_3^-) and carbonate (CO_3^{2-}) ion concentrations, known as alkalinity [Bauderet al., 2011]. The increase in the pH values for the treated grey water therefore may be attributed to the dissolution of oyster shells which

consist mainly of calcium carbonate ($CaCO_3$), in the grey water during treatment. As the oyster shells dissolved in the water, they might have released carbonate ions (CO_3^{2-}) which combined with the H+ ions present in the water to form hydrocarbonates. The formation of these hydrogen carbonates might have probably reduced the H+ ions levels in the water to raise the pH of the effluent water The difference in p-value ($p<0.05$) recorded for the pH levels of the untreated grey water and the treated grey water for weeks one and two were not statistically significant, but that of week three was statistically different. The permissible pH range for irrigation water as given by the FAO (1985) and USEPA (2004) is 6.5-8.5.

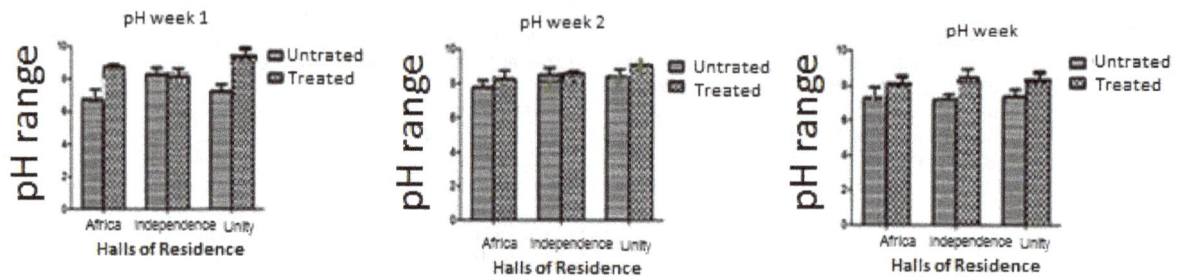

Fig. 3. pH levels for untreated and treated grey water from Africa, Independence and Unity Halls.

4.1.2. Total dissolved solids (TDS), salinity and conductivity

Water salinity, conductivity and total dissolved solids (TDS) are interrelated. The content of salts dissolved in water as well as the parameters derived from its presence in water such as conductivity and total dissolved solids (TDS) determine the chemical characteristics of the water. Water salinity refers to the total amount of salts dissolved in water but it does not indicate which salts are present in it. Salts contribute to the conductivity of water by dissolving to release positively charged ions (e.g. Ca^{++}, Na^+) and negatively charged ions(e.g. Cl^-, NO_3^-), which conduct electricity. The electrical conductivity of water on the other hand estimates the total amount of solids dissolved in water.

Results showed increasing trends in the salinity, conductivity, and TDS levels of the treated grey water relative to the untreated grey water. Values for TDS were highest at 331 mg/L and 429 mg/L for untreated grey water and treated grey water respectively (Figure 4). Both values were each recorded at Independence Hall. The difference in p-value ($p<0.05$) for the untreated grey water and the treated grey water samples over the three-week sampling period showed a significant variation. The permissible TDS level for irrigation water as given by Rowe and Abdel-Magid (1995) is in the range of 500 to 2000 mg/L.

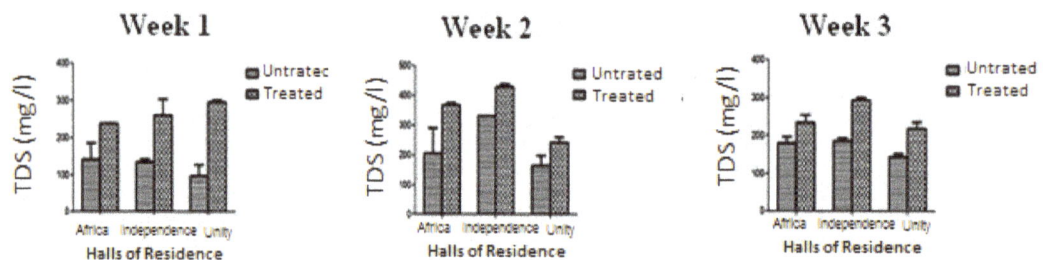

Fig. 4. TDS levels for untreated and treated grey water from Africa, Independence and Unity Halls.

Generally, the salinity levels of the treated grey water were higher relative to that of the untreated grey water. For the untreated grey water, Africa Hall recorded the least salt concentration (0.07 PSU), and Independence Hall recorded the highest salt concentration at 0.31 PSU. For the treated grey water, Africa Hall recorded the lowest salinity values; 0.23 PSU, 0.35 PSU, and 0.24 PSU for weeks 1, 2, and 3, respectively throughout the sampling period, while Independence Hall recorded the highest salt concentrations; 0.35 PSU, 0.60 PSU, and 0.333 PSU weeks 1, 2, and 3, respectively (Fig. 5). Salinity concentrations between the untreated grey water and treated grey water were significant. The permissible salinity level for irrigation water as given by the FAO (1985) ranges from 0.7 to 3.0 PSU.

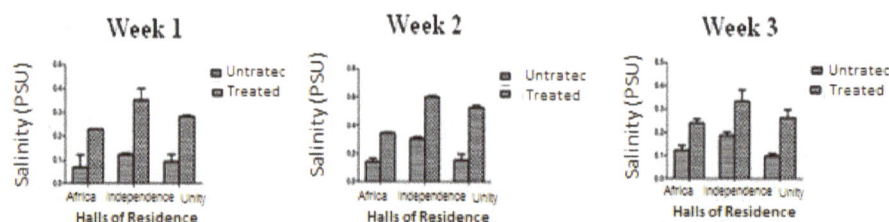

Fig. 5. Salinity levels for untreated and treated grey water from Africa, Independence and Unity Halls.

The conductivity levels of the untreated grey water varied from 196µS/cm to 652µS/cm, while that of the treated grey water varied from 363µs/cm to 1355µS/cm (Figure 6). Independence hall recorded the highest conductivity value in both untreated grey water and treated grey water. The difference in p-value ($p <0.05$) recorded for the conductivity levels of the untreated grey water and the treated grey water were statistically significant. The permissible conductivity level for irrigation water as given by the EPA (1995) is 800µS/cm. The relatively higher levels of salinity, conductivity, and TDS in the treated grey water samples than the untreated grey water samples may be due to the presence of inorganic soluble solids such as chloride, nitrate, sulfate, anions or sodium, magnesium, and calcium cations in the filter substrate through which the grey water percolated US.EPA (2012).

4.1.3. Nutrients

Grey water can contain nutrients including nitrogen and phosphorus Maimonet al., [2010]. Although wastewater can generally contain elevated concentrations of nitrogen and phosphorus (nutrients) and potassium not much nutrients (usually less than 5　　mg/L) occur in grey water (Surendran and Wheatley, 1998).

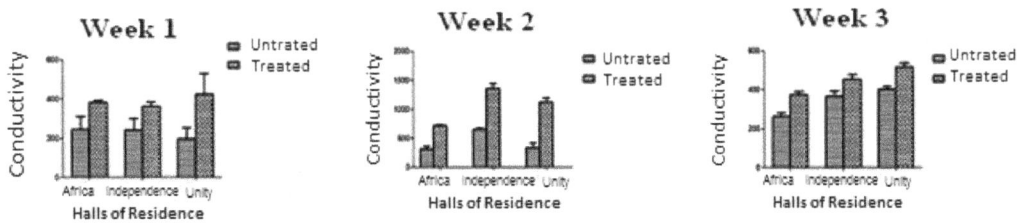

Fig. 6. Conductivity levels (μs/cm) for untreated and treated grey water from Africa, independence and Unity Halls.

4.1.4. Nitrate

Nitrate values recorded during the study period for untreated grey water ranged from 0.33mg/L to 1.5mg/L whereas nitrate values recorded for treated grey water relatively increased from 0.43mg/L to 2.19mg/L. Similar to pH, conductivity, TDS, and salinity, nitrate levels increased consistently in the filtered grey water throughout the study period. The highest increase was recorded in week three at 2.19mg/l. Bhumbla (1999), reported that nitrate in water is present as highly soluble salts. Therefore, when water moves on the surface of a soil, it dissolves some nitrates that are originally present in the surface layers of the soil and the dissolved nitrates move along with the water in the soil. This may have contributed to the consistent increase of nitrate levels in the filtered grey water.

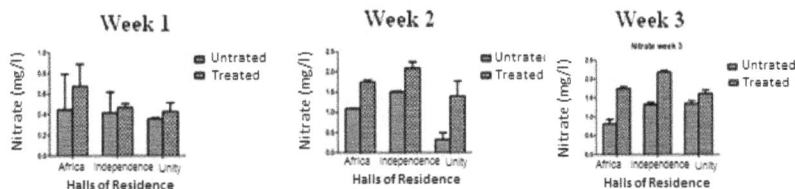

Fig. 7. Nitrate levels for untreated and treated grey water from Africa, Independence and Unity Halls.

The difference in p-value recorded for weeks one and three were not statistically different from the untreated grey water and the treated grey water, but that of week two were statistically different. The permissible nitrate level for irrigation water as given by the (EPA, 1995) is 10mg/L.

4.1.5. Phosphate

The levels of the effluent phosphate decreased at the end of each filtration process. Results indicated that reductions in phosphate levels was higher in week one (from 5.76 to 3.51mg/l) and week two (from 1.85 to 1.09mg/l). The level was higher in untreated grey water the treated (at 5.95mg/L and 5.25 mg/L respectively) (Figure 8). The high pH recorded may account for this observation. At high pH levels, phosphate may be locked up in the soil as insoluble forms of calcium and magnesium. Phosphate (PO_4^{3-}) is a triple negatively charged anion which means that it is strongly attracted to positively charged cations like calcium, magnesium, iron and aluminium. The fact is that when PO_4^{3-} forms a bond with these other minerals it becomes insoluble, locked-up, and does not travel along with the soil water.

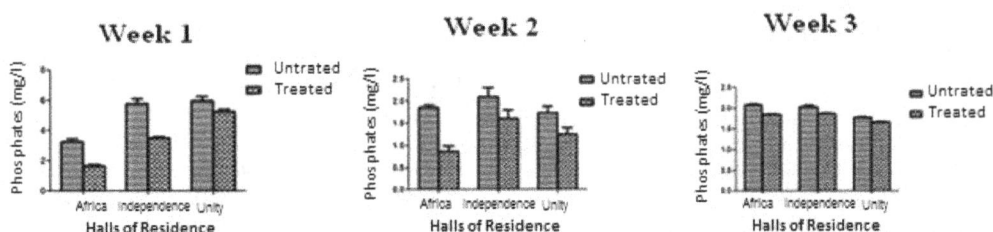

Fig. 8. Phosphates levels for untreated and treated grey water from Africa, Independence and Unity Halls.

The difference in p-value (p<0.05) recorded for weeks one and two showed no statistical difference between the untreated grey water and the treated grey water, but that of week three was statistically different. The permissible phosphate level for irrigation water as given by the EPA (1995) is 10mg/L.

4.1.6. Microbial quality

Microbial and chemical contamination of grey water poses a potential risk to human health [Dixon et al., 2000]. There were no recorded values for both total and faecal coliform for the treated grey water in weeks one and two. Filtration processes remove most microorganisms from waste water. Casanova et al. (2001), recounted that grey water contain relatively mildly less pathogenic microbial such as faecal coliforms, enterococci, and bacteriophages. In week two, no coliform count was recorded for both the untreated and treated grey water. In week three, Africa Hall and Independence Hall recorded equal faecal coliform counts of 2.35 x104CFU/100mL for the treated grey water. The highest total coliform count (2.35 x 105 CFU/100ml) in week three was recorded for the treated grey water samples with Unity Hall. There were no recorded coliform values for Unity Hall's treated grey water in week three. Also according to Huisman and

Wood (1974), it is essential to maintain a constant flow of water through a sand filter system which provides oxygen and food to the organisms that make up the 'schmutzdecke' layer and biological zone living within the top part of the sand and are responsible for much of the removal of disease-causing organisms. However, under stagnant conditions where the filter system is not constantly fed with

water to be treated, the biological zones can start to die off. This absence biological zone may inhibit the filter system from effectively removing microorganisms from the treated effluent.

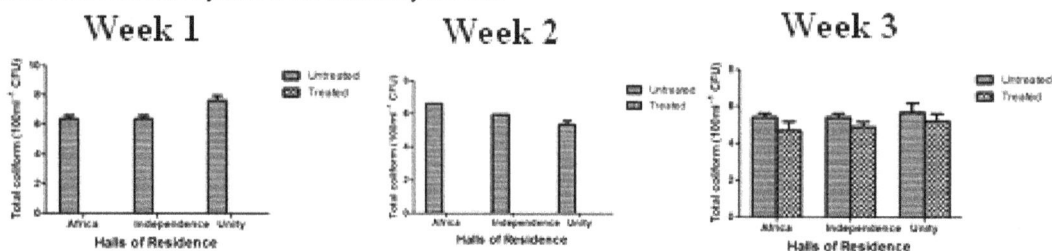

Fig. 9. Total coliform count for untreated and treated grey water from Africa, Independence and Unity Halls.

The difference in p-value (p<0.05) showed no statistical difference of the coliform count between the untreated grey water and the

treated grey water. The permissible coliform level for irrigation water as given by WHO (2006) is 1*103 CFU/100mL.

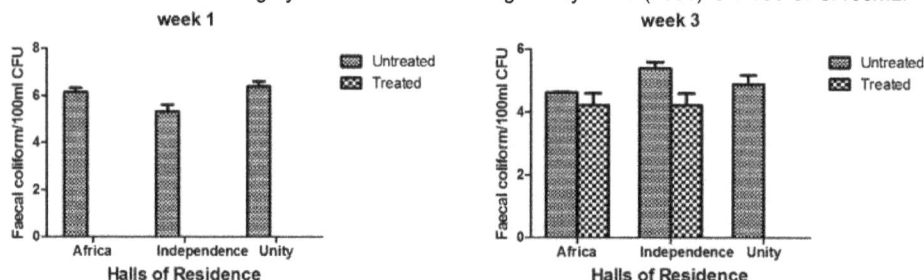

Fig. 10. Faecal coliform count for untreated and treated grey water from Africa, Independence and Unity Halls.

5. Conclusions

The study revealed a considerable decrease in phosphate levels in treated grey water over three-week study period and contains appreciable levels of TDS, salinity, conductivity, and nitrate at higher levels than that of untreated grey water. Most of the measured physico-chemical parameters in the treated water were below the

threshold for hazard as determined by globally acceptable irrigation water standards, although there were some increases. Also treated grey water was generally void of both total and faecal coliform. Therefore, treated grey water from KNUST campus does not appear to pose threat to crops as irrigation water suggesting that it can support irrigation.

References

Awuah E., Abrokwa K.A, Performance evaluation of the UASB sewage treatment plant at ames Town (Mudor), Accra. Reviewed paper, 33th WEDC International conference. Access to sanitation and safe water: global partnerships and local actions, (2008), pp. 20-26.

Bauder T.A., Colorado State University extension water quality specialist; Waskom R.M., director, Colorado Water Institute; Sutherland, P.L., USDA/NRCS area resource conservationist; and Davis, J.G., Extension soils specialist and professor, Soil and Crop Sciences (2011).

Bhumbla D.K., Agriculture Practices and Nitrate Pollution of Water, West Virginia University Extension Service (1999). Available at: http://www.caf.wvu.edu/wforage/-nitratepollution/nitrate.html, (Accessed October 2, 2012).

Brown R.R., Davies P., Understanding community receptivity to water re-use: Ku-ring-gai Council case study, Water Science and Technology 55 (2007) 283–290.

Dixon A., Butler D., Fewkes A., Robinson M., Measurement and modeling of quality changes in stored untreated grey water, Urban Water 1 (2000) 293-306.

Domenech L., Sauri D., Socio-technical transitions in water scarcity contexts: Public acceptance of grey water reuse technologies in the Metropolitan Area of Barcelona, Resources, Conservation and Recycling. (2010).

FAO, Water quality for agriculture. Irrigation and Drainage paper 29 Rev. 1. Food and Agriculture Organization of the United Nations,

Rome, (1985), pp. 174.

Finley S., Barrington S., Lyew D., Reuse of domestic grey water for the irrigation of food crops, Water, Air, & Soil Pollution 199 (2009) 235-245.

Friedler E., Quality of individual domestic grey water streams and its implication for on-site treatment and reuse possibilities. Environmental Technology, 25 (2004) 997-1008.

Gross A., Kaplan D., Baker K., Removal of chemical and microbiological contaminants from domestic grey water using a recycled vertical flow bioreactor (RVFB), Ecological Engineering 31 (2007) 107–114.

Health considerations, Available at: http://www.emro .who.int/ceha/pdf/Greywter% 20English%202006.pdf. (Accessed: October 11, 2012).

James (Mudor) T., Accra. Reviewed paper, 33th WEDC International Conference. Access to sanitation and safe water: global partnerships and local actions (2008) 20-26.

Jefferson B., Palmer A., Jeffrey P., Stuetz R., Judd S., Grey water characterization and its impact on the selection and operation of technologies for urban reuse. Water Science & Technology 50 (2004) 157-164.

Keraita B., Drechsel P., Amoah P., Influence of Urban Wastewater on Stream Water Quality and Agriculture in and around Kumasi, Ghana, Environment and Urbanization, 15 (2003) 171-178.

Khalaphallah R., Grey water treatment for reuse by sand filtration: study of pathogenic microorganisms and phage survival (2012).

Laban P., Can local people accept grey water technology? In: Grey water Use in the Middle East, McIlwaine and Redwood (eds). IDRC. (2010).

Morel A., Diener S., Grey water management in low and middle-income countries, review of different treatment systems for households or neighbourhoods-Sandec Report No. 14/06. Sandec (Water and Sanitation in Developing Countries) at Eawag (Swiss Federal Institute of Aquatic Science and Technology), Dübendorf, Switzerland (2006).

Protection Agency and U.S. Agency for International development, Washington D.C. World Health Organization, Overview of grey water management: Health considerations, (2006). Available at: http://www.emro.who.int/ceha/pdf/Greywter%20English%202006.pdf. (Accessed: October 11, 2012)

Rowe D.R., Abdel-Magid I.M., Handbook of Wastewater Reclamation and Reuse. CRC Press, Inc (1995), p. 550.

Toze S., Reuse of effluent water – benefits and risks. CSIRO Land and Water, CSIRO Centre for Environment and Life Sciences (2006). (Accessed: November 8, 2012).

USEPA, Guidelines for water reuse, U.S. Environmental Protection Agency, Washington DC, (2004).

USEPA, Manual Guideline for water reuse (1995), EPA/625/R-92/004, U.S. Environment

USEPA, Water: Monitoring and Assessment (2012), Conductivity. Available at: www.water.epa.gov. (Accessed: April 19, 2013).

WHO, Guidelines for the safe use of wastewater, excreta and grey water (2006). Available at: http://www.who.int/water_sanitation health/-wastewater/gsuww/en/index, (Accessed: September 18, 2012)

WHO, Regional Office for the Eastern Mediterranean Overview of grey water management: Health considerations (2006). Available at: http://www.emro.who.int/ceha/pdf/Greywter%20English%202006.pdf. (Accessed: October 11, 2012).

14

Synthesis and characterization of low-cost ion-exchange resins used for the removal of metal ions

R.Kannan Seenivasan[1], Veerasamy Maheshkumar[2,*], Palanikumar Selvapandian[2]

[1]Department of Chemistry, Government Arts College, Tamil Nadu, India.
[2]Post Graduate and Research Department of Chemistry, Raja Doraisingam Government Arts College,Tamil Nadu, India.

ARTICLE INFO

Keywords:
Phenol-formaldehyde resin
Sulphonated Abutilon indicumcarbon
Cation exchange capacity
Composites resin
Low cost ion exchangers
Endothermic

ABSTRACT

Phenol- formaldehyde resin (PFR) is used bad-tempered protecting agent in blending of different percentage by weight of Sulphonated Abutilon indicum, Linn.carbon (SAIC). we provide a synopsis of current developments in the use of ion exchange techniques in wastewater treatment. The prepared materials (PFR, composites, and SAIC) were characterized by FT-IR spectra, SEM and thermal (TGA) studies. The low cost ion exchangers (IERs) are used for the removal of some selective metal ions such as Na^+, K^+, Ca^{2+}, Cu^{2+}, Mg^{2+}, Zn^{2+} and Pb^{2+}. Thermodynamic equilibrium constants (ln K_c) are calculated for Zn^{2+}- H^+ ion exchanges using the composite resins and also the thermodynamic parameters such as ΔH°, ΔG° & ΔS° are evaluated from Van't Hoff plot. The cation exchange capacity (CEC) of the composites were found to decreased with the increasing the percentage of SAIC in PFR matrix. It was observed that the composites up to 20 % (w/w) blending of SAIC2 retain all the properties of original PFR. Hence, the blended composites could be used as low cost ion exchangers when SAIC partially replaces the original PFR up to 20 % (w/w) SAIC2 blending without affecting the physico-chemical, thermal, spectral properties and CEC values of PFR.

1. Introduction

The quality of water is of fundamental concern to mankind, since it is directly linked with human benefit. The water pollution and its impact on the environment are currently the focus of international attention. In developed nation are using dynamic methods to manage and prevent the environmental pollution caused by chemicals, but the developing countries are beginning to follow it. Most of the industries discharge the wastewater from some stage of their manufacturing processes. The indiscriminate disposal of untreated wastewater from various industries into the rivers or onto the domestic invariably pollutes surface/ground water and land. The contaminants in wastewater vary depending on the type of industry and nature of processes. The universal pollutants encountered in the industrial wastewater are soluble organics/inorganics, suspended solids, heavy metals, volatile materials, nitrogen, phosphorous, oil and greases. In the wastewater treatment, usually a decreasing the level of pollutants is achieved. A large number of methods have been used for the removal of harmful organic and inorganic constituents from wastewater. The available systems to treat wastewater can be broadly classified into physico-chemical and biological treatments. The selection of wastewater treatment process depends upon the nature of pollutants present in wastewater and quality of water required after the treatment process.

Ion exchange method is an appropriate technique, which could be employed under field condition for the removal of ionic materials from water and wastewater (Bolto et al. 1987). Ion-exchange process finds a valuable place in the treatment of metal wastes from plating and other industrial processes. Many viable resins are originated from petroleum products and there is a continual increase in their cost. Hence, there is an urgent need to find out the new low-cost ion exchange resins (IERs) and reduce its cost by blending it with sulphonated carbons (SCs) prepared from carbonaceous materials. Such types of low-cost ion

exchangers can be prepared by blending cheaper and freely available plant materials containing Phenolic groups. Attempts have also been made to prepare cheaper cationic resins from waste materials and natural products. The application of low-cost adsorbents obtained from agricultural by-products is widely used in the recent research as a replacement for costly conventional methods of removing heavy metal ions from wastewater (Homagai. 2012). Earlier studies have shown that the cheaper condensate ion exchangers could be prepared by partially blending the macroporous phenol-formaldehyde sulphonic acid resin (PFR or PFSAR) matrix by sulphonatedcarbon (SCs) prepared from coal (Sharma et al.1980), Egyptian corn cop (Metwallyet al. 1992), bagasse charcoal (Swamiappan et al. 1984), Egyptian bagasse pith (Metwally et al. 1994), ground nut shell (Chandrasekaran et al. 1987), wheat husk (Dheiveesanet al. 1988), spent tea and mangroves bark (Krishnamoorthy et al. 1997), starch (Farag et al. 1995), jute (Hassan. 2003), Accacianilotica (Kannan et al. 2003), cashew nut shell (Bato et al. 2003) and lignin (Zoumpoulakis et al. 2001). A careful literature survey revealed that no work has been done with Abutilon indicum. Abutilon indicum (local name; peelybooti or karandi) isa rigid, timbered, shrubby plant. It is widely distributed in the tropical countries (Kirtikar et al. 1980). It is known as "Atibala" in Hindi and found in the outer Himalayan tracts from Jammu to Bhutan up to an altitude of 1500m and extending through the whole of northern and central India (Rajurkar et al. 2009).

The present work is to synthesize andcharacterize (by IR, TGA, and SEM studies) the new condensate / low-cost ion exchangers of PhOH –HCHO type blended with sulphonatedAbutilon indicum, Linn.Carbon (SAIC) and to estimate the physico - chemical properties including the cation exchange capacity (CEC) for some selective metal ions.The effect of particle size of selective metal ions, regeneration level of Mg^{2+} ions loaded IERs by using NaCl and to find out thermodynamic parameters for the above ion exchange process.

Corresponding author Email: maheshmalar72@gmail.com

2. Materials and methods
2.1. Materials

Abutilon indicum, Linn. (AI) was used as a plant material and it is easily available in India. It belongs to the family Malvaceae and called 'Karandi' in Tamil and Atibalain Hindi. The plant material was nearby collected, cleaned, dried and cut into small pieces of about 0.5cm length. Phenol and formaldehyde (AR) were purchased from Fischer reagents (India). Concentrated sulphuric acid (Sp.gr. = 1.82) and other chemicals/reagents used were produced from SD fine chemicals, India.

2.2. Methods

Abutilonindicum, Linn.plant material (500 g) has been carbonized and sulphonated by using con. sulphuric acid, washed with distilled water to remove excess free acid (tested with $BaCl_2$ solution) and dried at 70 ^{0}C for 12 h. It was labeled as SAIC.
Pure phenol–formaldehyde resin was prepared according to the literature method (Ramachandran et al. 1984; Radhakrishnan et al. 1990). It was then ground, washed with distilled water and finally with double distilled (DD) water to remove the excess acid (tested with $BaCl_2$ solutions), dried, sieved (210 – 300μm) using Jayant sieves (India) and preserved for characterization (Kannan et al. 2003;Dheiveesan et al. 1988). It was labeled as PFR.

The condensates were obtained as per the method reported in the literature (Krishnamoorthy et al. 1997). The products with 10, 20, 30, 40 and 50 % (w/w) of SAIC in the condensates were labeled as SAIC1, SAIC2, SAIC3, SAIC4, and SAIC5 respectively. A separate sample of SAIC was also subjected to the characterization (instrumental and physico - chemical) studies.

2.3. Characterization of samples
2.3.1. Instrumental studies

FT-IR spectral data of pure resin (PFR), condensate with 20 % (w/w) of SAIC2 and pure SAIC were recorded with a JASCO FT-IR 460 plus FT-IR spectrophotometer by using KBr pellets. To establish the thermal degradation of the samples, TGA and DTA traces were obtained by using the thermal analyzer. SEM analysis of pure resin (PFR), condensate with 20 % (w/w) of SAIC2 and pure SAIC were obtained using Hitachi Scanning Electron Microscope (Model S-450), Japan.

2.3.2. Physico-chemical characteristics

Samples were crushed and sieved into a particle size of 210 – 300 μm using Jayant sieves (India). This was used for further characterization by using standard procedures (Kannan et al. 2003; Chandrasekaran et al. 1987) to find out the values of absolute density (wet and dry), the percentage of gravimetric swelling and percentage of attritional breaking. The solubility of these IERs was tested with various organic solvents and inorganic reagents.

The values of cation exchange capacity (CEC) were determined by using standard titration technique (Ramachandran et al. 1984), as per the reported method (Raghunathan et al. 1984; Bassett et al. 1989). The effects of initial concentration of metal ions, particle size, chemical and thermal stability of the resin on CEC ware determined (Kannan et al. 2003; Chandrasekaran et al. 1987; Krishnamoorthy et al. 1997).

After the exchange of H^+ ions by the Mg^{2+} ions, the regeneration level of the condensates loaded with Mg^{2+} ions were determined by using NaCl (brine) solution. The equilibrium constant (K_{eq}) of on exchange reaction between H^+ ions (bound resin) and Zn^{2+} ions (in aqueous solution) were obtained as per the literature method (Bato et al. 2003). Thermodynamic parameters ($\Delta G°$, $\Delta H°$ and $\Delta S°$) were computed by following the standard procedure and interpretation.

3. Results and discussion
3.1. Synthesis of composites

The experimental (observed) and theoretical (calculated) compositionof SAIC in the condensates (SAIC1 – SAIC5) is in good agreement with each other. The results are similar to those obtained by others (Vasudevan et al. 1978; Sharma et al. 1980). This indicates that the preparative methods adopted for the synthesis of PFR and its condensates (SAIC1 – SAIC5) are more dependable and reproducible. Exactly 11.5mL of formaldehyde, 12.5mL of conc. Sulphuric acid and 10mL of phenol are found to be optimum.

3.2. Characterization by instrumental studies

FT-IR studies were used to confirm the ion exchangeable groups present in the IERs based on various stretching frequencies. Fig.1 (a and b) indicates the appearance of absorption bands at 1033 - 1035 cm^{-1} (S = O str.), 1118 - 1186 cm^{-1} (SO_2 sym. str.) and 623 - 673 cm^{-1} (C-Sstr.) in pure resin (PFR), condensate resin with 20 % (w/w) of SAIC2 and pure SAIC, which confirm the presence of sulphonic acid group.

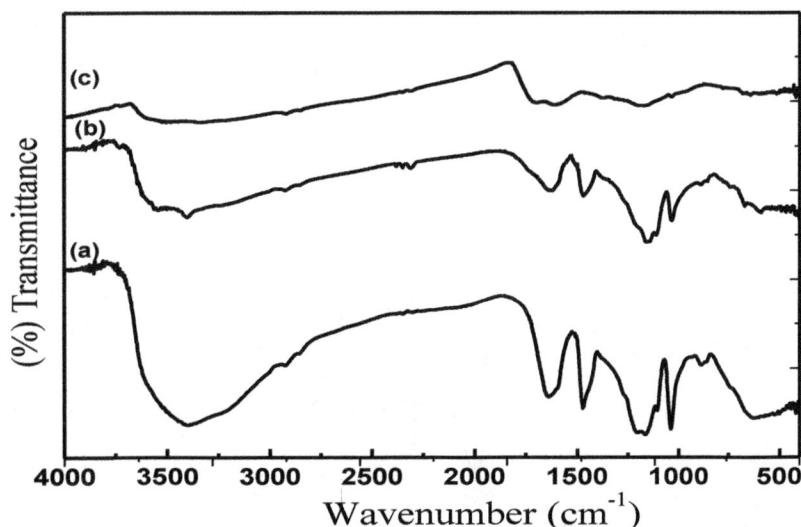

Fig.1. FT-IR spectra of (a) PFR (b) condensate resin with 20 % (w/w) of SAIC2 and (c) pure SAIC (100 %).

The appearance of broad absorption band at 3332 - 3404 cm^{-1} (H-bonded –OH str.) indicates the presence of phenolic and –OH group of in sulphonic acid group of the samples. The appearance of absorption band at 1610 – 1629 cm^{-1} (C-C str.) confirms the presence of aromatic ring in PFR, condensate resin with 20 % (w/w) of SAIC2 in PFR and pure SAIC. The absorption band at 1469 – 1581 cm^{-1} (-CH_2 def.)

concludes the presence of –CH$_2$ group in the samples. The weak absorption band at 887 – 894 cm^{-1}(-C-H def.) in the sample indicates that the phenols are tetra substituted.

Thermogravimetric analysis (TGA) is used for rapidly assessing the thermal stability of various substances (Sharma et al. 2001). The consequences of TGA curves are shown in Fig.2. From the TG data summary, the 16 % weight loss of PFR sample and the 14 % weight loss 20 % (w/w) of SAIC2 for up to 90 ^{0}C observed. It is clearly shown that the loss of moisture absorbed by the resin and composite SAIC2. The temperature between 100 ^{0}C-300 ^{0}C there is 27 % weight loss in PFR and 30 % weight loss for SAIC2. Up to 360 ^{0}C, approximately 44 % loss of weight was observed for both PFR and SAIC2. It is cleared observed that up to 100^{0} C weight loss of resin is due to the removal of water content present in the resin. Beyond this temperature, a gradual

chemical degradation of PFR and composite resin (PFR-SAIC2) occurs in 100 – 300 °C and up to 360 °C temperature range. Thermal studies indicate that the IERs are thermally stable up to 10 ^{0}C.

In Fig. 2a DTA curve shows that there are two peaks obtained in PFR, approximately at 60^{0}C and 280^{0}C. At 60 ^{0}C the presence of broad peak indicates the dehydration process of PFR. A peak at 280 ^{0}C shows the chemical changes, which occur due to thermal dehydration of PFR and reflects approximately 44 % weight loss in PFR. DTA curves of composite with 20 % (w/w) of SAIC2 (Fig. 2b) show that the two exothermic peaks are obtained at 70 ^{0}C and 290 ^{0}C respectively, which is comparable to PFR. The first peak shows the dehydration of SAIC2 and second sharp peak indicates the chemical changes arising because of thermal degradation of the condensate SAIC2.

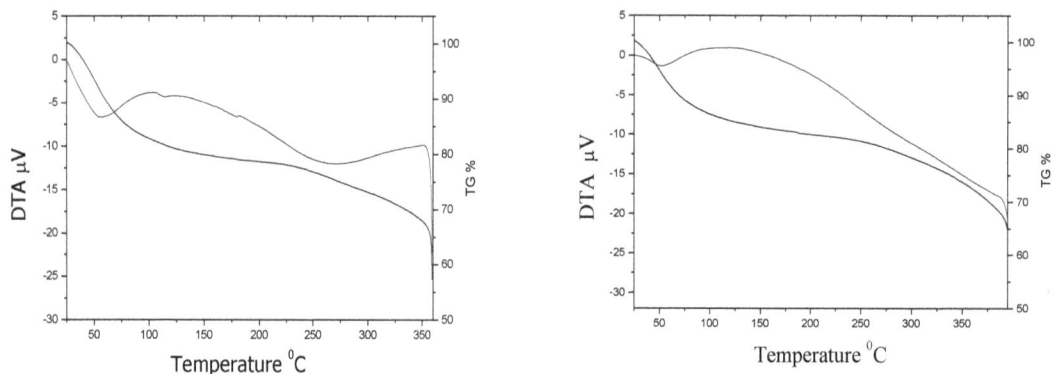

Fig. 2. Thermal studies of (a) PFR and (b) condensate with 20 % (w/w) of SAIC2.

SEM photos of PFR (Fig.3a and b), condensate SAIC2 (Fig. 3c and d) and pure SAIC (Fig.3e and f) with the magnification of 50 μm and 500 μm are given. SEM images that all the samples are macro permeable in nature. The high macro porous carbon obtained from Abutilon indicum, form the basin in which the phenol-formaldehyde

sulphonic acid particles are deposited. As a result, the pore diameter decreases in condensate SAIC2 as compared to pure SAIC. Therefore, the condensate resin SAIC2 has a great mechanical strength and little attritional breaking (Fig. 3) compare to pure SAIC.

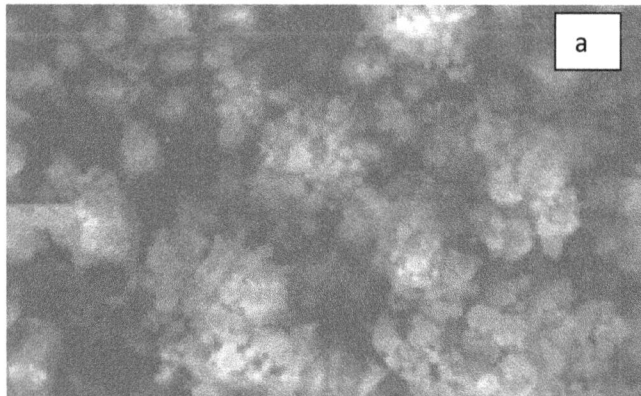

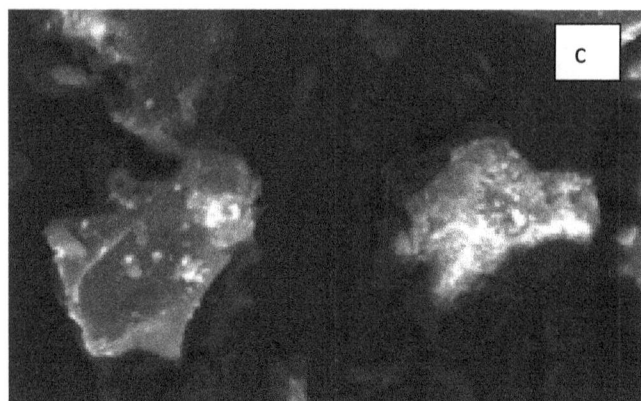

Fig. 3. SEM images of PFR (a & b), condensate resin with 20 % (w/w) ofSAIC2 (c & d) and pure 100 % SAIC (e &f).

3.3. Physico-chemical characteristics
3.3.1. Absolute density

Absolute density value (Table 1) decreases from pure resin(PFR) to the condensate containing highest percentage (w/w) of SAIC5 (50 %) and then finally to pure SAIC (100 %). The density of condensate resin in dry (dehydrated) and wet (hydrated) form depends upon the structure of resin, degree of cross-linking and its ionic nature (Sharma et al. 2001). As expected, the value of absolute density decreases with the increase in SAIC content (% w/w) in the condensate. The value of high absolute density (in both dry and wet condition) indicates a high degree of cross-linking and hence suitable for making columns for treating polar and nonpolar effluents liquids of high density. The values of absolute density of the resins in the dehydrated states are slightly higher than that of the hydrated state, but somewhat close to each other. Moreover, the values of absolute density in wet and dry states are close to each other indicating that the pores of the sample may be macroporous in nature (Krishnamoorthy et al. 1997) and close packing.

Table 1. Physicochemical properties of PFR, SAIC and condensatesSAIC1- SAIC5.

Sample	% of SAIC in IER	Density (g mL^{-1})		Percentage	
		Wet	Dry	Gravimetric Swelling	Attritional Breaking
PFR	0	2.015	2.080	80.37	9.00
SAIC1	10	1.778	1.734	65.68	18.25
SAIC2	20	1.604	1.605	55.82	19.52
SAIC3	30	1.435	1.426	54.29	24.36
SAIC4	40	1.257	1.222	44.71	31.25
SAIC5	50	1.195	1.187	42.20	38.65
SAIC	100	1.025	1.072	35.04	47.00

3.3.2. Gravimetric swelling

The value of gravimetric swelling percentage (Table 1) decreases from PFR (80.37 %) to SAIC (35.04 %). The average percentage of gravimetric swelling of the resin decreased with increasing SAIC content in the condensate. The values of gravimetric swelling percentage are found to be 80.37, 65.68 and 55.82 respectively, for 10, 20 and 30 % (w/w) blending of SAIC with the parent resin, viz., PFR. This indicates that up to 20 % (w/w) SAIC2 could be mixed with the PFR without affecting its property. The rigidity of the resin matrix is proved from the swelling measurements. Therefore, the cationic resin with higher SAIC content shows lower swelling, which reveals much lower rigid shape, and the rigidity decreases with the increase in the SAIC content in the condensate (Krishnamoorthy et al. 1997). It indicates that pure resin and condensates are rigid with non-gel macroporous structure (Bassett et al. 1989; Duraisamy et al.1987).

3.3.3. Attritional breaking

Attritional breaking value in Table 1 percentage increases with the increase in SAIC content in the resins, representing the stability of the resin, which increases from pure resin to SAIC. Therefore, the mechanical stability is good up to 20% (w/w) of SAIC2 with pure resin. This observation indicates that the capillary of the resin may be occupied by the sulphonated carbon (SAIC2) particles (Dheiveesan et al. 1988; Krishnamoorthy et al. 1997).

3.3.4. Solubility of IERs

The chemical stability of the samples in terms of its solubility in various solvents and reagents was determined. It reveals that PFR, condensates, and SAIC are practically insoluble in almost all the solvents and reagents. Therefore, these samples could be used as ion exchangers for treating non-aqueous effluents. However, the samples are found to be partially soluble in 20 % (w/w) NaOH solution, which indicates the presence of phenolic groups. Hence, these ion exchange materials cannot be used for the treatment of industrial effluents having high alkalinity. The insolubility of the samples even in the trichloroacetic acid express the rigidity, i.e., the high degree of cross-linking.

3.3.5. Cation exchange capacity

CEC data are given in Table 2. The CEC values decrease with the increase in the percentage (w/w) of SAIC in PFR for 0.1M solution of metal ions. CEC range (in m. mol. g^{-1}) for condensates 0-100 % (w/w) of SAIC with PFR: 1.4864 - 0.1791 for Na$^+$ ion; 1.6216- 0.2786 for K$^+$ ion; 1.6875 - 0.8484 for Ca^{2+} ion; 1.6934 – 1.1968 for Cu^{2+} ion; 1.8111 -0.6464 for Mg^{2+} ion; 1.7052 – 1.01 for Pb^{2+} ion and 1.8343 – 1.1312 for Zn^{2+} ion. Fig.4 shows the percentage CEC for various IERs-metal ion systems.

The Value of CEC for individual metal ion depends upon its atomic radius or atomic number. At the same time, the CEC value depends upon the anionic part of the metal salt. i.e., inter ionic forces of attraction between anions and cations, which plays a vital role in deciding the value of CEC for a particular metal salt solution (Natarajan et al. 1993; Dimov et al. 1990; Son et al. 2001). From the CEC data given in Table 2, the CEC of the samples is found to decrease in the following order.

$$Zn^{2+}>Pb^{2+}> Mg^{2+}> Cu^{2+}>Ca^{2+}>K^+> Na^+$$

3.3.6. Selectivity of metal ions

The selectivity order of metal ions, i.e., orders of CEC value also depend upon the ionic potential and the hydrated atomic radius of the metal ions in solution (Dimov et al. 1990). The order of exchange affinities of various metal ions is not unique for all the ion exchange systems. Only under dilute conditions, Hofmeister series is applicable. But, under high concentration the order is different (Kunin, et al. 1958). It is equally important to note that the relative behavior of these ions for other ionic phenomena is different from the attraction order under the

same experimental condition (Bonner et al.1995). The observed order in the present study is different from that of the Hofmeister or lyotropic series. This may be due to the concentration of the metal ion in effluent solution, which is relatively high and also due to the selectivity of the metal ions. Therefore, each composite resin (prepared from plant material) system has its own specific selectivity of the particular metal ion.

Table 2. Cation exchange capacity of PFR, pure SAIC and condensate SAIC1 – SAIC5 (H$^+$) form of 0.1 M solution of selective metal ions at 303K.

Sample	% of SAIC in IER	Cation exchange capacity, in m. mol. g^{-1} 0.1 M solution						
		Na$^+$	K$^+$	Ca^{2+}	Cu^{2+}	Mg^{2+}	Pb^{2+}	Zn^{2+}
PFR	0	1.4864	1.6216	1.6875	1.6934	1.8111	1.7052	1.8343
SAIC1	10	0.8509	0.9552	1.5352	1.4039	1.5009	1.5655	1.6008
SAIC2	20	0.8109	0.9253	1.515	1.3736	1.515	1.515	1.5099
SAIC3	30	0.7014	0.7711	1.3584	1.3029	1.2625	1.4443	1.4796
SAIC4	40	0.6865	0.7462	1.1059	1.2625	1.111	1.3988	1.4392
SAIC5	50	0.6169	0.597	1.0049	1.2322	1.0504	1.2019	1.313
SAIC	100	0.1791	0.2786	0.8484	1.1968	0.6464	1.01	1.1312

Also, the CEC data given in Fig 4, conclude that the condensate up to 20 % (w/w) mixing of SAIC with PFR (SAIC2) retains nearly 81.11–90.95 % of CEC bivalent metal ion and 54.45-58.90 for monovalent metal ions. Hence, 20% (w/w) blending of SAIC2 in PFR will reduce the cost of original resin. It is observed that the CEC decreases as the percentage of SAIC content in the condensate increases. Hence, any chemical method requiring ion exchangers of the low value of CEC, 20 % (w/w) blended SAIC –PFR resin could be used. SAIC can be inexpensively prepared from the corresponding plant material (Abutilon

indicum), which is freely available in plenty, in this study area viz., Tamil Nadu, India.

The percentage values of CEC for exchange of H$^+$ ions with Na$^+$, K$^+$, Cu^{2+}, Ca^{2+}, Mg^{2+}, Zn^{2+} and Pb^{2+} ions in 0.1M solution are about 56-65 % for   SAIC1 - SAIC5 compared to pure commercial resin(CRs) (100 %). As an average, the condensates SAIC1 - SAIC5 retain nearly 60 % of CEC of CRs. This indicates that the condensates can partially replace commercial resins (CRs) in making the ion exchangers for industrial applications.

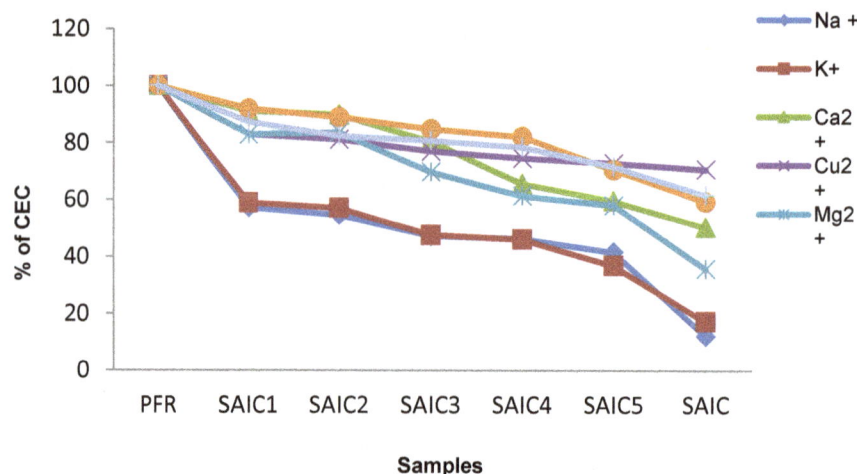

Fig. 4. Cation Exchange Capacities of H$^+$ form of PFR, condensates and Pure SAIC.

3.3.7. Effect of stability of IERs on CEC

The effect of different reagents and heat on the values of CEC for Mg^{2+} ions for various cationic resins is shown in Table 5. On treatment with 20 % (w/w) NaOH solution, 1.0 – 2.2 % reduction in CEC value is noted. Upon treatment of the resins with an organic solvent like benzene, the loss in CEC is noted to be 1.0 – 3.0 %. The decrease in

CEC value on treatment with boiling water is 0.4 – 1.5 %, for condensates with different amounts of SAIC in the resin. When the resins are heated for 10 h at 100 ^{0}C the value of CEC decreases (8 – 18 %) compared to the resins, which are not heated. All these observations reveal that the condensates have good thermal and chemical stability.

Table 3.   Chemical and thermal effect on CEC of PFR and condensates for exchange with   0.1 M Mg^{2+} ions at 303 K.

Reagents	Cation exchange capacity, in m. mol. g^{-1} 0.1M solution					
	PFR	SAIC1	SAIC2	SAIC3	SAIC4	SAIC5
CEC (of untreated)	1.855	1.714	1.535	1.413	1.345	1.287
20% (w/v) NaOH	1.745	1.613	1.514	1.343	1.243	1.285
Benzene	1.696	1.656	1.525	1.315	1.287	1.225
1M HCl	1.768	1.635	1.498	1.356	1.298	1.204
Water	1.808	1.723	1.541	1.334	1.302	1.214
Thermal treatment	1.616	1.532	1.386	1.132	1.102	1.006

CEC data given in Table 3, describe that the particle size of IERs < 210 μm are fine, 300 – 500μm and > 500 μm are common as to cause very low value of CEC compared to 210 – 300μm particle size. Hence, in order to have the effective CEC, the bed size and particle size of IER should be maintained and the recommended particle size of IER is 210 – 300μm for preparing columns for in ion exchange studies.

3.3.8. Regeneration of IERs

The regeneration data obtained with 40 mL of 0.2 M NaCl (brine) solution conclude that it effectively regenerates PFR, condensate resins, and pure SAIC. Exactly 40 mL of 0.2M brine solution effectively acts as a regenerating agent for original PFR (Fig. 5). Most of the CRs are in Na^+ form and hence 40mL of 0.2 M NaCl is to be used as a regenerator for every 2 g of the resin.

Table 4. Effect of particle size on CEC of PFR and condensate obtained by blending PFR with 20 % (w/w) of SAIC at 30 °C.

Sample	Particle Size(micron)	Cation exchange capacity,in m.mol. g^{-1} 0.1M solution			
		Na^+	Ca^{2+}	Mg^{2+}	Zn^{2+}
PFR	<200	1.4170	1.6228	1.7698	1.7816
	200-300	1.4864	1.6875	1.8333	1.8345
	300-500	1.3288	1.5934	1.7522	1.7934
	>500	1.2936	1.4994	1.6346	1.4935
SAIC2	<200	0.8080	1.3994	1.1413	1.4998
	200-300	0.8109	1.5150	1.5150	1.5099
	300-500	0.7977	1.4347	1.4847	1.4897
	>500	0.7770	1.3641	1.4594	1.4746

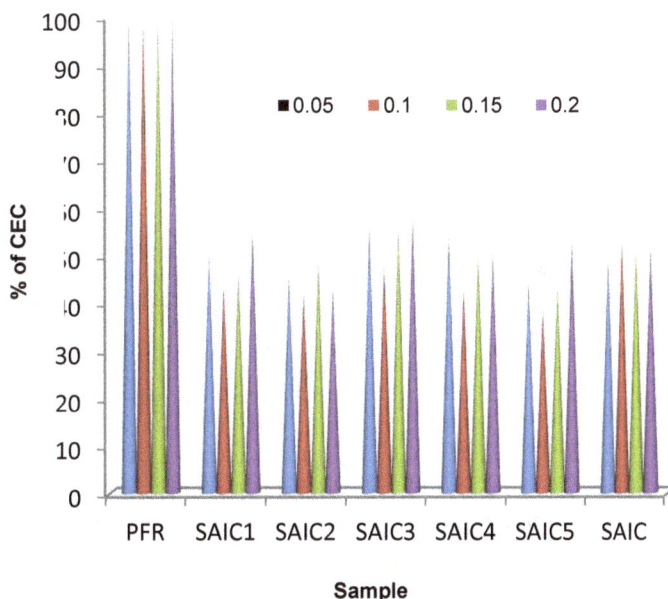

Fig. 5. Regeneration level for PFR condensates and SAIC by using NaCl after the exchange with Mg^{2+} ions.

3.3.9. Equilibrium Constant and Thermodynamic Parameters

In order to determine the effect of temperature on the exchange of Zn^{2+} by SAIC2, experiments were carried between 303 K and 333 K. The thermodynamic parameters for the removal of Zn^{2+} by IER (SAIC2) were calculated using the following basic thermodynamic equations.

$$\Delta G^\circ = - RT \ln K_c \qquad (1)$$

$$\ln K_c = \Delta S^\circ/R - \Delta H^\circ/RT \qquad (2)$$

Where R is the universal gas constant, 8.314 J/mol./K and T, the absolute temperature (K) and ΔH°, ΔG° & ΔS° are the changes in enthalpy (J/mol.), Gibb's free energy (J/mol.) and entropy (J/K/mol.) respectively. The values of ΔH° and ΔS° could be determined from the slope and intercept of the linear plot of $\ln K_c$ versus 1/T. where R is the gas constant, K_c is the equilibrium constant and T is the solution temperature in Kelvin. The thermodynamic parameters can be calculated from Vant Hoff plot (Fig. 6). The positive value of ΔH° (6350J/mol.) for the adsorption of Zn^{2+} by SAIC2 showed endothermic nature of the overall process. Negative ΔG^0 value (-882 J mol⁻¹)

indicates that the ion exchange reaction is spontaneous (Selvapandian et al. 2015).

4. Conclusions

From the current study, it is concluded that the PFR could be blended with upto 20 % (w/w) of SAIC2, without affecting its spectral, thermal and physico-chemical properties and also retain the CEC values of PFR. The effect of particle size and initial concentration of Zn^{2+} ions on CEC, its regeneration level by NaCl solution was studied. To have the effective CEC, the particle size of IER should be maintained between 210 and 300μm. CEC values of various metal ions of blends up to 20 % (w/w) SAIC2 were found to be very close to that of PFR. Equilibrium studies for the removal of Zn^{2+} ion reveal that the process is spontaneous, endothermic and occur with an increased randomness. SEM images have well defined micrometric structures. Therefore, the composites obtained from cationic matrix blending of PFR with 20 % (w/w) of SAIC2 will absolutely lower the cost which can be used in wastewater treatment, especially for the removal of metal ions.

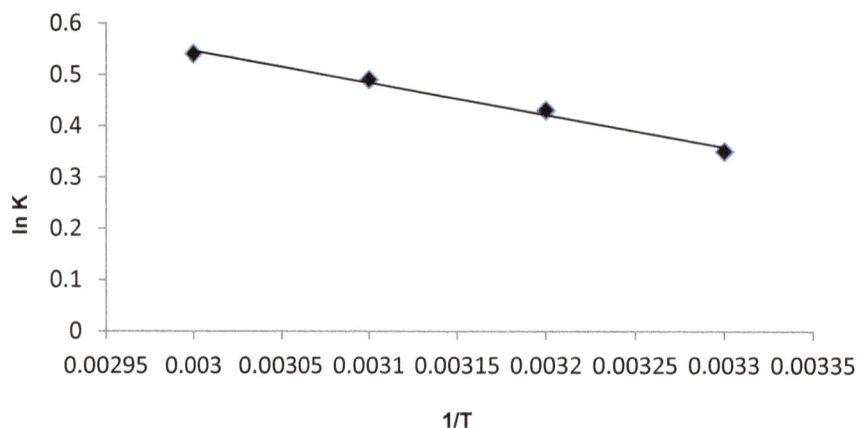

Fig. 6. Vant Hoff plot for the removal of Zn^{2+} by SCAC2.

Acknowledgments

The authors wish to thank the Principal and Head, Department of chemistry, R.D.Govt. Arts College, Sivagangai, for providing research facilities and encouragement. The authors also thank Department of chemistry, Thiagarajar college of Engineering and Technology, Madurai for recording TGA studies.

References

Bassett G. H., Jefferry J., Mendham J., R.C. Denney (eds.)., Vogel's Text Book ofQuantitative Chemical Analysis, 5thEdn., Longman Group Ltd., London, UK, (1989) pp. 186 – 192.

Bato R. C., Tanayo J. P., Jennifer J., MidredM., FPRDI J., 25 (2003)115.

BoltoB. A., Pawlowski L. (eds.), Waste Water Treatment by Ion Exchange, Oxford & IBH Publ. Co., New Delhi, (1987).

Bonner O., Easterling G., Weit D., Holland V., J Am. Chem. Soc., 77 (1955)

Chandrasekaran M. B., Krishnamoorthy S.,Studies on the removal of metal ions by various cashewnutshell-groundnutshell, Journal of the Indian Chemical Society, 64 (1987)134.

Dheiveesan T., Krishnamoorthy S., J.Indian Chem. SocLXV (10) (1988) 731.

Dimov D.K., Petrova E.B., Panayotov I.M., Tsvetanov Ch.B., Macromolecules, 21 (1990) 2733.

Duraisamy N., Krishnamoorthy S., Journal of the Indian Chemical Society, 64 (1987) 701.

Farag S., Synthesis and Physicochemical Studies of Starch-Sulphonated Phenol Formaldehyde Cationic Exchangers, Starch 47 (1995) 196.

Hassan M. L., Egyptian J. Chem 46 (2003) 329.

Homagai P.L., Development of Natural Cation Exchanger for the Treatment of Lead ions from Aqueous Solution, Journal of Nepal Chemical Society 29 (2012) 34-43.

Kannan N., SeenivasanR. K., Mayilmurugan R., Phenol-formaldehyde cationic matrices substituted by sulphonated acacia niloticacharcoal, Indian Journal of Chemical Technology 10 (2003) 623.

Kirtikar K.R., Basu B.D., Indian Medicinal Plants, 2nd Ed., BishenSingh, Mahendra Pal Singh, India, 1 (1980) 314–15.

KrishnamoorthyS., Mariamichel A., Synthesis and characterization of new composite ion exchangers, Asian Journal of Chemistry 9 (1997)136.

Krishnamoorthy S., Mariamichel A., synthesis and characterization of new composite ion-exchangers, journal of scientific & industrial research, 56 (1997) 680-685.

Kunin R., Ion Exchange Resin, Wiley, Newyork and London, 2nd edn., (1958).

MetwallyM. B., MetwallyN. E., SamyT .M., Synthesis and studies of Egyptian bagasse pith Phenol-formaldehyde cationic exchangers, Journal of Applied Polymer Science 52 (1) (1994) 61.

MetwallyM S., Metwally N.S., Synthesis and properties of synthetic Egyptian corncob-phenol formaldehyde cationic exchangers, Polymer-Plastics Technology and Engineering 31(9-10) (1992) 773.

Natarajan M., Krishnamoorthy S., studies on p-cresol-formaldehyde cationic resins substituted by coconut shell charcoal, Research and Industry 38 (1993) 278-281.

Radhakrishnan S., Krishnamoorthy S., Research and Industry 35 (1990) 188.

Raghunathan J., Krishnamoorthy S., Journal of the Indian Chemical Society 61 (1984) 911.

Rajurkar R., Jain R., Matake N., Aswar P., Khadbadi S.S., Anti-inflammatory Action of Abutilon indicum (L.) Sweet Leaves by HRBC Membrane Stabilization, Research Journal of Pharmacy and Technology 2 (2009) 415-416.

Ramachandran.S., Krishnamoorthy S., Resorcinol-formaldehyde cationic resins, Indian journal of technology 22(1984) 355-358.

Selvapandian P., Ananthakumar K., Removal of methylene blue from aquous solutions by Cynodondactylon leaf powder 5 (2015) 71-72.

Sharma N. L .N., Vasudevan P., Indian journal of technology 17 (1979) 450.

Sharma N. L.N., Vasudevan P., Journal of the Indian Chemical Society 57 (1980)191.

Sharma N.L.N., Vasudevan P., Journal of Macromolecular Science A12 (1978) 1401.

Sharma Y.R., Elementary Organic Spectroscopy, Chand S., Co.Ltd., India, Ch. 3 (2001) 100 – 104.

Son W. K., Kim S.H., Park S. G., Synthesis and exchange properties of sulfonated poly(phenylene sulfide) with alkali metal ions in organic solvents, Bulletin of the Korean Chemical Society 22 (2001) 53-58.

Swamiappan N., Krishnamoorthy S., Res. Ind., 29 (1984) 293.

Vasudevan P., Sing M., Satish N., Sharma N.L.N., J. Poly. Sci., Polmery Chemistry, 16 (1978) 2545.

Zoumpoulakis L., Simitzis J., Ion exchange resins from phenol/formaldehyde resin-modified lignin, Polymer International 50 (2001) 277-283.

15

Preparation and characterization of PES nanofiltration membrane embedded with modified graphene oxide for dye removal from algal wastewater

Negin Shaabani, Sirus Zinadini*, Ali Akbar Zinatizadeh

Environmental Research Center, Department of Applied Chemistry, Razi University, Kermanshah, Iran.

ARTICLE INFO	ABSTRACT

Keywords:
Nanofiltration
Antifouling membrane
Algal wastewater
Color removal

The present work was concentrated to study the ability of nanofiltration membrane as a treatment method of algal colored wastewater discharge from Islamabad refinery, Kermanshah, Iran. The polyether sulfone nanofiltration membrane was modified with sodium dodecyl sulfate (SDS) as an anionic surfactant and applied for treatment of colored wastewater. Water contact angle Scanning electron microscopy (SEM) and were applied to characterization of prepared membranes. The pure water flux, relative flux reduction as a parameter that represents antifouling property of membrane and also dye rejection were studied by dead-end and cross-flow filtration system in the present research. The period of the filtration time was extended about 6 hours to evaluate the stability and flux reduction of membrane. The results indicated 23.26% flux reduction was observed for modified membrane that confirms the antifouling property of prepared membrane. The results demonstrated that the permeate was completely transparent (100% dye removal, 98.2% turbidity removal), and the pure water flux was enhanced for modified membrane to 27.21 (Kg/m².h). In the present research nanocomposite polymeric membrane are introduced as an appropriate option for the treatment of natural colored wastewater.

1. Introduction

There are some reasons that can make the treatment of water as an essential issue. The increasing population of the world and also the increase in water consumption, make a water reuse as a necessity of the today's world (Nawaz et al .2013). Also the industrial wastewater discharge into environment has a negative effect on that (Qadir et al .2008) . It is harmful to discharge in river water and known as a hinder of water treatment for the different application (Azarian et al .2007). So the water shortage and water pollution are become as one of the big challenge in the world and the water reuse defined as a solution to overcome that (Cosgrove et al .2015). There are different ways for water and wastewater treatment such as physical, chemical and biological treatment. Among them, membrane processes have been used extensively as a wastewater treatment method for dye removal from wastewater (Nawaz et al .2014). Despite all the advantages of the membrane, fouling is a severe problem of them, that affects membrane performance (Mosqueda-Jimenez et al .2006). Membrane fouling can be reduced by increasing of hydrophilicity and it can be possible by different way of membrane modification (sun et al .2017).

In the last years, the properties and performance of membranes was improved by the incorporation of carbon nanofillers into the polymer matrix (Bhattacharya et al .2016). The small amount of these kind of nanoparticles can make an improvement in the properties of their composite materials (Bhattacharya et al .2016). Among different methods for membrane fabrication, blending has some benefits such as: Convenient preparation conditions and sufficiency for industrialization (Yang et al .2017). Due to the hydrophilic nature of graphene oxide, it known as a suitable nanoadditive to enhance hydrophilicity of membrane (Chang et al .2014). Zinadini and his coworkers fabricated an antifouling mix matrix PES membrane by embedding graphene oxide nanoplates (Zinadini et al .2014). The prepared membrane shown high pure water flux and hydrophilicity. Yang et al (2017) prepared a graphene oxide modified poly(m-phenylene isophthalamide) nanofiltration membrane with improved water flux, antifouling property and dye rejection. Shukla et al (2017) fabricated polyphenyl sulfone/GO nanocomposite membrane with improved in an antifouling property. Also the prepared membrane demonstrated an improvement in mechanical property. Ganesh et al (2013) prepared Psf/GO composite membrane with significant improvement in pure water flux and salt rejection. Most of these researches reported agglomeration of nanoparticles as a major problem of nanocomposite polymeric membranes (Zinadini et al.2014-Yang et al.2017). The use of surfactant suggested as a solution of these problem (Morsy et al.2014). Therefore, at the present work graphene oxide was incorporated into PES polymeric matrix and sodium dodecyl sulfate (SDS) was selected as an anionic surfactant to play an important role as dispersing agents. This ionic surfactant contains anionic functional groups at their head. It has been able to coating nanoparticles due to the hydrophilic sulfonated group in the molecules and also increase hydrophilicity. The performance of prepared membrane was evaluated by pure water flux measurement, dye rejection and flux reduction as a parameter to indicate antifouling. The algal colored wastewater was applied to examine the performance of prepared membrane in dye removal experiments.

2. Materials and methods
2.1. Materials

Polyethersulfone (ultrason E6020p, Mw=58000 g/mol and glass transition temperature Tg=225˚c) and dimethyleacetamide (DMAC) was used as solvent from BASF co, Germany. Poly vinylpirolidone

(PVP) with 25000g/mol molecular weight was provided from Merck as pore former agent. The synthetic Graphene oxide was used as a carbon nanofiller. Sodium dodecyl sulfate (SDS) as an anionic surfactatant was used from Merck for dispersion of carbon nanofillers in the casting solution and increase hydrophilicity of polymer. Algal wastewater (total COD=305.7, turbidity=56 NTU, dye concentration= 800-1000 mg/lit) from Islamabad refinery, Kermanshah, Iran was applied as a natural colored wastewater. Distilled water was used throughout the experiments.

2.2. Preparation of the membrane

The composition of casting solution for prepared membranes are listed in Table 1. The phase inversion techniques were used for preparation of membranes. In the first step 0.5 wt.% of GO and certain amount of SDS were dispersed into the DMAC and was sonicated for 30 minutes using DT 102H Bandelin ultrasonic(Germany). Then, PES and PVP were dissolved in the dope solution by continues stirring for 24 h. After that, casting solution was cast by a film applicator on to a smooth glass plate with 150μm thickness. Then the glass plate was submerged into the non-solvent bath (distilled water at a temperature of 15 °C). After membrane formation, the prepared membranes were kept in water for 24 h to allow complete phase separation. Then membranes were drying between two sheets of filter paper for 24h at room temperature.

Table1. The compositions of casting solutions.

Membrane type	PES (wt. %)	PVP (wt. %)	G.O (wt. %)	SDS (wt. %)
M1	20	1.0	0.5	-
M2	20	1.0	0.5	0.5

2.3. Characterization of prepared membranes

The hydrophilicity of the membrane surface was examined by contact angle goniometer (G10, KRUSS, Germany) at 25 °C and a relative humidity of 50%. All contact angles are measured by 2μL of de-ionized water. To minimize the experimental error value, the contact angle values in five different locations are randomly measured and reported. The morphology of prepared membrane was studied by a Philips scanning electron microscope (Philips X100). The samples of the prepared membrane were cut into small pieces and were submerged in nitrogen liquid for one minute and were frozen. Then the frozen membrane was broken and kept in air for drying. After sputtering with gold, they were viewed in very high vacuum conditions at 25kV.

2.4. Membrane performance measurements

To evaluation of prepared membranes, pure water flux, dye removal and flux reduction were examined. To measurement of the pure water flux, the dead-end stirred cell (200 ml volume) was used with a membrane surface area of 12.56 cm² conjunct to a nitrogen gas line. Pressurized nitrogen was used to force the liquid through the membrane. The operational pressure was 4 bar .The water flux $J_{w,1}$(kg/m².h) was calculated as follow:

$$J_{w,1} = \frac{M}{A \cdot \Delta t}$$ (1)

where, M (kg) is the weight of the permeates , A (m²) is the membrane effective area and Δt (h) is the permeation time. The experiments were done at 25 and the average of three times was represented in Tables. Cross-flow filtration cell was applied for measure of relative flux reduction and evaluation of membrane stability at the pressure of 5 bar and flowrate of 130 Kg/h. The fig 1, showed the schematic of cross-flow filtration cell that used for examination of the membrane modified with anionic surfactant.

2.5. Antifouling performance

The flux of the membrane may be reduced during the algal wastewater filtration due to the fouling. To investigate the antifouling property of modified membrane, the relative flux reduction (RFR) parameter was measured. That was calculated by the following equation:

$$RFR(\%)= (\frac{J_f - J_i}{J_i}) \times 100$$ (2)

where J_f is the final flux and J_i is the initial flux. The lower relative flux reduction indicates the higher antifouling property.

Also to examination of antifouling performance of prepared membrane, the flux recovery ratio (FRR) of prepared membrane calculated based on water flux before and after the passing milk powder solution with 800 ppm concentration from the membrane. That was measured according following equation:

$$FRR(\%)=(\frac{J_{w,2}}{J_{w,1}}) \times 100$$ (3)

Where $J_{w,1}$ is the initial water flux and $J_{w,2}$ is the water flux after milk powder filtration (kg/m².h).

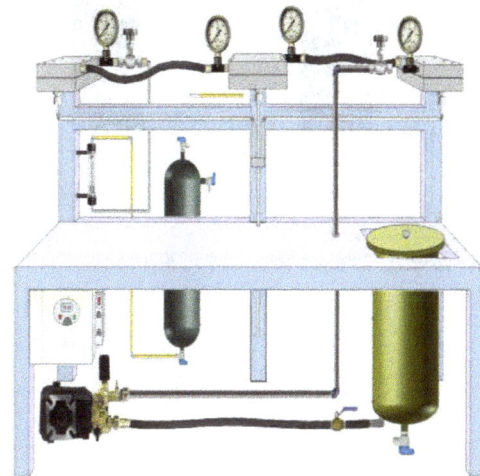

Fig. 1. The schematic of cross-flow filtration system.

3. Results and discussion
3.1. Morphology of the prepared membranes

The SEM technique was used for examine the structure and morphology of the nanocomposite polymeric membranes. The cross-sectional SEM images are shown in Fig 2. Both of the membrane show similar structure with dense top layer and finger-like sub layer. Finger like pores for M2 is significantly wider than M1. This is due to the hydrophilic nature of GO and SDS that can enhance the mass transfer during the phase inversion and make a wider pore channels (Vatanpour et al.2011).

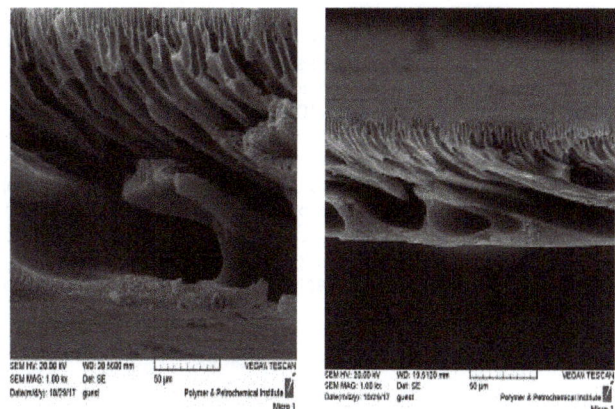

Fig. 2. Cross-section SEM images of the membrane modified with GO-SDS(left) and membrane modified with GO (right).

3.2. Pure water flux and membrane hydrophilicity

Water contact angle measurement was applied to examine the hydrophilicity of the prepared membrane. More hydrophilic membranes have lower contact angle. As demonstrated in in Fig. 3, the contact angle of membrane modified with GO and SDS was 52.8°, which is lower than the membrane modified with just graphene oxide (55.3°). By improving hydrophilicity, water molecules pass more easily through the membrane and the pure water flux will be increase. Fig. 4 showed the pure water flux data of the prepared membrane that indicates higher performance of M_2 membranes.

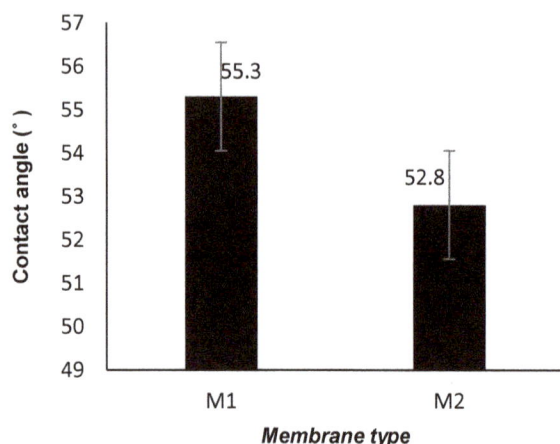

Fig. 3. Static contact angle of the prepared membranes.

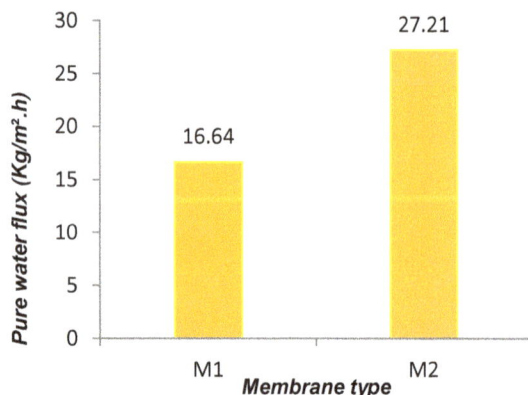

Fig. 4. Pure water flux of the prepared membrane (after 30 min).

3.3. Nanofiltration performance

To investigate the nanofiltration performance of the prepared membrane, the algal wastewater was selected and the removal of algal colored was examined. After filtration, the permeate flow was quite transparent and free of any dyes (100% algal colored removed). Also, the turbidity was measured about 56 and 1 NTU for influent and filtrated respectively that shows 98.2% turbidity removal. It can be attributed to the electrostatic charge of membrane surface and dye. When the pH value is in the range of 2-9, the graphene oxide (Yang et al.2017) and anionic surfactant are negatively charged Which causes repulsion between dye and the negative charge of the membrane surface. The image of influent and permeation solution before and after the filtration was shown in Fig 5.

3.4. Antifouling properties of the membranes

To evaluate the stability and antifouling property of membrane modified with GO and SDS, the cross-flow filtration cell was applied at the flow rate of 130 Kg/h and the pressure of 5 bar. During the long term filtration of the algal wastewater, the pores of the membrane are fouled and the flux decreased. The period of the filtration time was extended about 6 hours and the amount of flux reduction was measured. The relative flux reduction for M2 was calculated about 23.26 %, that shows

the antifouling property of the modified membrane. Fig. 6, demonstrated the graph of flux (Kg/m².h) versus time (hour) for M2.

Fig. 5. Comparison of the wastewater appearance before (right) and after membrane treatment (left).

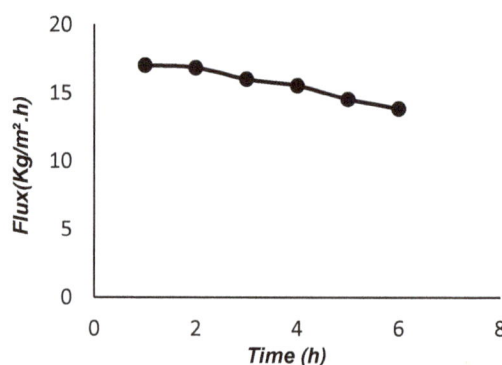

Fig. 6. The flux of algal wastewater during the 6 hours filtration by cross-flow filtration cell.

Also the results of FRR are demonstrated in Fig. 7, and determined as a parameter to indicates the antifouling property of prepared membrane. As shown in Fig 7, the FRR parameter for modified membrane with GO and SDS is 83.79 %, that is higher than bare PES membrane. These results shows that the modified membrane with hydrophilic nanoparticles such as graphene oxide and surfactants such as SDS shows more antifouling properties.

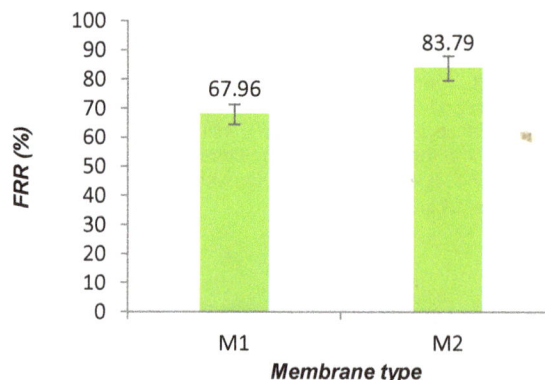

Fig. 7. Flux recovery ratio of the unfilled and modified PES membranes after 90 min milk powder solution filtration at 4 bar by dead-end cell.

4. Conclusions

In this research, polyether sulfone nanofiltration membrane modified with GO and SDS by blending method. The addition of constant concentration (0.5 wt.%) of graphene oxide and anionic surfactant (SDS) into the PES nanofiltration membrane, decrease the contact angle and increase hydrophilicity and pure water flux and improved antifouling property. The morphology of prepared membranes studied by SEM images. The modified membrane with GO and SDS showed the wider finger-like structure. Also the modified membrane showed significant nanofiltration performance and eliminate near 100 % algal waste dye. Also after 6 hours the flux reduction of modified membrane was 23.26% that shows the good antifouling property of that. All the results demonstrated that, PES membrane modified with GO and SDS considered as a suitable option for dye removal from algal wastewater.

References

Azarian G.H., Mesdaghinia A.R., Vaezi F., Nabizadeh R., Nematollahi D., Algae removal by electro-coagulation process, application for treatment of the effluent from an industrial wastewater treatment plant, Iranian Journal of Public Health 36(4) (2007) 57-64.

Bhattacharya M., Polymer nanocomposites—a comparison between carbon nanotubes, graphene, and clay as nanofillers, Materials 9(4) (2016) 262.

Chang X., Wang Z., Quan S., Xu Y., Jiang Z., Shao L., Exploring the synergetic effects of graphene oxide (GO) and polyvinylpyrrodione (PVP) on poly (vinylylidenefluoride)(PVDF) ultrafiltration membrane performance, Applied Surface Science 316 (2014) 537-548.

Cosgrove W.J., Loucks D.P., Water management: Current and future challenges and research directions, Water Resources Research 51(6) (2015) 4823-4839.

Ganesh B., Isloor A.M., Ismail A.F., Enhanced hydrophilicity and salt rejection study of graphene oxide-polysulfone mixed matrix membrane, Desalination 313 (2013) 199-207.

Morsy S.M., Role of surfactants in nanotechnology and their applications, Int. J. Curr. Microbiol. App. Sci 3(5) (2014) 237-260.

Mosqueda-Jimenez D.B., Huck P.M., Characterization of membrane foulants in drinking water treatment, Desalination 198(1-3) (2006) 173-182.

Nawaz M.S., Ahsan M., Comparison of physico-chemical, advanced oxidation and biological techniques for the textile wastewater treatment, Alexandria Engineering Journal 53(3) (2014) 717-722.

Nawaz M.S., Gadelha G., Khan S.J., Hankins N., Microbial toxicity effects of reverse transported draw solute in the forward osmosis membrane bioreactor (FO-MBR), Journal of membrane science 429 (2013) 323-329.

Qadir A., Malik R.N., Husain S.Z., Spatio-temporal variations in water quality of Nullah Aik-tributary of the river Chenab, Pakistan, Environmental Monitoring and Assessment 140(1-3) (2008) 43-59.

Shukla A.K., Alam J., Alhoshan M., Dass L.A., Muthumareeswaran M., Development of a nanocomposite ultrafiltration membrane based on polyphenylsulfone blended with graphene oxide, Scientific Reports 7 (2017) 41976.

Sun W., Liu J., Chu H., Dong B., Pretreatment and membrane hydrophilic modification to reduce membrane fouling, Membranes 3(3) (2013) 226-241.

Vatanpour V., Madaeni S.S., Moradian R., Zinadini S., Astinchap B., Fabrication and characterization of novel antifouling nanofiltration membrane prepared from oxidized multiwalled carbon nanotube/polyethersulfone nanocomposite, Journal of Membrane Science 375(1-2) (2011) 284-294.

Yang M., Zhao C., Zhang S., Hou P. Li, D., Preparation of graphene oxide modified poly (m-phenylene isophthalamide) nanofiltration membrane with improved water flux and antifouling property, Applied Surface Science 394 (2017) 149-159.

Zinadini S., Zinatizadeh A.A., Rahimi M., Vatanpour V., Zangeneh H., Preparation of a novel antifouling mixed matrix PES membrane by embedding graphene oxide nanoplates, Journal of Membrane Science 453 (2014) 292-301.

16

Short review on membrane distillation techniques for removal of dissolved ammonia

Batool Shahroie[1], Laleh Rajabi[1], Ali Ashraf Derakhshan[2,*]

[1]Polymer Research Center, Department of Chemical Engineering, Razi University, Kermanshah, Iran.
[2]Environmental Research Center, Faculty of Chemistry, Razi University, Kermanshah, Iran.

ARTICLE INFO	ABSTRACT
Keywords: Ammonia Membrane distillation Wastewater	When ammonia discharged into water resources, it has a negative effect on aquatic life as a major water pollutant. Therefore, removing of ammonia from wastewaters has become an essential need for last decades concurrent with developing in the industry and agriculture. Hence there are emerged various techniques for removing the solvated ammonia which among them membrane distillation (MD) is the powerful technique for wastewater treatment. In the thermally process of membrane distillation, only volatile molecules are transferred through hydrophobic membrane. The microporous membrane is a barrier for separation of permeate (cool side-liquid or gas phase) from feed (hot side-liquid or gas phase). The vapor pressure gradient is a propulsion force for migration volatile molecules into the permeate side. In this short review paper, we summarized the surveys about membrane distillation techniques in removal of solvated ammonia.

Contents

1. Introduction

Ammonia (NH_3) penetrated into the natural waters by industrial, domestic and agricultural waste water discharges have become a major environmental problem. Small amounts of discharged NH_3 without any purification can have harmful effects on aquatic life. Due to the toxic nature of ammonia, the use of biological processes to purify wastewater from ammonia is not so simple.

The removing and recovery of NH_3 and its derivatives from wastewaters can be performed by biological, physical, chemical, or a combination of them such as adsorption, chemical precipitation, membrane filtration, reverse osmosis, ion exchange, air stripping, breakpoint chlorination and biological nitrification (Degermenci et al. 2012; Tchobanoglous et al. 1991). Recently there are much attention to membrane distillation (MD) for separation of volatile pollutants from wastewaters because of its potentially low energy necessity. The MD process has capability for recycling of industrial wastewaters, and can be advantageous for high-temperature wastewater streams with relatively low levels of volatile compounds (Xie et al. 2009).

In the thermally process of membrane distillation, only volatile molecules are transferred through hydrophobic membrane. The microporous membrane is a barrier for separation of permeate (cool side-liquid or gas phase) from feed (hot side-liquid or gas phase). The vapor pressure gradient is a propulsion force for migration volatile molecules into the permeate side (Xie et al. 2009; Banat et al. 1998). Finally, migrated volatile compounds are either condensed or removed in the vapor phase, depending on the configuration (Xie et al. 2009; El-

Bourawi et al. 2006; Lawson et al. 1997). In this paper we try to review the membrane distillation techniques in the wastewater treatment in order to remove ammonia and considering their advantages and disadvantages.

2. Membrane distillation techniques
2.1. Direct contact membrane distillation

Structure of a Direct contact membrane distillation (DCMD) are illustrated schematically in Fig. 1. In the membrane, Evaporator and permeate sides are charged with liquid hot-feed water and cooled permeate, respectively. The vapors passing through the membrane condense directly inside the liquid phase at the membrane surface. The single membrane layer has the low insulating properties hence a disadvantage of DCMD is the high sensible heat loss between condenser and evaporator sides.

Hollow fiber membrane contactors nominate a suitable alternative to remove various volatile contaminants (Tan et al. 2006; Ozturk et al. 2003; Zhang et al. 1985). These membranes provide a barrier between liquid phase and volatile contaminants, and these volatile molecules penetrate to membrane pores in order to reach liquid phase. To achieve less mass transfer resistances, it is necessary that the membranes used to remove volatile pollutants usually have a hydrophobic structure. Because of good hydrophobicity and feasibility, Polyvinylidene fluoride (PVDF) is an attractive membrane material to form asymmetric membranes (Tan et al. 2006; Jian et al. 1997; Deshmukh et al. 1998).

---

*Corresponding author Email: derakhshan.ali.a@gmail.com

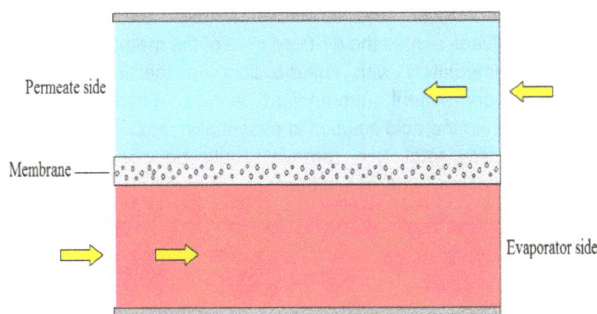

Fig. 1. Structure of direct contact membrane distillation (DCMD).

Hollow fiber membranes of PVDF with different morphological structures (Fig. 2) were prepared by Xiaoyao Tan et al. (Tan et al. 2006), to tailor for NH_3 separation from water. In order to accelerate ammonia removing, the aqueous solution of H_2SO_4 was utilized as stripping solution. The results revealed that increasing the pH is capable of promoting the NH_3 elimination. Post-treatment of PVDF membrane with ethanol was improved both the hydrobicity and the effective surface porosity, and subsequently improved the NH_3 removal. In this process, the feed velocity of acid solution and initial concentration of NH_3 had little impacts on the NH_3 elimination.

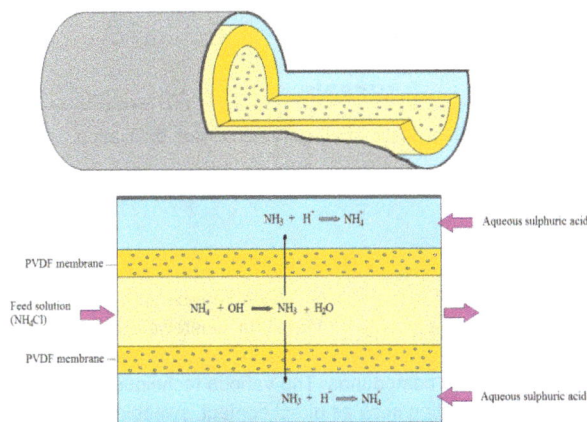

Fig. 2. PVDF hollow fiber membranes with different morphological structures.

In the other attempt, polypropylene hollow-fiber membranes were utilized to attain effective removal of dissolved ammonia (Ashrafizadeh et al. 2010). In order to accelerate ammonia removing, the aqueous solution of H_2SO_4 was utilized as stripping solution. Polypropylene membrane was shown to be very efficient in separating NH_3 from the wastewaters, in the best conditions, NH_3 removal of over 99 % was achieved. Attained results indicate that the velocities and initial concentrations of the NH_3 and H_2SO_4 solutions had insignificant effects on the NH_3 elimination. Increasing the pH of feed solution up to 10 enhanced the elimination of NH_3 meaningfully while insignificant improvements attained in upper than 10 value.

Increasing the feed velocity of NH_3 solution enhanced its removal in the range studied (Ashrafizadeh et al. 2010). Highly promising results can be attain using a submerged membrane contactor for NH_3 extraction. The direct NH_3 removal from particle rich substrates and less consuming input energy are the advantages of this method. B. Lauterböck et al. (2012) were utilized a hollow-fiber membrane contactor module for continuous NH_3 elimination in an anaerobic digestion process. The hollow-fiber membranes were directly immersed into the digestate of the anaerobic reactors.

The wastewater of slaughterhouse was used as feed for reactors with NH_4^+ concentrations ranging from 6-7.4 g/L. In this membrane reactor, the ammonia level was significantly decreased by about 70 %. The continuous ammonia removal causes to improve substrate conversion rates, a more stable process performance and an increased biogas yield (Lauterböck et al. 2012).

2.2. Sweeping gas membrane distillation (SGMD)

Sweeping gas membrane distillation (Fig. 3), uses a channel structure with an empty gap on the permeate side. The volatile compounds can be distilled with a low surface tension and an inert gas removes these vapor from the permeate side. Then condensation of vapors takes place outside the module by an external condenser.

The lower conductive heat loss and reduced mass transfer resistance are the advantages of sweep gas MD towards other configurations (El-Bourawi et al. 2006). In addition, this module provides a superior permeate flux and evaporation efficiency (Xie et al. 2009). Therefore, among various membrane distillation methods, the SGMD was indicated to be prominent method for the removing volatile components from wastewaters (Xie et al. 2009; Khayet et al. 2003; Rivier et al. 2002). Also membrane wetting is minimized when SGMD has less condensation of water droplets in the membrane pores (Franken et al. 1987). The ammonia elimination from wastewaters with high NH_3 concentration (500-10,000 mg/L) has been studied (Ding et al. 2006; Zhu et al. 2005). The mass transfer coefficient for SGMD was indicated to be similar to vacuum membrane distillation (VCMD) at NH_3 concentrations of up to 3200 mg/L while the selectivity was found 27-100 % higher (Ding et al. 2006). Some industries discharge the wastewater containing lower ammonia concentrations and we know that the feed concentration has influence on MD performance. The ammonia elimination from wastewater containing low value of NH_3 (100 mg/L) has been simulated in experiments with SGMD (at pH 11.5) by Zongli Xie et al (Xie et al. 2009). It has found that the raising of feed temperature causes to enhance in the permeate flux meaningfully, but reducing the selectivity. Also increasing in flow rates of feed and sweep gas promoted NH_3 removal efficiency and permeate flux. Up to 97 % ammonia removal could be achieved in the best of conditions, to give a purified water containing only 3.3 mg/L of NH_3 (Xie et al. 2009).

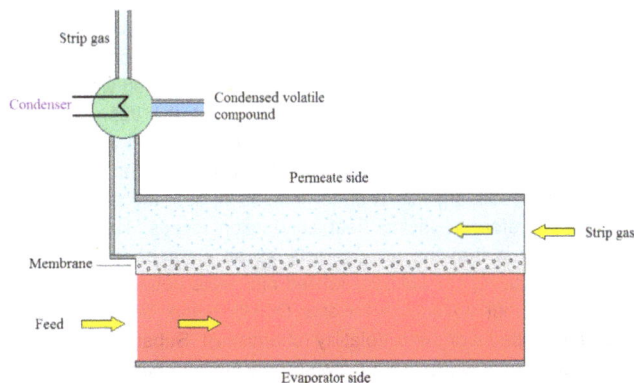

Fig. 3. Schematic of sweeping gas membrane distillation.

2.3. Vacuum membrane distillation (VCMD)

In Vacuum membrane distillation includes an air gap channel configuration (Fig. 4). The volatile compounds that have transferred through the membrane, are sucked out by the vacuum from permeate channel and condenses outside the module. The advantages of VCMD are that leaving a larger effective membrane surface active and a reduction of the boiling point. However, providing the technical equipments for generation of a vacuum is a disadvantage to this method (https://en.wikipedia.org/wiki/Membrane_distillation). pH is a critical factor for NH_3 removal applications by VCMD when increasing the feed pH caused to enhancing ammonia removal efficiencies. EL-Bourawi et al. (2007) were investigated the applicability of VCMD for NH_3 removal from its aqueous solutions. The results showed that higher value for feed temperatures, pH and initial feed concentrations and lower value for downstream pressures promote NH_3 removal efficiency.

This is found that the pH value is to be a most effective factor. Mass transfer significantly affected by temperature and concentration polarization between feed border layers. Increasing in feed flow velocity is caused to decreasing in temperature and concentration polarizations. The resistance to mass transfer is shown to change from being mainly located in the feed side at low flow velocities and feed temperatures to be closely located through the membrane pores at 55.7 °C and logically higher feed flow velocity of 0.84 m/s. Although higher feed temperatures and lower downstream pressures increase remarkably the total trans membrane flux and the NH_3 removal rate, the corresponding ammonia separation factors were decreased. Ammonia removal efficiencies higher than 90 % with separation factors of more than 8 were achieved by El-Bourawi team (El-Bourawi et al. 2007).

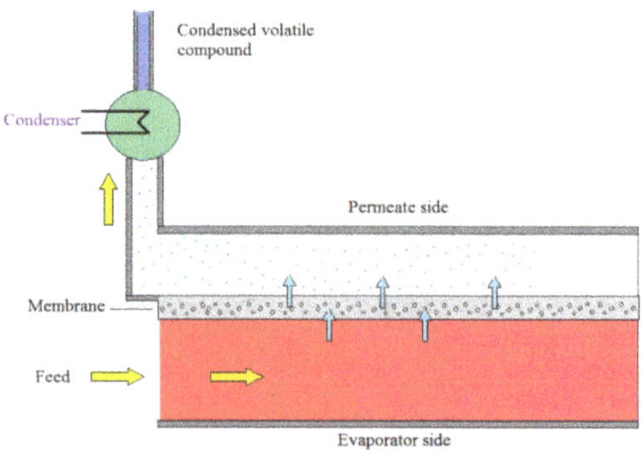

membrane interface. NH_3 volatilizes through the feed-membrane interface, diffuses across the air-filled pore of the membrane, and finally it reacts immediately with sulfuric acid on the interface to form nonvolatile component, ammonium sulfate. Therefore, the NH_3 concentration in the acid solution is essentially zero.

Theoretically total NH_3 removal could be possible under this separation system, whereas difference in ammonia partial pressure between the feed and the receiving solution is a driving force for this membrane contactor process (Hasanoğlu et al. 2010).

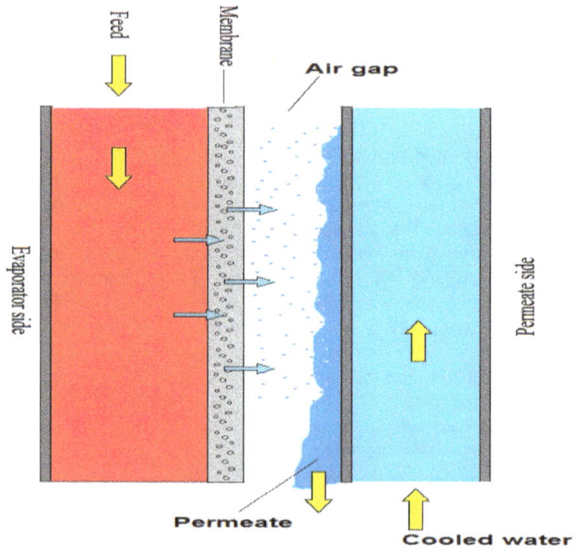

Fig. 4. Vacuum membrane distillation includes an air gap channel configuration.

2.4. Air gap membrane distillation (AGMD)

In air-gap MD (http://en.wikipedia.org/wiki/Evaporator), the evaporator channel resembles that in DCMD, while the permeate gap filled with air exists between the membrane and a cooled wall (Fig. 5). Before condensation on the cooler wall surface, the vapor diffusing through the membrane must additionally overcome this air gap. The advantage of this method is the high thermal insulation (http://en.wikipedia.org/wiki/Thermal_insulation) near the condensation channel, therefore reducing heat conduction losses. However, the disadvantage is that the air gap acts as an extra barrier for mass transport, reducing the surface- related permeate output compared to DCMD. A further advantage towards DCMD is the fact, that volatile (http://en.wikipedia.org/wiki/Volatility (chemistry)). Substances such as alcohol or other solvents (with a low surface tension) can be separated from diluted solutions, because there is no contact between the liquid permeate and the membrane with AGMD (https://en.wikipedia.org/wiki/Membrane_distillation). Hasanoğlu et al. (2010) were used polypropylene (PP) and polytetrafluoroethylene (PTFE) membranes so that contact the NH_3 solutions and the receiving solution (Diluted solutions of H_2SO_4). The hydrophobic hollow-fiber separates the feed including aqueous ammonia on the shell side and the receiving solution on the lumen side. The pores of hydrophobic membrane filled by an air gap which is not wetted by the aqueous solutions. First, NH_3 molecules penetrates from the feed into the feed-

Fig. 5. Schematic of air-gap membrane distillation.

3. Conclusions

In this short review paper we summarized the surveys about membrane distillation in removal of solvated ammonia. The MD process has capability for recycling of industrial wastewaters, and can be advantageous for high-temperature wastewater streams with relatively low levels of volatile compounds. The various techniques have utilized in membrane distillation such as direct contact, sweeping gas, vacuum and air gap membrane distillation which can be led to ammonia treatment from wastewater. The single membrane layer has the low insulating properties hence a disadvantage of DCMD is the high sensible heat loss between condenser and evaporator sides. The lower conductive heat loss and reduced mass transfer resistance are the advantages of sweep gas MD towards other configurations. The advantages of VCMD are that leaving a larger effective membrane surface active and a reduction of the boiling point. The advantage of AGMD is the high thermal insulation near the condensation channel, therefore reducing heat conduction losses.

References

Ashrafizadeh S.N., Khorasani Z., Ammonia removal from aqueous solutions using hollow-fiber membrane contactors, Chemical Engineering Journal 162 (2010) 242-249.

Banat F.A., Simandi J., Desalination by membrane distillation, Separation Science and Technology 33 (1998) 201-206.

Ahmed S., Rasul M.G., Martens W.N., Brown R., Hashib M.A., Heterogeneous photocatalytic degradation of phenols in wastewater: a review on current status and developments, Desalination 261 (2010) 3–18.

Degermenci N., Nuri Ata O., Yildız E., Ammonia removal by air stripping in a semi-batch jet loop reactor, Journal of Industrial and Engineering Chemistry 18 (2012) 399–404.

Deshmukh S.P., Li K., Effect of ethanol composition in water coagulation bath on morphology of PVDF hollow fibre membranes, Journal of Membrane Science 150 (1998) 75-85.

Ding Z., Liu L., Li Z., Ma R., Yang Z., Experimental study of ammonia removal from water by membrane distillation (MD): the comparison of three configurations, Journal of Membrane Science 286 (2006) 93-103.

El-Bourawi M.S., Ding Z., Ma R., Khayet M., A framework for better understanding membrane distillation process, Journal of Membrane Science 285 (2006) 4-29.

EL-Bourawi M.S., Khayet M., Ma R., Ding Z., Li Z., Zhang X., Application of vacuum membrane distillation for ammonia removal, Journal of Membrane Science 301 (2007) 200-209.

Franken A.C.M., Nolten J.A.M., Mulder M.H.V., Bargeman D., Smolders C.A., Wetting criteria for the applicability of membrane distillation, Journal of Membrane Science 33 (1987) 93-103.

Hasanoğlu A., Romero J., Pérez B., Plaza A., Ammonia removal from wastewater streams through membrane contactors: Experimental and theoretical analysis of operation parameters and configuration, Chemical Engineering Journal 160 (2010) 530-537.

Jian K., Pintauro P.N., Asymmetric PVDF hollow-fiber membranes for organic/water pervaporation separations, Journal of Membrane Science 135 (1997) 41-53.

Khayet M., Godino, M.P., Mengual, J.I., Possibility of nuclear desalination through various membrane distillation configurations: a comparative study, International Journal Nuclear Desalination 1 (2003) 30-46.

Lauterböck B., Ortner M., Haider R., Fuchs W., Counteracting ammonia inhibition in anaerobic digestion by removal with a hollow fiber membrane contactor, Water Research 46 (2012) 4861-4869.

Lawson K.W., Lloyd D.R., Membrane distillation. Journal of Membrane Science 124 (1997) 1-25.

Ozturk I., Altinbas M., Koyuncu I., Arikan O., Yangin C.G., Advanced physico-chemical treatment experiences on young municipal landfill leachates, Waste Management 23 (2003) 441-446.

Rivier C.A., Garcia-Paya M.C., Marison, I.W., Stockar U.V., Separation of binary mixtures by thermostatic sweeping gas membrane distillation, Journal of Membrane Science 201 (2002) 1-16.

Tan X., Tan S.P., Teo W.K., Li K., Polyvinylidene fluoride (PVDF) hollow fiber membranes for ammonia removal from water, Journal of Membrane Science 271 (2006) 59-68.

Tchobanoglous G., Burton F.L., Wastewater Engineering, 3rd edition, 1991, 1178.

Zhang Q., Cussler E.L., Microporous hollow fibers for gas absorption. I. Mass transfer in the liquid, Journal of Membrane Science 23 (1985) 321-332.

Zhu Z., Hao Z., Shen Z., Che J., Modified modelling of the effect of pH and viscosity on the mass transfer in hydrophobic hollow fiber membrane contactors, Journal of Membrane Science 250 (2005) 269-276.

17

Water pollution management in wells of zawar village for investigation of effects of nitrogen fertilizers in nitrate entry into groundwater

Ali Roholamin Kasmaei[1,*], Mehdi Nezhad Naderi[2], Zaynab Bahrami[2]

[1]Member of Water and Sewer Company of Gilan, Rasht, Iran.
[2]Department of Civil Engineering, Tonekabon Branch, Islamic Azad University, Tonekabon, Iran.

ARTICLE INFO	ABSTRACT
Keywords: Chemical parameters Fertilizer Groundwater Nitrate Water quality in rural wells	Application of N fertilizers in agricultural operations is one of the important sources of nitrate entry into groundwater. In Iran, especially in coastal areas with a high groundwater level, in agricultural areas, there is a risk of pollution of groundwater and surface water to nitrates. This research was conducted with the aim of determining the concentration of chemical parameters of drinking water wells in a Tonekabon village and comparing with acceptable standards. The present study was carried out on groundwater resources of Zawar village of Tonekabon city for six months and then data were analyzed to determine the concentration of chemical parameters and water resources type based on anions and cations in water. The results of this study during the investigation of different wells showed that the total number of samples tested from a drinking water well in Zawar, Tonekabon, the range of nitrate concentration from 8 to 33.7 mg / L, TDS from 233 to 435 mg in liters and the total hardness varied from 211 to 372 mg/L. According to the definition of pollution, the wells were classified in the permitted class in terms of nitrate. However, the approach of nitrate levels in some wells to 20 mg per liter (a sign of the impact of human activities) is also worthy of serious consideration. determining the quality of the area reduces the amount of nitrate in the groundwater and thus increases the quality of groundwater.

1. Introduction

For a long time, groundwater has been considered as one of the vital sources of water supply and meeting the needs of human societies. In fact, this water supply has led to the establishment of civilizations in the field of plains and lands away from freshwater rivers, and today it also plays an important role in the economic growth of various societies.

In recent years, with the increase in population in the country and the lack of attention to how scientific exploitation of groundwater, this vital source, like other resources, has undergone dramatic changes and the amount of groundwater extraction from aquifers has been increasing day by day. In Iran, with increasing population, the use of groundwater has become more prevalent, so that agricultural water supply, drinking water and water requirements are mostly met by the underground water resources.

Water is not found purely in nature, but it always contains amounts of solutes, suspended matter and soluble gases, so water resources in different regions have different characteristics.

One of the most important issues in hydrology is the quality of water. Because most of the hydrological activities are for supplying water for agricultural or drinking and industrial purposes, each of them qualitatively and qualitatively must have qualitative characteristics and certain criteria, and if such a water supply is not possible, these activities are ineffective. Today, water quality surveys have become widespread and include issues related to surface water and groundwater pollution.

The issue of pollution is not only in industrialized countries, but also in developing and developing countries. For example, in most cities of Iran where drinking water is supplied from underground resources, the problem of contamination of these resources with nitrates and other toxic elements that may be used by sewage wells or fertilizers and pesticides used in agriculture and associated with penetrating water into aquifers comes to notice. Therefore, in hydrologic studies along with a quantitative study of water content, its qualitative criteria are also examined.

The purpose of water quality control is to be aware of the status and the process of changing the physical, chemical and microbial properties of water resources at the site of use. Qualitative control processes confirm the possibility of continuing to supply water from a specified range or source. Therefore, conducting chemical tests and determining the main parameters for each source of drinking water, is one of the essential and essential activities of quality control in all water and wastewater companies. Be After analyzing, for each water supply source, a series of data and results are obtained that in most cases these data are abandoned raw and it is difficult to achieve a correct and comprehensive understanding of these data, but also requires time There is a lot. On the other hand, given the high cost of the laboratory in each water and wastewater company, the importance of qualitative analysis of the desirable use of these data is necessary, to examine the status of the source of water supply, as well as the status of resources at the village level, City, province and in each region. By doing statistical analysis of the results, comprehensive knowledge and knowledge can be obtained that this knowledge improves the productivity and aims of investment and innovation. It is also designed to control the pollution of groundwater. Chemical parameters of water samples are necessary. Studies in this area have long history due to the importance of the subject. The World Health Organization published the first guidelines for the quality of drinking water from 1984 to 1985. And in 1988, the revision of the guidelines began until 1988, when revisions were made (Nabihzadeh Nodehi and Faezi Razi 2007). In Iran, the standards for

*Corresponding author E-mail: Rooholaminali@yahoo.com

physical and chemical drinking water were first prepared in 1954 and after four times the revision of the one hundred and ninety-fifth session of the National Committee of the Standard was published as the official standard of Iran (Imandal et al. 1997). In this area, studies have also been conducted in different parts of the country. In Safari studies on mineral resources in the city of Mianeh, the main problem of water resources was the total hardness, TDS and bicarbonate ions, and other parameters were good to acceptable (Safari and Vaezei 2003). According to a study by (Kacaroglu and Gunay 1997), seasonal fluctuations in nitrate concentration (10-10 mg/lit) were observed in underground water samples, so that during the wet season, low concentrations and in the dry season, the concentration of the measured Made in this study, seasonal variations in nitrate concentration in wells were observed, these changes may be due to: 1. groundwater recharge 2. changes in the concentration of pollution sources 3. changes in meteorological conditions (fall, evaporation) 4- Underground water level fluctuations and agricultural activities. Farshad et al. (1998) reported that nitrate and nitrite ions were 51.96 and 16.16 mg/lit, respectively, in the study of nitrate and nitrite ions in the industrial units of Tehran region of Karaj. In a comprehensive study on nitrate in drinking water in Qazvin, from 2000 to 2001, about 31% of wells had nitrate levels above the limit (Farshad and Imandel 2002). In a study by Dindarloo et al. (2006) on drinking water quality in Bandar Abbas, it was found that fluoride ion was 2.47 mg/lit, while nitrite, chlorine, sulfate, sodium and TDS also exceeded Reported a standard limit and also placed drinking water in the area in a very difficult water type and reported that the city's underground water resources are among the sources of water cannot be used alone. In a study on the groundwater quality of Hamedan plain by Rahmani and Shokohi (2007), the results of this study showed that due to the low table recession, the quality changes of groundwater reservoir have not yet reached the acute state, but Majority in 18 stations is more polluted than other stations, and 43.3% of these stations have high concentration of nitrate and one third of stations with TDS above the standard level. Water quality studies in some of cities, including Tehran, Arak and Mashhad, have also shown that the concentration of nitrate in water of some of the wells is more than standard, so that these chains use the operating circuit for consumption drinking has been delegated to municipalities for agricultural use. Poor irrigation management and agronomic activities, along with undesirable hydrodynamic conditions, are among the most important factors in groundwater contamination. Latif et al. (2005) conducted a study to determine the pollution level of groundwater nitrate in Mashhad plain and to identify the causes and source of contamination. For this purpose, 40 wells in the drinking, agricultural and industrial parts of the plain for the period. They sampled 6 months from July to December and compared their chemical and microbial parameters to international standards. The results showed that the polluted areas of the population and the good quality of water in the agricultural and industrial sectors. Also, the high concentration of nitrate in parts of Mashhad was due to leakage of domestic sewage to groundwater. Ehteshami and Sharifi (2007) measured the nitrate pollution in the groundwater resources of the city of Ray, which consists of 40 to 50 percent of their drinking water, measured nitrate in the central and eastern parts of the city to 65 mg /lit. They showed that the sewage collection network could reduce the nitrate concentration to 30 mg/lit. Society studied.

2. Study method

The method of study in this research is field trial. Sampling is carried out to determine the concentration of chemical parameters of groundwater resources in the drinking water of the villages of Tonekabon city. The city of Tonekabon is located between the geographical coordinates of North 's 45º and Eastern 12º51. The height of this city is from the sea level of 20 meters and its area is 2140 square kilometers. For this purpose, a drinking water well ring was selected in the study area of Tonekabon. In the village of Zawar, it was sampled monthly for 6 months. Due to the location of the well in the vicinity of agricultural land, this area was selected for studying chemical parameters. Characteristics of well is shown in Table 1.

Table 1. Location and characteristics of well sampled.

Well	x	Y	depth (m)	Rate of well discharge (liters per second)
Zawar	36.74091	50.97175	72	15

3. Periodic analysis of drinking water quality

For periodic analysis of the quality of drinking water sources, with a view to ensuring quality during consumption, as well as a review of quality change over time, a plan has been developed so that with the available facilities and constraints, there is a high degree of certainty and a better estimate of the annual change in resources. Chemical analysis of resources in all seasons of a year requires the use of more force and budget and higher costs. Therefore, a plan has been developed that takes into account the existing conditions for the desired management objectives. In this project, a chemical analysis period for all water supplies will be carried out in the first six months of the year. Then seasonal control will be based on measuring one of the main criteria for examining TDS changes, electrical conductivity, and changes to it. Nitrite, nitrate and ammonium changes can also be considered. Bringing on-site analytical results for electrical conductivity and laboratory analysis results, in addition to making water resource quality changes possible, helps control the processes of both systems (sampling, measuring, and recording results). Investigating and investigating cases where there is a discrepancy is necessary by the laboratory's quality control officer.

3.1. Steps to implement the plan

The stages of implementation of the project are as follows:
1. An equal list of water supplies for periodic laboratory and seasonal measurements is provided.
2-The same standard instructions for sampling method and location are provided and taught.
3. Planned to sample these groundwater resources according to the standard instructions and all chemical water variables are measured.
4. For sources with no significant difference in data, the results of a complete analysis of qualitative variables in the laboratory are considered valid data.
5- It is planned to analyze and analyze the results of the control samples. From the analysis of periodic results and control with seasonal data, considering the validity of the data, reliable and valid results for each water source are determined.

3.2. Review the results of the analysis

The results of the analysis are recorded in mg/lit (electrical conductivity in micro siemens), along with the resource specifications of the city in Excel, and the items are selected based on their effects and their role in determining the water quality, including: turbidity, TDS, electrical conductivity, total hardness Total alkalinity, temporary hardness, permanent stiffness, calcium hardness and magnesium hardness, sulfate, chlorine, fluorine, nitrate, nitrite, phosphate, ammonia, bicarbonate, sodium, magnesium, calcium, iron, potassium and magnesium. American Public Health Association recommended limit for nitrate in drinking water in terms of nitrogen 10 mg/lit (in terms of nitrate 50 mg /lit) and permissible nitrite level of 1 mg/lit (in terms of nitrite 3 mg/lit) recommendation has given. Water sources are classified into three groups of nitrate concentration: highly contaminated (more than 50 mg/lit), contaminated (20-50 mg/lit) and slightly contaminated (less than 20 mg/lit).

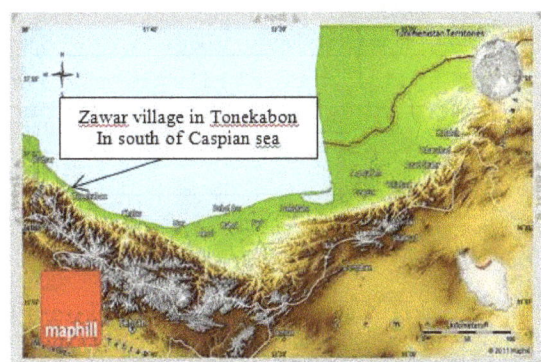

Fig. 1. Zawar village in Tonekabon, in south of Caspian Sea (http://www.maphill.com/iran/mazandaran/maps/physical-map/).

Table 1. Results of experiments on chemical parameters of Zawar (March 2016).

Unit	Results (mg/lit)	Test method	Test title	Rows
NTU	2.55	Odometry	Opacity	1
S/Cmμ	568	Conductivity measurement	Electric conductivity (E.C)	2
mg/lit	276	Conductivity measurement	Dry residue (TDS)	3
Caco3	278	Titration	Total hardness	4
Caco3	214	Titration	Total alkalinity	5
Caco3	170	Titration	Temporary difficulty	6
Caco3	108	Titration	Permanent hardship	7
Caco3	182	Titration	Calcium hardness	8
Caco3	96	Titration	Magnesium hardness	9
F^-	0.1	Spectrophotometer	Fluorine	10
CL^-	6.1	Spectrophotometer	Chlorine	11
SO_4^{2-}	68	Spectrophotometer	Sulfate	12
NO_3^-	24	Spectrophotometer	Nitrate	13
NO_2^-	0.21	Spectrophotometer	Nitrite	14
PO_4^{3-}	0.15	Spectrophotometer	Phosphate	15
Ammonium ion	0.09	Spectrophotometer	Ammonia	16
I^-	1.24	Spectrophotometer	Iodor	17
HCO_3^-	260.97	Titration	Bicarbonate	18
$CO3^{-2}$	–	Titration	Carbonate	19
OH^-	–	Titration	Hydroxide	20

Table 2. Results of experiments on chemical parameters of Zawar (May 2016).

Unit	Results (mg/lit)	Test method	Test title	Rows
NTU	3.18	Odometry	Opacity	1
S/Cmμ	567	Conductivity measurement	Electric conductivity (E.C)	2
mg/lit	275	Conductivity measurement	Dry residue (TDS)	3
$CaCO_3$	274	Titration	Total hardness	4
$CaCO_3$	210	Titration	Total alkalinity	5
$CaCO_3$	172	Titration	Temporary difficulty	6
$CaCO_3$	102	Titration	Permanent hardship	7
$CaCO_3$	171	Titration	Calcium hardness	8
$CaCO_3$	103	Titration	Magnesium hardness	9
F^-	0.09	Spectrophotometer	Fluorine	10
CL^-	4.5	Spectrophotometer	Chlorine	11
SO_4^{2-}	65	Spectrophotometer	Sulfate	12
NO_3^-	20.4	Spectrophotometer	Nitrate	13
NO_2^-	0.213	Spectrophotometer	Nitrite	14
PO_4^{3-}	0.15	Spectrophotometer	Phosphate	15
Ammonium ion	0.06	Spectrophotometer	Ammonia	16
I^-	1.17	Spectrophotometer	Iodor	17
HCO_3^-	256.097	Titration	Bicarbonate	18
$CO3^{-2}$	–	Titration	Carbonate	19
OH^-	–	Titration	Hydroxide	20

4. Conclusions

The chemical parameters of rural water resources in Tonekabon are not problematic from a health point of view, but because of the lower quality of extracted water, some of the underground resources can be consumed if necessary, with other waters such as surface water to the extent that it provides the desired range of standards, mixing and then using.

Although the concentration of nitrite and nitrate in all cases is lower than the standard drinking water in Iran. But according to the definition of water pollution in terms of nitrate wells are located on the "infected" and "slightly contaminated" floors. In addition, the difference in concentration of nitrate in the wells of the area, as well as the approximation of some nitrogen nitrates to 20 mg/lit (which is a sign of the effect of human activities), is also important and deserves serious attention.

According to the above, it is suggested that studies be repeated over time to understand the impact of human activities on nitrate and nitrite contamination. Nitrogen nitrate concentration at different stages was not affected by the short time interval between stages. Our findings indicate that nitrate concentrations are higher in winter and spring compared to the rest of the seasons. The highest concentration of nitrate is usually in the late winter (March of 2015) and mid-spring (May 2015). It is observed that the amount of nitrate in March is the maximum that can be attributed to nitrate washing due to winter rainfall and the beginning of the growing season and fertilization Be Plant growth in April and May can result in the use of soil nitrate and reduced its leaching. Therefore, there is no possibility of nitrate leaching to zero, and under any circumstances some nitrogen is leached as nitrate from the soil, but it can be minimized by proper management. Of course, the beginning of the wintering season at the end of winter at this time caused the maximum concentration of nitrate, although by diluting the concentration of nitrate decreased, but did not return to the initial

amount and increased by fertilization. This principle is not always the case, and other factors such as rainfall can change it. This research is an underlying study to investigate the distribution of chemical parameters such as nitrate and to identify the cause of contamination, its origin and further studies on the effects of different factors such as soil characteristics, geology, fertilization management, type and planting system, Irrigation water and fluctuations in groundwater aquifers are necessary in each region. Investigations showed that nitrate and nitrite concentration in one well in the village of Zawar is a sign that there is a risk of increasing nitrate concentration. However, it is recommended that water quality control of the city be carried out on a regular basis and the sanitary protection of wells should be fully observed and, if not possible, new sources should be replaced.

References

Dindarloo K., AliPour V., Farshidfar Gh., Chemical quality of drinking water in Bandar Abbas, Hormozgan Medical Journal 10 (2006) 57-62.

Ehteshami M., Sharifi A., Modeling and assessment of Rey's groundwater quality, Journal of Environmental Science and Technology 31 (2007) 125-135.

Farshad A., Imandel K., Nitrate and nitrite in wells of industrial units in west of Tehran, Journal of School of Public Health and Institute of Public Health Research1 (2002) 33–44.

Imandal k., Physical and chemical properties of drinking water. Standard Booklet, No. 1052, Institute of Standards and Industrial Research of Iran (1997).

Kacaroglu F., Gunay G. Groundwater nitrate pollution in an alluvium aquifer, Eskisehir urban area in its vicinity, Turkey, Journal of

Environmental Geology 3 (1997) 178-184.

Latif M., Mousavi S.F., Afyoni M., Velayatei S., Determination of nitrate pollution and its origin in groundwater of Mashhad Plain, Journal of Agricultural Sciences and Natural Resources 12 (2005) 32 -21.

Mazandaran map. From available site:http://www.maphill.com/iran/mazandaran/maps/physical-map/.

Nabihzadeh Nodehi R., Faezi Razi D., Guidelines for the quality of drinking water. First Edition. Tehran: Nasat Publishing (1997) 224.

Rahmani A., Shokohi R., Evaluation of groundwater quality in Bahar plain of Hamedan. The 10th National Conference on Environmental Health, Hamadan, 8-10 (2007).

Safari G., Vaezei F., Evaluation of the quality of drinking water in the city of Mianeh, The 6th National Conference on Environmental Health, Sari, (2003).

18

Preparation and characterization of high flux PES nanofiltration membrane using hydrophilic nanoparticles by phase inversion method for application in advanced wastewater treatment

Sirus Zinadini*, Foad Gholami

Environmental research center, Department of Applied Chemistry, Faculty of Chemistry, Razi University, Kermanshah, Iran.

ARTICLE INFO

ABSTRACT

Keywords:
Membrane
Nanofiltration
Antifouling
Dye rejection
Nanoparticles

In this research, in order to application of polymeric membrane for high quality treatment of wastewater, the synthesis, characterization, antifouling properties and performance of blended nanofiltration membranes were investigated. The chemical and physical characteristic influence of embedded hydrophilic dendrimer polycitrate-Alumoxane nanoparticles in membranes matrix was investigated by measuring permeability, filtration of fouling agent, water contact angle and the performance was assessed by calculating of Flux recovery ratio (FRR) and pure water flux. Also, to visual evaluation of thick of skin layer and pores shape, scanning electron microscopy (SEM) techniques was applied. The membrane surface hydrophilicity was improved by adding polycitrate-Alumoxane nanoparticles that can be attributed to the presence of hydrophilic functional groups on surface that was confirmed by contact angle experiments. The modified poly ether sulfone (PES) NF membrane revealed high resistance against fouling and high dye removal efficiency compared with that of the pristine PES. The FRR value of the PES membrane was increased from 39 to 98 % by blending 0.5 wt. % hydrophilic dendrimer polycitrate-Alumoxane nanoparticles. Also, Direct red 16 removal percentage was obtained 82 and 99 for unfilled and modified membrane, respectively.

1. Introduction

The hydrophobic nature of poly ether sulfone (PES) polymer caused a high interfacial energy with water-rich media among other membrane materials such as polyacrylonitrile (PAN), cellulose acetate (CA), polyamide (PA) and polyamide-imide (PAI) that makes the unmodified PES membrane have a great desire to fouling during wastewater filtration. Therefore, worldwide usage of PES membranes is still greatly restricted by membrane fouling, which decreases the flux and enhances the operation cost by requiring extra process of cleaning (Zhu et al. 2013).

Many strategies include pretreatment of wastewater, chemical surface alteration (like hydrophobic or hydrophilic and negatively or positively charged surface), optimization of process variables, module arrangement optimization have been done to control membrane fouling that among these methods, improvement of surface hydrophilicity seems to be a capable way to reduce membrane fouling (Rana and Matsuura. 2010).

Several techniques have been performed for development of membrane hydrophilicity such as bonding of hydrophilic monomers onto the membrane surface, blending an amphiphilic terpolymer or phthalate plasticizers in the polymer matrix, functionalization of the polymer, coating of membrane surface with hydrophilic polymer, and embedding hydrophilic nanoparticles (Han et al. 2011; Wang et al. 2008; Rahimpour et al. 2009; Chang et al. 2009; Na et al. 2000).

Introducing hydrophilic inorganic Nps in the membrane matrix (embedding method) such as TiO_2 (Rahimpour et al. 2011), Al_2O_3 (Liu et al. 2011), SiO_2 (Yu et al. 2009), Fe_3O_4 (Zinadini et al. 2014), graphene oxide and carbon nanotubes can make better the hydrophilicity, the water permeability and the antifouling characteristic of synthesized

membranes. Boehmite is an aluminum oxide hydroxide (AlOOH) particle, containing OH groups bound to its surface. The formula of the boehmite exhibits an excess water from the properly crystallized boehmite form (AlOOH), principally due to physically adsorbed water on the crystallite surface (15 wt. % H_2O) (Deer et al. 1992). Therefore, using these nanoparticles as a base for preparation of hydrophilic nano materials could be promising.

Citric acid (CA) is a multifunctional chemical compound that supplies effective functionality contributed to the ester bond-crosslink formation and balance of the polymer network hydrophilicity. CA-derived biomaterials developments are dependent upon the significant requirements for many applications. Dendrimer polycitrate-Alumoxane is a derivation of CA which has abundant hydrophilic OH and COOH functional groups and make them strongly hydrophilic. In this paper, the successful fabrication of novel dendrimer polycitrate-Alumoxane-PES nanocomposite membranes by the phase inversion method was reported. The effect of nanoparticle concentrations on the hydrophilicity of membranes surface, permeability, morphology, and antifouling performance is investigated.

2. Materials and methods
2.1. Materials

Analytical grade dimethylacetamide (DMAc) as solvent and polyethersulfone (PES ultrason E6020P with MW= 58,000 g/mol) as a polymer were provided by BASF Company (Germany). Polyvinyl pyrrolidone (PVP) with molecular weight (PVP K30) of 25,000 g/mol was purchased from Mowiol, Germany. Aluminum nitrate [Al $(NO_3)_3.9H_2O$], sodium hydroxide, citric acid, and decane were provided from Merck. Distilled water was utilized during this study.

Corresponding author Email: sirus.zeinaddini@gmail.com

2.2. Fabrication of asymmetric mixed matrix PES nanofiltration membranes

The immersion precipitation phase inversion technique was used for the preparation of asymmetric dendrimer polycitrate-Alumoxane-PES NF membranes. The components of casting solutions for unfilled and modified membranes are summarized in Table 1.

Table 1. The compositions of casting solutions.

Membrane type	PES (wt. %)	PVP (wt. %)	PC-Nanoparticle (wt. %)
Unfilled PES	20	1.0	-
Modified PES	20	1.0	0.5

The definite percentage of nanoparticles (0.5 wt. %) were added into DMAc and dispersed using DT 102H Bandelin ultrasonic (Germany) for 15 min to make better the homogeneity. After dispersing hydrophilic NPs in the solvent, PES and PVP were dissolved in the dope solution accompanied by continuous stirring for 24 h. After air bubbles removal, the membrane with the thickness of 150 μm was casted on glass plate. Then, the casted membrane was horizontally immersed in coagulation bath (distilled water) at room temperature for membrane solidification. Finally, to assure the perfect phase inversion the membranes were immersed in fresh distilled water about one day. Afterward, the membranes were dried at room temperature.

2.3. Characterization of the prepared membranes

The wettability and hydrophilicity of solid membrane surface was quantified using the sessile drop contact angle that is measured by a contact angle goniometer (G10, KRUSS, Germany) at 25 °C and a relative humidity of 50 %. Also, to minimize the experimental error, the measurement of contact angle was done at least five random locations on each surface.

The cross sectional structure of the freeze-dried membranes was studied using Philips-X130 and Cambridge SEM. To provide electrical conductivity of the fractured membranes, the samples were sputtered with gold. The SEM analyze were performed in very high vacuum conditions operating at 20 kV.

2.4. Membrane performance measurements

In order to investigate performance of the prepared nanofiltration membranes, tests of the permeate flux, dye rejection and powder milk fouling in a homemade stirred dead-end system with a volume capacity of 150 ml and an effective membrane surface area of 12.56 cm^2 connected with a nitrogen gas line were done (Fig. 1).

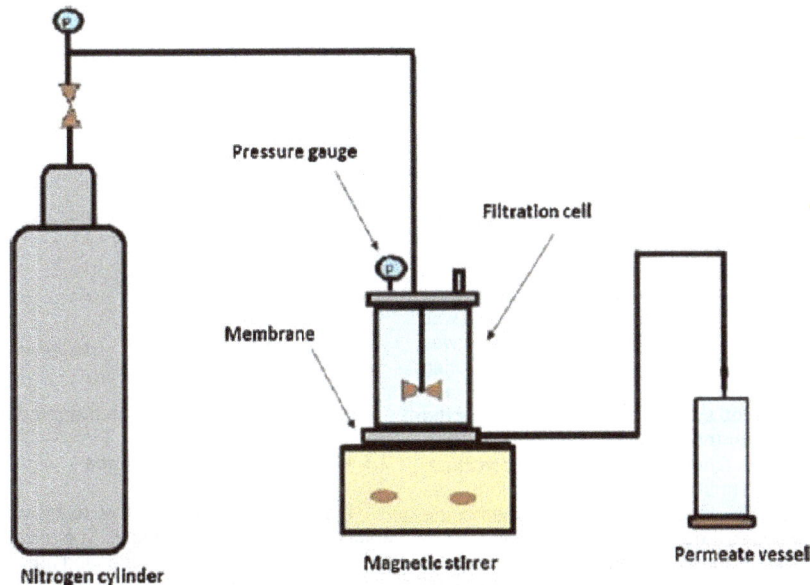

Fig. 1. Schematic of dead end system.

The stirring rate and operating pressure were adjusted to 500 rpm and 4 bar, respectively. After reaching steady state permeation, the water flux, $J_{w,1}$ (kg/m^2 h), was computed using the equation (1), where M is the weight of the water permeates gathered (kg), A is the effective membrane area (m^2) and Δt is the sampling time (h).

$$J_{w,1} = \frac{M}{A\,\Delta t} \tag{1}$$

The prepared membranes were subjected to the synthetic colored feed with 30 mg/L concentration (typical value of Direct Red16 in wastewaters). For assessment of nanofiltration performance and dye removal capability, rejection (R) is determined as follow:

$$R\,(\%) = \left(1 - \frac{C_p}{C_f}\right) \times 100 \tag{2}$$

where C_P is the concentration of a particular component of permeate and C_f is the feed concentration.

2.5. Antifouling experiments

After performing water flux filtration, the dead end cell was refilled immediately with 8000 ppm concentration of powder milk solution as a strong foulant and permeation flux Jp (kg/m^2 h) was measured by collection of the permeate at transmembrane pressure of 4 bar for 1 h. Then, the fouled membranes were rinsed with distilled water for 20 min, then the water flux of regenerated membrane for the second time, $J_{w,2}$ (kg/m^2 h), was measured. The FRR as a suitable index of antifouling characteristic was computed as follow:

$$FRR\,(\%) = \left(\frac{J_{w,2}}{J_{w,1}}\right) \times 100 \tag{3}$$

Generally, higher FRR indicates better antifouling property of the prepared membrane.

3. Results and discussion
3.1. Morphology of the prepared membranes

In order to evaluate the changes induced in the skin-layer and sub-layer of the prepared membranes, the cross-sectional SEM images of the unfilled and modified nanocomposite membranes are displayed in Fig. 2.

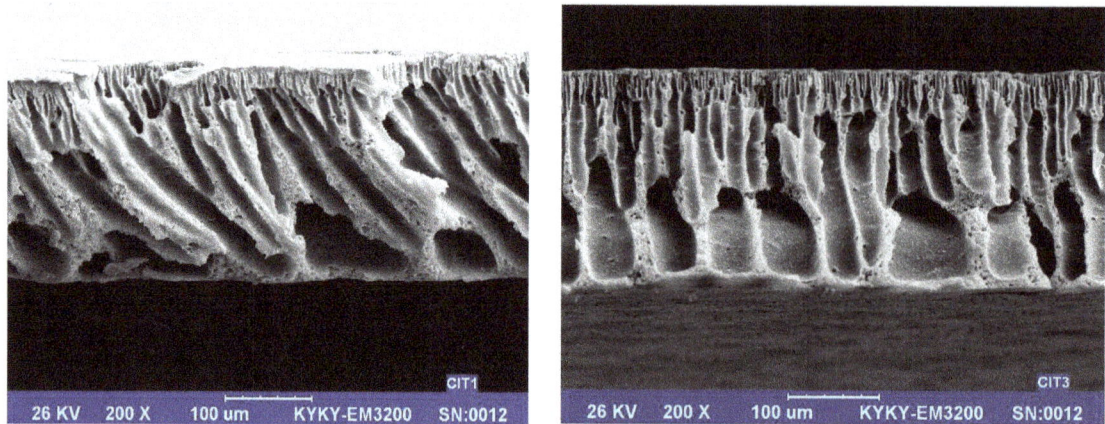

Fig. 2. Cross-section SEM images of the prepared membranes (left) Unfilled PES, (right) NPs 0.5 wt. %.

The membranes exhibited a typical asymmetric structure composed of a thin skin-layer and a porous bulk with a finger-like structure. From the Fig. seen, the addition of NPs caused a significant decrease in the top-layer thickness of the membranes. The addition of the dendrimer polycitrate-Alumoxane NPs increases the solution thermodynamic instability in the gelation bath (nonsolvent), and hydrophilic nature of the NPs increases the mass transfer rate between the solvent and the non-solvent in coagulation bath, which promotes a rapid phase inversion and results in large pore formation in low amount of the nanoparticles at the membrane skin-layer (Vatanpour et al. 2012).

3.2. Pure water flux and membrane hydrophilicity

The hydrophilicity of the membrane surface can be investigated by water contact angle measurement. Lower contact angle indicates that the membrane surface is more hydrophilic in nature. As shown in Fig. 3, by addition of polycitrate-Alumoxane nanoparticles to the casting solutions, the hydrophilicity of the mixed matrix membranes was improved. The contact angle of the 0.5 wt. % blended membrane was 54.2°, which is much lower than the unfilled membrane (62.8°). During membrane formation, the hydrophilic polycitrate-Alumoxane nanoparticles migrate towards the top surface of the membrane as the top layer was more exposed to water (non-solvent).

This migration decorates the functional groups of polycitrate-Alumoxane nanoparticles on the membrane top surface and improves the membrane hydrophilicity. By increasing of the membrane hydrophilicity with the nanoparticles addition, water molecules were attracted into the membrane matrix and promoted to pass through the membrane, thus enhancing the membrane flux. Fig. 4 reveals the pure water flux of the prepared membranes. As shown in this Fig., the trend of increasing in pure water flux is well matched with hydrophilicity improvement.

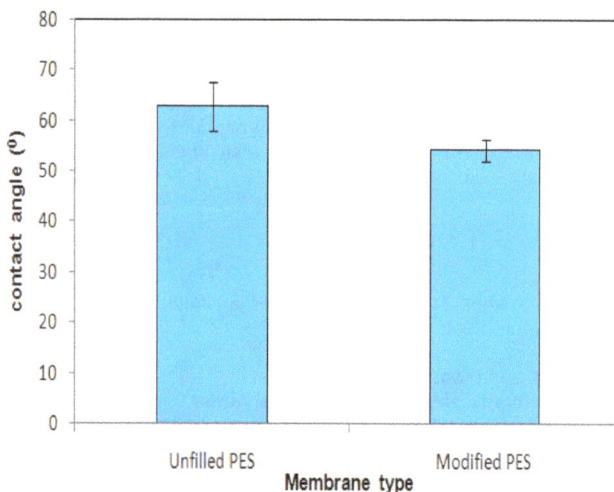

Fig. 4. Pure water flux of the prepared membranes (after 60 min).

3.3. Nanofiltration performance

The retention results of dye rejection after 60 min filtration of dye solution are shown in the Fig. 5. The rejection capability of the prepared polycitrate-Alumoxane NPs blended membrane was higher than that of unfilled PES membrane. Due to acidic functional groups of dendrimer polycitrate-Alumoxane, it can induce negative charge on the surface of the prepared membrane, causing high retention between negative dye and negative surface. The results showed that the behavior of the prepared nanofiltration membrane could be classified as Donnan exclusion mechanism (Sarkar et al. 2010), which is negatively charged. The membrane in contact with an aqueous solution gets an electric charge by dissociation of surface functional groups, causing electrostatic repulsion of the dye. Fig. 5. Dye retention performance of the prepared unfilled and modified PES membrane (4 bar, pH= 6.0 ± 0.1, 30 mg/L Direct Red 16, after 60 min filtration).

3.4. Antifouling properties of the membranes

Results of the fouling parameters are shown in Fig. 6. Comparing FRR as the most demonstrative factor in the antifouling capability of the prepared membranes indicates that addition of polycitrate-Alumoxane nanoparticles might be influential on the fouling reduction. The FRR for the unfilled PES membrane (39 %) was lower than the FRR for the membranes prepared by embedding nanoparticles (81 %). This indicates the high antibiofouling property of the modified membrane was induced by polycitrate-Alumoxane nanoparticles. The smoother and high hydrophilic surface of membrane indicates reversible attachment of the foulants on the membrane top surface and a higher flux recovery ratio. The higher FRR is an index of better antifouling property for the membrane.

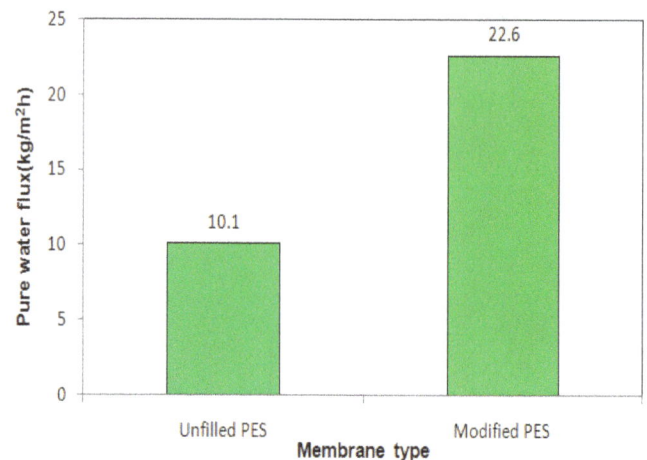

Fig. 3. Static contact angle of the prepared membranes.

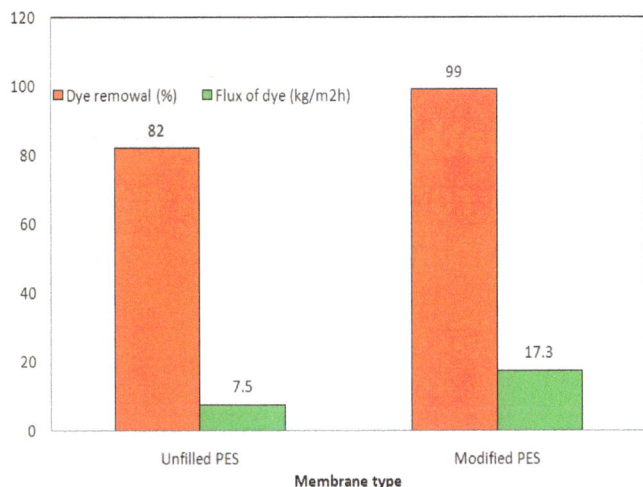

Fig. 5. Dye retention performance of the prepared unfilled and modified PES membrane (4 bar, pH= 6.0 ± 0.1, 30 mg/L Direct Red 16, after 60 min filtration).

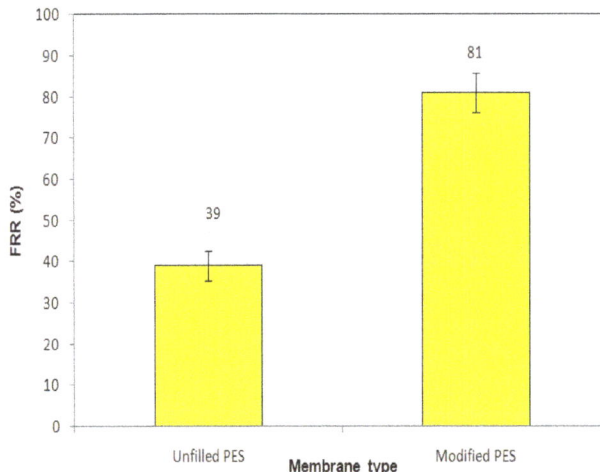

Fig. 6. Water flux recovery ratio of the unfilled and modified PES membranes after milk powder fouling (average of three replicates was reported).

Addition of nanoparticles results in partial blockage of membrane surface pores and reduces the pore radius of membrane surface. Since the roughness of membrane surface depends on pore size and pore density, the reduction of pore size may lead to creation of smoother surface. The trend of FRR is matched by hydrophilicity of the membranes. Hydrophilic surface of membranes can form a water layer by adsorption of water molecules, which retards the adsorption of protein and other fouling agents (Zinadini et al. 2014).

4. Conclusions

This was an attempt to study the effectiveness of the dendrimer polycitrate-Alumoxane nanoparticles on the characteristics of the PES mixed matrix membranes. The influences of blended nanoparticles on the morphology and performance of the fabricated NF membranes were investigated by pure water flux, dye removal and fouling measurements. The results indicated that the hydrophilic branch of dendrimer formed on the nanoparticles is located on the surface of membrane by migration of the nanoparticles to the surface that enhanced the membrane hydrophilicity as well as the surface properties. A significant improvement was observed in fouling prevention and dye removal in the prepared modified membrane compared with that of the unfilled PES.

References

Chang Y., Ko C.-Y., Shih Y.-J., Quémener D., Deratani A., Wei T.-C., Wang D.-M., Lai J.-Y., Surface grafting control of PEGylated poly (vinylidene fluoride) antifouling membrane via surface-initiated radical graft copolymerization, Journal of Membrane Science 345 (2009) 160–169.

Deer W.A., Howie R.A., Zussman J., An Introduction to the Rock-Forming Minerals, John Wiley Sons Inc., New York, 1992.

F. Liu, M.R. Moghareh Abed, K. Li, Preparation and characterization of poly (vinylidene fluoride) (PVDF) based ultrafiltration membranes using nano Al_2O_3, J. Membr. Sci. 366 (2011) 97–103.

Han M.J., Baro˝na G.B., Jung B., Effect of surface charge on hydrophilically modified poly (vinylidene fluoride) membrane for microfiltration, Desalination 270 (2011) 76–83.

Na L., Zhongzhou L., Shuguang X., Dynamically formed poly (vinyl alcohol) ultrafiltration membranes with good anti-fouling characteristics, Journal of Membrane Science 169 (2000) 17–28.

Rahimpour A., Jahanshahi M., Rajaeian B., Rahimnejad M., TiO_2 entrapped nano-composite PVDF/SPES membranes: Preparation, characterization, antifouling and antibacterial properties, Desalination 278 (2011) 343–353.

Rahimpour A., Madaeni S.S., Zereshki S., Mansourpanah Y., Preparation and characterization of modified nano-porous PVDF membrane with high antifouling property using UV photo-grafting, Applied Surface Science 255 (2009) 7455–7461.

Rana D., Matsuura T., Surface modifications for antifouling membranes, Chemical Reviews 110 (2010) 2448–2471.

Sarkar S., Sengupta A.K., Prakash P., The Donnanmembrane principle: opportunities for sustainable engineered processes and materials, Environmental Science & Technology 44 (2010) 1161–1166.

Vatanpour V., Madaeni S.S., Rajabi L., Zinadini S., Derakhshan A.A., Boehmite nanoparticles as a new nanofiller for preparation of antifouling mixed matrix membranes, Journal of Membrane Science 401–402 (2012) 132–143.

Wang X., Chen C., Liu H., Ma J., Preparation and characterization of PAA/PVDF membrane-immobilized Pd/Fe nanoparticles for dechlorination of trichloroacetic acid, Water Research 42 (2008) 4656–4664.

Yu L.-Y., Xu Z.-L., Shen H.-M., Yang H., Preparation and characterization of PVDF–SiO_2 composite hollow fiber UF membrane by sol–gel method, Journal of Membrane Science 337 (2009) 257–265.

Zhu X., Loo H.E., Bai R., A novel membrane showing both hydrophilic and hydrophobic surface properties and its non-fouling performances for potential water treatment applications, Journal of Membrane Science 436 (2013) 47–56.

Zinadini S., Zinatizadeh A.A., Rahimi M., Vatanpour V., Zangeneh, H., Beygzadeh M., Novel high flux antifouling nanofiltration membranes for dye removal containing carboxymethyl chitosan coated Fe_3O_4 nanoparticles, Desalination 349 (2014) 145–154.

Application of response surface methodology (RSM) for optimization of ammoniacal nitrogen removal from palm oil mill wastewater using limestone roughing filter

Arezoo Fereidonian Dashti*, Mohd Nordin Adlan, Hamidi Abdul Aziz, Ali Huddin Ibrahim

School of Civil Engineering, Engineering Campus, Universiti Sains Malaysia, Nibong Tebal, Penang, Malaysia.

ARTICLE INFO

Keywords:
Adsorption
Ammonia nitrogen
Filtration rate
Limestone (LS)
Horizontal Roughing Filter (HRF)

ABSTRACT

The creation of very pollute palm oil mill waste water has resulted in semiserious environmental hazards. The reason for the current study is to test the optimal removal of ammonia nitrogen (NH_3-N) from palm oil mill waste water by filtration using inexpensive filters media in place of current methods, to remove ammonia nitrogen from palm oil mill effluent. A series of batch and column studies were conducted using a different particle size of limestone (4, 12 and 20 mm) at various filtration rates of 20 ml/min, 60 ml/min and 100 ml/min. An experimental model design was conducted using Central Composite Design (CCD) in Response Surface Methodology (RSM). RSM was used to calculate the outcomes of process variables and their role in reaching ideal conditions. Equilibrium isotherms in this study were evaluated using the Langmuir and Freundlich isotherm. Using statistical analysis, the NH_3-N removal model proved to be very significant with very low probability values (0.0001). The column study showed that ideal NH_3-N removal was attained using a lower flow rate and smaller sized limestone (LS). The ideal conditions found when using 4 mm limestone and a 20 ml/min flow rate. This resulted in 45.3% removal of NH_3–N which was seen in the predicted model, and fit well with the laboratory results (45%). The adsorption isotherm data fit the Langmuir isotherm.

1. Introduction

Palm oil mill waste water is a natural effluent from palm oil industry (Hassan et al. 1996). Palm oil mill effluent (POME) is rich in natural carbon with a nitrogen content around 0.2 g/L, ammonia nitrogen 0.5 g/L and biochemical oxygen demand (BOD) higher than 20 g/L (Ma et al. 2001). In 1993, world palm oil production was 13.7 million tons. Malaysia alone created 7.4 million tons (Yusof 1994). Presently, there are over 270 palm oil mills in Malaysia. The average factory, produces about double the amount of POME compared to the amount of crude palm oil produced (Ma 1982; Qush 1991). Currently, approximately 85% of POME treatment is based on anaerobic and facultative ponding systems used by Malaysian palm oil mills (Rahim and Raj 1982; Wong 1980; Chan and Chooi 1982), which are known for long hydraulic retention time (HRT), which is regularly over of 20 d, and requires big plots of land or digesters (Chin et al. 1996). Currently POME is predominantly treated anaerobically in lagoons that release bio-gas into the atmosphere (Madaki and Seng 2013). In addition to adding to the greenhouse effect, a carbonaceous matter is created, which could be utilized for the profitable product. Unfortunately, the quality of after-treatment wastewater did not meet the discharge requirement fixed by the Malaysian Department of Environment, therefore further treatment is needed before it can be used (Zhang et al. 2008). Nutrient compounds such as nitrate (No_3^-), nitrite (No_2^-) and ammonia nitrogen (NH_3-N), that are regularly found numerous kinds of wastewater and water, can find their way to rivers, drinking water reservoirs and lakes. Nitrogen is a vital nutrient to all living creatures. It is a fundamental building block of plants as well, it can be found in

animal proteins. Changes in microbes releases ammonia and if the concentration of NH_3-N exceeds 0.3–0.5 mg/L it can reduce the dissolved oxygen aquatic life needs in order to promote the growth of algae (AWWA 1990). Treatment of wastewater NH_3-N is necessary to alleviate environmental problems such as polluting, eutrophication and decomposition (Rozic et al. 2000). The removal of NH_3-N can be done chemically, physically, biologically or by using a combination of these methods. Air stripping, ion exchange, membrane filtration, adsorption, chemical precipitation, denitrification, biological nitrification, reverse osmosis and breakpoint chlorination are all available technologies used to carry out these processes (Metcalf and Eddy 2004). Standard wastewater treatment methods, are plagued by maintenance problems, operational issues and are costly to build. Easy maintenance, constant and dependable physiochemical treatment is more desirable than biological systems. Aguilar et al. (2002) studied physio-chemical removal of NH_3-N using activated silica, powdered activated carbon and precipitated calcium carbonate. They discovered that ammonia removal was lowered by 3–17%. Ion exchange often utilizes natural resins, which are extremely discerning but costly, though, less-expensive alternate natural and waste materials can be used instead. Numerous researchers have studied the efficacy of using different inexpensive materials for ammonia removal such as clay and zeolite (Sarioglu, 2005; Demir et al. 2002; Celik et al. 2001and Rozic et al. 2000); limestone (Aziz et al. 2004a); organic and waste materials such as waste cement, discarded paper and concrete (Ahsan et al. 2001), activated sepiolite and sepiolite (Ozturk and Bektas 2004; Balci and Dincel 2002). Roughing filtration technology is a filtration process that utilizes a medium filter that has low filtration rates. It is chiefly used as pretreatment to lessen solid matter prior to slow filtration (Wegelin 1988; Boller 1993). This paper discusses laboratory investigations on the use of various sizes of limestone filter media through varied filtration rates for the removal of NH_3-N using a roughing filter. Water samples tested in the experiment were taken from the wastewater of a palm oil mill in NibongTebal, Pulau Pinang.

2. Materials and methods
2.1. Wastewater analysis

POME and palm oil mill effluent which were obtained from the United Palm Oil Mill Sdn. Bhd, Nibong Tebal, Pulau Pinang were selected as the case study of the present research. The pounding POME treatment system has been employed to treat wastewater. This study involves two main data collection methods including field measurement and laboratory experiment. Field measurement included tests for pH and temperature, while laboratory experiments involved tests for Chemical Oxygen Demand (COD), Biological Oxygen demand (BOD_5), Suspended Solids (SS), Colour, turbidity, and Ammonia Nitrogen. All of the tests were conducted in accordance with the Standard Methods for the Examination of Water and Wastewater (APHA 2005). The typical characteristics of raw POME are illustrated in Table 1.

Table 1. Palm oil mill effluent characteristics from a polishing pond.

Parameter	Unit	Min	Max	Average	*Standard
pH	-	7.89	8.77	8.33	5.0-9.0
Turbidity	NTU	200	650	425	-
COD	mg/l	2200	3300	2750	100
BOD_5	mg/l	120	210	165	50
BOD/COD	-	0.054	0.063	0.058	-
Colour	PtCo	3000	5000	4000	-
SS	mg/l	400	730	652	400
NH_3-N	mg/l	190	300	245	150
Temperature	C˚	35	37	36	-

* Department of Environmental Standards

2.2. Materials

Natural limestone media, provided from Perak, Malaysia, was used in this study. Prior to the experiment, the limestone was crushed into three media size: small size (4 mm), medium size (12mm) and large size (20mm). They were washed with water, and then dried. The composition of limestone used in this study was tested in the XRD Laboratory, School of Material and Mineral Resources Engineering, Engineering Campus, Universiti Sains Malaysia, as presented in Table 2.

Table 2. Composition of limestone.

Component	Percentage (%)
MgO	1.722
SiO_2	0.5212
CaO	56.5931
Fe_2O_3	0.0707
SrO	0.0376
NiO	0.0199
K_2O	0.0243
CuO	0.0112
LOI	41

2.3. Column study

In this study, the horizontal roughing filter was used for the treatment. The laboratory filter made of a special Perspex material with 750 mm width, 200 mm length and 90 mm height, was filled with limestone in various sizes of 4 mm, 12 mm, 20 mm. The experiment was carried out at the Environmental Laboratory Two, School of Civil Engineering, Universiti Sains Malaysia. Filtration process was run with 3 different flow rates which were 20 ml/min, 60 ml/min and 100 ml/min (Table 3). Inlet and outlet samples were taken for each flow rate before and after the retention time for seven days consecutively. The schematic diagram of laboratory-scale roughing filter is shown in Fig. 1 Removal percentage, percentage=$(c_1-c_2)/(c_1) \times 100$. Where C_1 and C_2 indicated untreated water and treated water concentration respectively.

Table 3. Operational parameters of the filter are shown.

Operational parameters	Column
Column heights	90 mm
Column lengthen	200 mm
Column material	Perspex
Width	750 mm
Volume of filter	13500 ml
Area of hols	3817 mm²
Adsorbent	Limestone
Particle size range	4 mm, 12 mm, 20 mm
Mode of flow	Horizontal
Flow rate	20 ml/min, 60 ml/min, 100 ml/min
Type of pump	Masterflex LS
Retention time	317 min, 260min, 106 min, 64 min, 88min

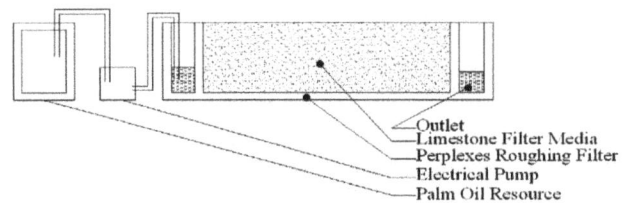

Fig. 1. Schematic of laboratory-scale roughing filter.

2.4. Equilibrium studies

In this study, adsorption isotherms were used to portray the media performance, and the relationship between the effects of dosage and settling times prior to the main experiment. Settling could examine the effect of particle sedimentation after shaking (Hussain et al. 2011). Different masses of media were used in 8 conical flasks from 25 g to 105 g, and the settling rate of the POME was observed for every 1 hour in 6 conical flasks for 6 hours, and shook at 350 rpm for 2 hours. Data for Freundlich and Langmuir isotherms generation were obtained by varying the amount of adsorbent dosage and different settling time. The isotherm constants and least squares correlation coefficients (R^2) of both models were compared together. The amount of adsorption at equilibrium, q_e (mg/g), was calculated by using the equation as shown below.

$$q_e = \frac{(c_0 - c_e)V}{M} \qquad (1)$$

where C_e and C_0 (mg/L) are the liquid-phase concentrations of equilibrium conditions and NH_3–N at initial, respectively. V (L) is the volume of the POME and M (g) is the mass of limestone used.

2.5. Experimental design and analysis

Experimental design of the process for optimum removal of NH_3–N from palm oil mill waste was carried out using RSM. The RSM is a collection of mathematical and measurable systems that are valuable for the advancement of synthetic responses and mechanical procedures, and are regularly utilized for experimental designs (Bas and Boyac 2007). In this study, RSM was used to evaluate the connection between independent variables and response (ammoniacal nitrogen removal, %), as well as to enhance the applicable circumstances of variables in order to forecast the best value of responses. Central Composite Design (CCD), which is the most commonly used type of RSM, was used to define the effects of operational variables on ammoniacal nitrogen removal efficiencies. According to Guven et al. (2008), CCD is a useful design that is appropriate for sequential experimentation because it enables a realistic amount of information to be used to examine lack of fit when a suitable number of experimental values exist. RSM and CCD were generated from the Design Expert 10.02 software program. The two significant independent variables considered in this study were flow rate (A) and size of limestone (B) as shown in Table 4. Based on some initial experiments, each independent variable was different over three levels between −1 and +1 at the determined ranges. Thirteen experiments were improved with six duplications to assess the pure error. The applicable model is the quadratic model characterized by Eq. (2).

$$Y = \beta_0 + \sum_{i=1}^{k} \beta_i X_i + \sum_{i=1}^{k} \beta_{ii} X_i^2 + \sum_{i<1}^{k} \sum_j \beta_{ij} X_i X_j + ..\ell \quad (2)$$

where, Y is the response; β_0 is a constant coefficient; β_j, β_{ij} and β_{ij} are the interaction coefficients of quadratic terms, respectively; k is the number of studied factors; X_j and X_i are the variables and e_i is the error. The value of correlation coefficient (R^2) represents the quality of the fit of the quadratic model. The key indicators that show the significance and fitness of the model used include the Adequate Precision, F-value model (Fisher variation ratio) and probability value (Prob>F) (Arslan 2009; Meyrs and Montgomery 2002). Immediate consideration of multiple responses contains the initial creation of a suitable response surface model. Then, after identifying a set of operational conditions at the maximizes the targeted or minimum maintains response, in the ranges (Bas and Boyac 2007; Meyrs and Montgomery 2002).

3. Results and discussion

A total of 13 runs of the CCD experimental design and response are shown in Table 4. The observed percent removal efficiencies of NH$_3$–N were varied between 4 and 45%.

Table 4: Experimental design for treatment of POME using roughing filter limestone

Std	Run	Factor 1 Flow rate (ml/min)	Factor 2 Limestone (mm)	Response NH$_3$-N (mg/l)
11	1	60	12	30
8	2	60	20	13
12	3	60	12	34
6	4	100	12	22
5	5	20	12	38
10	6	60	12	33
9	7	60	12	32
2	8	100	4	37
1	9	20	4	45
4	10	100	20	4
7	11	60	4	39
3	12	20	20	11
13	13	60	12	33

3.1. Analysis of variance (ANOVA)

Table 5 illustrates the analysis of variance (ANOVA) of the variable for the predicted response surface quadratic model for ammonia nitrogen removal efficiency. According to Table 5, a low probability value (F<0.0001) and the F-value of 47.28; showed that the model was significant for NH$_3$–N removal. The model is not significant when values of P>F are greater than 0.1000 while values of P>F that are less than 0.0500 signify that the model terms are significant (Körbahti and Tanyolaç 2008). The "Adequate Precision" ratio of the model was 22.25 (Adequate Precision >4), which was an adequate signal for the model (Ölmez 2009). The Lack of Fit (LOF) for response was insignificant relative to pure error. An insignificant (LOF) is preferable because to create a model that fits the experimental data is the primary objective (Hosseini 2012). The value of correlation coefficient (R^2=0.97) obtained in the present study for NH$_3$–N removal was higher than 0.80, indicated that only 3.78% of the total dissimilarity might not be explained by the empirical model. Joglekar and May (1987), mentioned that for a good fit of the model the correlation coefficient should be at a minimum of 0.80. High R^2 value demonstrated a decent agreement between the observed results and calculated within the range of the experiment. The coefficient of variance (CV) is determined as a ratio of the standard error of assessment to the average of the observed response explained by the reproducibility of the model. CV more than 10%, shows model is considered reproducible (Aziz et al. 2011). In this study, A, B, B^2, were significant model terms. Insignificant model terms, which have limited influence such as A^2 and AB, were kept out from the study to recuperate the model. The final regression model in terms of their coded factors is explicit by the following second-order equation below.

NH$_3$–N removal, %=35.41- 5.22A – 9.23B – 2.09A^2- 3.14B^2+ 0.18 AB   (3)

An empirical relationship between NH$_3$–N removal efficiency and process variables in terms of actual factors can show by the following second-order equation below.

NH$_3$–N removal, %=45.48491+0.017924A+ 0.29795B -1.30388E-003A^2- 0.095097B^2-+7.81250E-004AB   (4)

Ultimately, it is important to verify that the selected model provides adequate conjecture of the real system. The model adequacy is evaluated by applying the diagnostic plots provided by the Design Expert 10.0.2 software, such as predicted versus actual value plots and normal probability plots of the residuals. As presented in Fig. 2, the predicted values of NH3–N removal efficiency attained from actual experimental data and the model were in good agreement (Myers et al. 2016). Perturbation plot (Fig. 3) exhibited the comparative effects of two independent variables on NH$_3$–N removal efficiency. In Fig. 3, a sharp curvature in flow rate (A) and size of limestone (B) showed that the response of NH$_3$–N removal efficiency was slightly sensitive to these two process variables

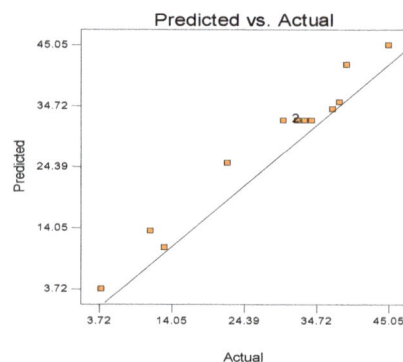
Fig. 2. Design Expert plot; predicted versus actual.

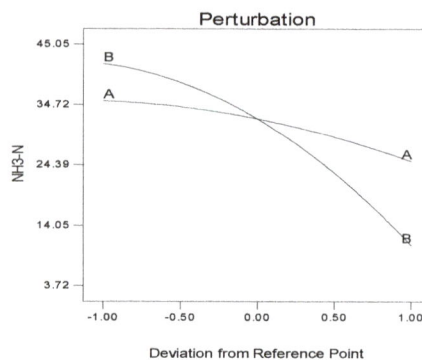
Fig. 3. Perturbation plot for NH$_3$–N removal values plot for NH$_3$– N removal.

Table 5. ANOVA for analysis of variance and adequacy of the quadratic model.

Source	Sum of Squares	Degree of freedom	Mean square	F-value	Prob>F
Model	1766.19	5	353.38	47.28	0.0001
A	152.62	1	152.62	20.42	0.0027
B	698.36	1	698.38	93.43	0.0001
A^2	12.02	1	12.02	1.61	0.2453
B^2	102.31	1	102.31	13.69	0.0077
AB	0.25	1	0.25	0.033	0.8601
Residual	52.32	7	7.47		
Lack of Fit	43.12	3	14.37	6.25	0.0544
Pure error	9.20	4	2.30		

SD = 2.73, R^2 = 0.97, R^2adj= 0.95, Adeq Precision=22.25, CV = 9.58

2.2. Ammoniacal nitrogen removal efficiency

The 3D surface responses of the quadratic model and contour plots were determined by utilized to assess the relationships between response and independent variables and design 10.02 software. (Fig. 4), illustrates that the NH$_3$-N removal increased with the decrease of limestone size and flow rate. The maximum and minimum of NH₃-N removal efficiencies obtained by the roughing filter were 45% (limestone size = 4 mm, flow rate = 20 ml/min) and 4% (Limestone size= 20 mm, flow rate = 100 ml/min), respectively. Faster flow rate causes lesser time required for the particle to travel along the settling distance and stick onto the media layer or be absorbed (Muhammad et al. 1996; Nwonta and Ochieng 2009). The results showed that when the filter media was small and the flow rate was low, the NH₃-N removal was more effective. These findings are in agreement with research reported by Maung (2006).

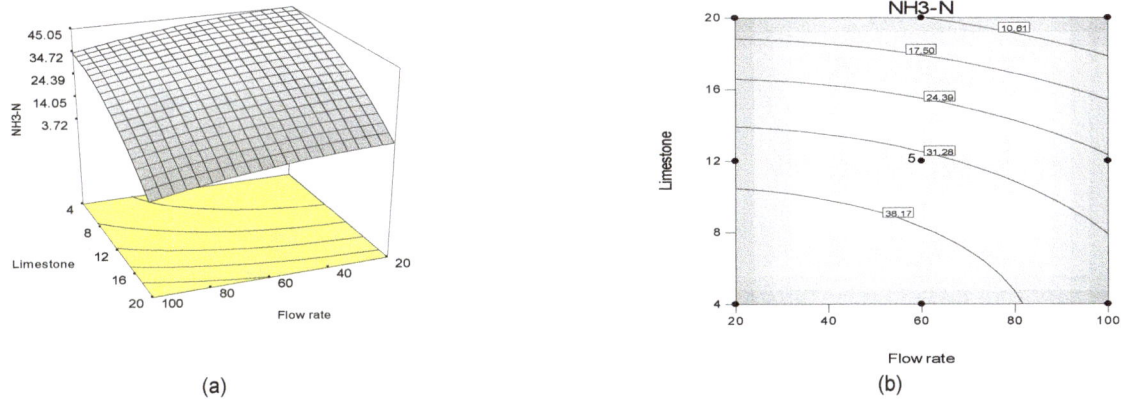

(a)

(b)

Fig. 4. Response surface (a) and contour plots (b) for NH₃–N removal efficiency as a flow rate (ml/min) and size of limestone (mm).

3.3. Optimization of operational conditions

For determining the optimum value of ammonia nitrogen removal efficiency using the Design Expert 10.02 software. According to the software optimization step, the desired goal the response (NH₃–N removal efficiency) was defined as "maximum" while for each operational condition (limestone size mm and flow rate ml/min) was chosen "within the range" to achieve the high performance. The program brings together the individual desirability into a single number, and then searches to maximize this function. Consequently, the optimal operating conditions and respective percent of removal efficiencies were established, and the results are presented in Table 6. As shown in Table 6, 45.3% NH₃–N removal was predicted according to the model under optimized operational conditions (limestone size mm and flow rate ml/min). The desirability function value was found to be 1.0 for these optimum conditions. The optimum results were then confirmed by an additional experiment. Forty-five percent removal of NH₃–N was obtained from the laboratory experiment which fits well with the predicted response value.

Table 6. The responses at an optimum condition for maximum removal.

Parameters	Flow rate (ml/min)	Limestone (mm)	Removal	Desirability
NH₃_N	20.40	4.01	45.03	1.000

3.4. Adsorption isotherms

Equilibrium isotherms in this study were examined using the Langmuir and Freundlich isotherms. Langmuir and Freundlich's isotherms are the isotherm models used most of often for describing adsorption characteristics of the adsorbents used in water and wastewater treatments (Isa et al. 2007; Chabani et al. 2007).The Langmuir isotherm theory assumes monolayer coverage of adsorbate over a homogeneous adsorbent surface (Langmuir 1918). The linear form of Langmuir isotherm equations is given as:

$$\frac{1}{x/m} = \frac{1}{QbC_e} + \frac{1}{Q} \quad (5)$$

where, b (L/mg) and Q (mg/g) are the Langmuir constants related to rate of adsorption and adsorption capacity, separately, Ce (mg/L) is defined as the equilibrium concentration of the adsorbate and x/m (qe) is the quantity of adsorbate per unit of adsorbent (mg/g). A straight line was obtained when 1/(q$_e$) was plotted against 1/Ce. While b was deduced from the intercept Q was calculated from the slope (Fig. 5 (a) and it shows that equilibrium data were fit with the Langmuir model. The R² value and constants are listed in Table 7. The features of the

Langmuir isotherm can be explicit in terms of equilibrium parameter RL (Weber and Chakravorti 1974), which is shown by:

$$R_l = \frac{1}{1+bC_e} \quad (6)$$

where, C₀ (mg/L) is the initial nitrogen ammonia concentration and b is the Langmuir constant. The value of RL exhibit the type of isotherm to be unfavorable (RL>1), linear (RL=1), and favorable (RL<1) (Isa et al. 2007). As shown in Table 7, the RL value for adsorption of NH₃-N onto limestone was 0.154, indicating that the adsorption process was favorable. The Freundlich model is an empirical equation, which assumes that the adsorption process takes place on heterogeneous surfaces (Freundlich, 1906). The Freundlich isotherm equations is given as:

$$\log q_e = \log K + \frac{1}{n} \log C_e \quad (7)$$

where, q$_e$ (mg/g) is the amount of ammonia nitrogen adsorbed per unit mass of adsorbent; C$_e$ (mg/L) is defined as the equilibrium concentration of the adsorbate; n and K are Freundlich constants with exhibit the favorability of the adsorption process; K (mg/g (L/mg)1/n) shows the adsorption capacity of the adsorbent. The value of 1/n<1 demonstrates that the adsorption capacity rises and the adsorption mechanism is favorable. The value of 1/n>1 indicates that the adsorption process is weak and the adsorption mechanism is unfavorable (Hussain et al. 2007; Aziz et al. 2008; Aziz et al. 2011). The linear correlation-regression (R²) and values of K for Freundlich isotherm are given in Table 8. It can be seen from Table 7 that the Langmuir isotherm fit the data better than the Freundlich. This is also confirmed that the higher value of R² for Langmuir (0.888) compared to Freundlich (0.838), exhibited that the adsorption of NH₃-N occur as monolayer adsorption on a homogenous surface in adsorption attraction. Accordingly, the results show, the Langmuir isotherm described the adsorption isotherm data and the adsorption capacity obtained was 0.231× 10⁻³ mg/g. This result is thought to be low when compared to others who also used limestone to remove NH₃–N from wastewater.

Table 7. Langmuir isotherm parameters for NH₃-N adsorption.

Media	B(L/mg)	Q (mg/g)	R²	R$_L$
Limestone	0.023	4.658×10⁻⁵	0.888	0.154

Table 8. Freundlich isotherm parameters for NH₃ -N adsorption.

Media	K$_F$	1/n	R²
Limestone	0.231× 10⁻³	-0.256	0.838

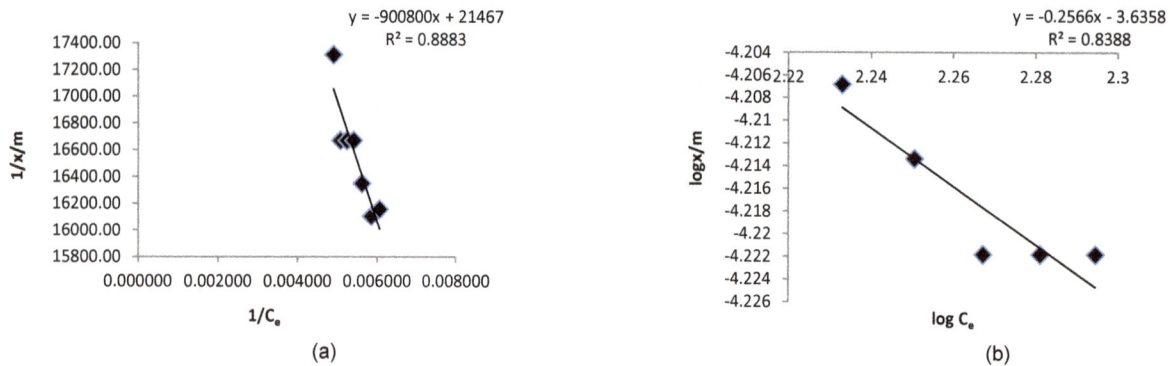

$$y = -900800x + 21467$$
$$R^2 = 0.8883$$

$$y = -0.2566x - 3.6358$$
$$R^2 = 0.8388$$

(a)   (b)

Fig. 5. (a) Langmuir isotherm and (b) Freundlich isotherm for NH_3–N adsorption.

4. Conclusions

In the present study, optimization of ammoniacal nitrogen removal from palm oil mill effluent using horizontal roughing filter and limestone as a filter media was investigated. Optimization of the treatment process was concentrated on the influence of operating variables and interaction among all the parameters, such as flow rate and size of limestone using CCD in RSM. The multiple correlation coefficients of determination R^2 was 0.97, showing that the actual data fitted well with the predicted data. The adsorption data fitted well with the Langmuir isotherm at R^2=0.888. The optimum results attained from the model indicated that 45% of NH_3-N removal was achieved at limestone size of 4 mm, and flow rate of 20 ml/min, while and 4% removal was obtained at limestone size of 20 mm, and flowrate of 100 ml/min. Faster flow rate requires lesser time for the particle to travel along the settling distance and stick onto the media layer or be absorbed. The results showed that when the filter media was small and the flow rate was low, the NH_3-N removal was more efficient.

References

Aguilar M.I., Saez J., Llorens M., Soler A., Ortuno J.F., Nutrient removal and sludge production in the coagulation-flocculation process, Water Research 36 (2002) 2910–2919.

Ahsan S., Kaneco S., Ohta K., Mizuno T., Kani K., Use of some natural and waste materials for wastewater treatment, Water Research 35 (2001) 3738–3742.

Arslan A.I.T.G., Olmez-Hanci T., Treatment of azo dye production wastewaters using Photo-Fenton-like advanced oxidation processes: Optimization by response surface methodology, Journal of photochemistry and Photobiology A: Chemistry 202 (2009) 142-153.

AWWA. Water Quality and Treatment. McGraw-Hill, New York (1990).

Aziz H.A., Adlan M.N., Ariffin K.S., Heavy metals (Cd, Pb, Zn, Ni, Cu and Cr (III)) removal from water in Malaysia: Post treatment by high quality limestone, Bioresource technology 99 (2008) 1578-1583.

Aziz H.A., Adlan M.N., Zahari M.S.M., Alias S., Removal of ammoniacal nitrogen (N-NH3) from municipal solid waste leachate by using activated carbon and limestone, Waste Management & Research 22 (2004a) 371–375.

Aziz S.Q., Aziz H.A., Yusoff M.S., Bashir M.J., Landfill leachate treatment using powdered activated carbon augmented sequencing batch reactor (SBR) process: Optimization by response surface methodology, Journal of Hazardous Materials 189 (2011) 404-413.

Balci S., Dincel Y., Ammonium ion adsorption with sepiolite: use of transient uptake method, Chemical Engineering and Processing: Process Intensification 41 (2002) 79–85.

Baş D., Boyacı İ.H., Modeling and optimization I: Usability of response surface methodology, Journal of Food Engineering 78 (2007) 836-845.

Boller M., Filter mechanisms in roughing filters, Journal of Water Supply Research and Technology - Aqua (1993) 42, 174–185.

Celik MS., Ozdemir B., Turan M., Koyuncu I., Atesok G., Sarikaya H.Z., Removal of ammonia by natural clay minerals using fixed and fluidized bed column reactors, Water Supply 1 (2001) 81–99.

Chabani M., Amrane A., Bensmaili A., Kinetics of nitrates adsorption on Amberlite IRA 400 resin, Desalination 206 (2007) 560-567.

Chan K.S., Chooi C.F., Ponding system for palm oil mill effluent treatment. In: Proceedings of Regional Workshop on Palm Oil Mill Technology and Effluent Treatment. PORIM, Malaysia, (1982). pp. 185‾192.

Chin K., Lee S., Mohammad H., A study of palm oil mill effluent treatment using a pond system. Water Science & Technology 34 (1996) 119–123.

Demir A., Gunay A., Debik E., Ammonium removal from aqueous solution by ion exchange using packed bed natural zeolite. Water S.A, (2002). 28 (3), 329–335.

Freundlich H.M.F., Over the adsorption in solution, Journal of Physical Chemistry 57 (1906) 1100-1107.

Güven G., Perendeci A., Tanyolaç A., Electrochemical treatment of deproteinated whey wastewater and optimization of treatment conditions with response surface methodology, Journal of Hazardous Materials 157 (2008) 69-78.

Hussain S., Aziz H.A., Isa M.H., Ahmad A., Van Leeuwen J., Zou L., Umar M., Orthophosphate removal from domestic wastewater using limestone and granular activated carbon, Desalination 271 (2011) 265-272.

Isa, M.H., Lang L.S., Asaari F.A.H., Aziz H.A., Ramli N.A., Dhas J.P.A., Low cost removal of disperse dyes from aqueous solution using palm ash, Dyes and Pigments 74 (2007) 446-453.

Joglekar A.M., May A.T., Product excellence through design of experiments, Cereal Foods World 32 (1987) 857–868.

Körbahti B.K., Tanyolaç A., Electrochemical treatment of simulated textile wastewater with industrial components and Levafix Blue CA reactive dye: Optimization through response surface methodologym Journal of Hazardous Materials 151 (2008) 422-431.

Langmuir I., The adsorption of gases on plane surfaces of glass, mica and platinum, Journal of the American Chemical society 40 (1918) 1361-1403.

Ma A.N., Chow C.S., John C.K., Ibrahim A., Isa Z., Palm oil mill effluent-a survey, In Proceedings PORIM regional workshop on palm oil mill technology and effluent treatment. Palm Oil Research Institute of Malaysia (PORIM), Serdang, Malaysia (1982) pp. 155-168.

Ma A.N., Extraction of crude palm oil and palm kernel oil, In Selected readings on palm oil and its uses, PORIM report. Palm Oil Research Institute of Malaysia (PORIM), Serdang, Malaysia. (1994) pp. 24-34.

Ma AN., Chow C.S., John C.K., Ibrahim A., Isa Z., Palm oil mill effluent a survey. In: Proceedings of the PORIM Regional Workshop on Palm Oil Mill Technology and Effluent Treatment. Palm Oil Research Institute of Malaysia (PORIM) Serdang, Malaysia, (2001) pp. 233–269.

Madaki Y.S., Seng L., Palm oil mill effluent (POME) from Malaysia palm oil mills: waste or resource, International Journal of Science, Environment and Technology 2 (2013) 1138-1155.

Metcalf and Eddy I., Wastewater Engineering: Treatment, Disposal and Reuse, fourth ed. McGraw-Hill Co. New York (2004).

Montgomery C.D. Design and Analysis of Experiments6th Ed, John Wiley and Sons, USA (2005).

Myers R.H., Montgomery D.C., Response Surface Methodology: Process and Product Optimization using Designed Experiments2nd Ed, John Wiley and Sons, USA (2002).

Myers R.H., Anderson-Cook C.M., Montgomery D.C., Wiley Series in Probability and Statistics: Response Surface Methodology: Process and Product Optimization Using Designed Experiments (2016).

Ölmez T., The optimization of Cr (VI) reduction and removal by electrocoagulation using response surface methodology, Journal of Hazardous Materials 162 (2009) 1371-1378.

Ozturk N., Bektas T.E., Nitrate removal from aqueous solution by adsorption onto various materials, Journal of Hazardous Materials 112 (2004) 155–162.

Qush S.K., Resources recovery and by-product utilisation in anaerobic treatment of palm oil mill effluent, In Proceedings workshop on anaerobic digestion technology in pollution control. Palm Oil Research Institute of Malaysia (PORIM), Serdang, Malaysia (1991) pp. 105-121.

Rahim B.A., Raj R., Pilot plant study of a biological treatment system for palm oil mill effluent, In: Proceedings of Regional Workshop on Palm Oil Mill Technology and Effluent Treatment. PORIM, Malaysia, (1982) pp. 163–170.

Rozic M., Cerjan-Stefanovic S., Kurajica S., Vancina V., Hodzic E., Ammoniacal nitrogen removal from water by treatment with clays and zeolites, Water Research 34 (2000) 3675–3681.

Sarioglu M., Removal of ammonium from municipal wastewater using natural Turkish (Dogantepe) zeolite, Separation and Purification Technology 41 (2005) 1–11.

Weber T.W., Chakravorti R.K., Pore and solid diffusion models for fixed-bed adsorbers, AIChE Journal (1974) 228-238.

Wegelin M., Horizontal roughing filtration: a design, construction and operation manual, CEPIS (1988).

Wong F.M., A review on the progress of compliance with the palm oil control regulations. Seminar on Advances in Palm Oil Effluent Control Technology, Kuala Lumpur, (1980) pp. 142–149.

Yusof B., The palm oil industry, export trade and future trends. In Selected readings on palm oil and its uses. PORIM report. Palm Oil Research Institute of Malaysia (PORIM), Serdang, Malaysia (1994). pp. 212-221.

Zakaria S.N.F.B. Peat water treatment using Limestone and Natural Zeolite. M.Sc, Universiti Sains Malaysia (2013).

Zhang Y., Li Y.X., Qiao L., Xiangjun N., Zhijian M., Zjang Z., Integration of biological method and membrane technology in treating palm oil mill effluent, Journal of Environmental Sciences 20 (2008) 558-564.

20

Advanced oxidation processes treating of Tire Cord production plant effluent

Maryam Habibi[1], Ali Akbar Zinatizadeh[1,*], Mandana Akia[2]

[1]Water and Wastewater Research Center (WWRC), Department of Applied Chemistry, Faculty of Chemistry, Razi University, Kermanshah, Iran.

[2]Department of Mechanical Engineering, University of Texas Rio Grande Valley, Edinburg, USA.

ARTICLE INFO

Keywords:
Tire Cord wastewater degradation
Advanced oxidation process
$O_3/UV/H_2O_2$

ABSTRACT

The degradation of an industrial wastewater (Tire Cord factory) with low BOD_5/COD ratio (0.1-0.2) was investigated using advanced oxidation processes (AOPs) (i.e. hydrogen peroxide, UV/H_2O_2, O_3/H_2O_2 and $UV/O_3/H_2O_2$ treatments). In order to investigate the effects of influential variables on the process performance, four independent factors involving two numerical factors (initial H_2O_2 concentration and initial pH) and two categorical factors (ozonation and UV irradiation) were selected. The process was modeled and analyzed using response surface methodology (RSM). The region of exploration for the process was taken as the area enclosed by initial H_2O_2 concentration (0-20 mM) and initial pH (3-11) boundaries at three levels. For two categorical factors (ozonation and UV irradiation), the experiments were performed at two levels (with and without application of each factor). Two dependent parameters (TCOD removal and BOD_5/COD ratio) were studied as the process responses. As a result, initial H_2O_2 concentration showed a reverse impact on the responses; an increasing effect at low concentrations (0-10 m mol/l) and a decreeing effect at higher concentrations (10-20 m mol/l). The maximum and minimum the responses were obtained at H_2O_2 concentration of 10 and 20 mmol/l and initial pH 3 and 11, respectively. $O_3/UV/H_2O_2$ system showed better performance with 32 % for TCOD removal efficiency and 0.41 for BOD_5/COD ratio.

1. Introduction

Industrial effluent is one of the important pollution sources for water resources with unpredictable toxicological and ecotoxicological effects (Shi 2003; Di Marzio et al. 2005). Post treatment systems currently used in wastewater treatment plants include membrane technology, activated carbon adsorption, and sand filters. However, none of these treatment methods is effective enough to generate water with acceptable levels of the most recalcitrant pollutants (e.g., phenols, pesticides, solvents, household chemicals and drugs, etc.). Therefore, a further treatment stage is often necessary to achieve standard levels for the removing persistent chemicals. This stage can entail the application of advanced oxidation processes (AOPs), which are recommended when wastewater components are stable and not biologically degraded. In the AOPs, very reactive species–hydroxyl radicals (OH') is produced which cause a complete mineralization of pollutants to CO_2, water, and inorganic compounds, or at least their transformation into more innocuous products (Shi 2003). There are several different processes which produce hydroxyl radicals; e.g. ozonation (Latifoglu and Gurol 2003; Sanchez-Polo et al. 2007; Lucas et al. 2010), Fenton process (Ghaly et al. 2001; Kusic et al. 2006; Horsing et al. 2012), UV/O_3 (Shen and Ku 1999; Garoma et al. 2008), O_3/H_2O_2 (Qiang et al. 2010), photocatalytic process (Keller et al. 2003; Lin et al. 2012) and etc. Ozone has an ability to oxidize various organic contaminants in water and wastewater such as naphthalene, anthracine and hydroxylated aromatics like phenols (Latifoglu and Gurol 2003). However, ozone reaction with saturated organic compounds is very slow. In a research work, using direct ozonation, 12% COD reduction after 180 min has been reported for treatment of

winery wastewater with COD content of 4650 mg/l (Lucas et al. 2010). Ozonation process combined with H_2O_2 has been also examined and found to be one of the most practical AOPs because of its simplicity (Katsoyiannis et al. 2011). It is noted that an appropriate range of H_2O_2 concentration must be applied because H_2O_2 acts not only as a HO generator, but also as a HO' Scavenger through generating peroxyl radical (HO_2') during the O_3/H_2O_2 process as follows:

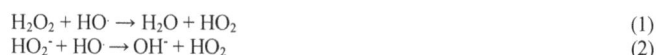

$$H_2O_2 + HO^- \rightarrow H_2O + HO_2 \tag{1}$$
$$HO_2^- + HO^- \rightarrow OH^- + HO_2 \tag{2}$$

It was also reported that the appropriate range of H_2O_2 concentration depends upon the type of solutes and their concentrations (Crittenden et al. 1999). O_3/H_2O_2 process was used for degradation of various compounds such as methyl-tert-butyl ether (MTBE) (Safarzadeh-Amiri 2001), 17β-estradiol and 17α-ethinylestradiol (Guedes Maniero et al. 2008) and humic acid Kosaka et al. 2001).

Simultaneous application of ozone with ultraviolet radiation (O_3/UV) is another AOP tested. It showed to be more effective compared with ozone alone. Application of the O_3/UV for the treatment of winery wastewater with COD content of 4650 mg/l, could remove 21% of COD after 180 min of reaction (Lucas et al. 2010). A very effective mineralization of TOC (94%) with a UV radiation of 96 W and an ozone dosage of 3.8 g/h was obtained (Poyatos et al. 2010). Although UV radiation itself has ability to destroy organic molecules, the efficiency of direct photolysis of organic matter is somewhat difficult and it depends on the compound's reactivity and photosensitivity. Moreover, most of the commercially used dyes are usually designed to be light resistant

*Corresponding author E-mail: zinatizadeh@gmail.com

(Kusic et al. 2006). The direct photolytic effect on the compounds dissolved in the winery wastewater was insignificant (Lucas et al. 2010). However, when UV irradiation is combined with some powerful oxidant, such as H_2O_2, organic matter degradation efficiency can be significantly enhanced because H_2O_2 absorbs UV light, and breaks down into OH radicals, degrading the contaminant via OH radical oxidation (Kusic et al. 2006; Rosenfeldt et al. 2006). UV/H_2O_2 process has been used for degradation of various organic compounds such as p-chlorophenol (Ghaly et al. 2001), methyl orange (Haji et al. 2001), naphthalenesulfonic acids (Sanchez-Polo et al. 2007) and cyclohexanoic acid (Afzal et al. 2012). In a study, the UV/H_2O_2 process examined for removing naproxen from surface water and more than 93% removal efficiency was obtained at pH 6 after 3 min (Poyatos et al. 2010).

Another combined AOP is $O_3/UV/H_2O_2$ process which is able to oxidize organic matters faster than ozone, showing an enhanced photochemical oxidation effect. This is principally due to the photolysis of ozone and the enhanced mass transfer of ozone for generation of hydroxyl radicals that react rapidly with the organic matter in the target wastewater. Treatment of winery wastewater with COD content 4650 mg/l using the $O_3/UV/H_2O_2$ process showed a COD removal of 35% after 180 min (Lucas et al. 2010). An $O_3/UV/H_2O_2$ process under UV irradiation of 6.4 mW cm^{-2} and an initial H_2O_2 concentration of 7 mg L^{-1} has been used for removal of carbamazepine (CBZ) and 96.5% removal efficiency achieved (Im et al. 2012).

Tire cord manufacturing plant uses polyesters and polyamides as the basic material for production of required textiles to be used in tire. The compounds found in Tire Cord wastewater (TCW) are mostly recalcitrant and non-biodegradable. It is noted that some of the chemicals, i.e. pyridine compounds, are not even detected in COD test. Up to the date, no study on degradation of TCW by the AOP processes has been reported. The present study was therefore undertaken to examine the degradation of TCW by combination of advanced oxidation processes.

2. Materials and methods
2.1. Wastewater characteristics

Tire Cord production wastewater (TCW) was taken from a working Co. Producing Tire Cord, Kermanshah, Iran. The characteristics of the TCW are shown in Table 1.

Table 1. Characteristics of Tire Cord manufacturing wastewater

Parameters	Unit	Amount
TCOD	(mg/l)	450-500
BOD$_5$	(mg/l)	80-100
TSS	(mg/l)	120-360
pH	-	7-7.8

2.2. Experimental set-up

The experimental set-up used for treatment of tire cord wastewater (TCW) by the advanced oxidation processes consists of an air compressor, air dryer, ozone generator, photo reactor and a washing bottle (Figs. 1a and b). Fig. 1a represents an image of the experimental setup used in this study. The air dryer consisted of a column which was filled with a high adsorptive molecular sieve (silica). Ozone was generated using a laboratory ozonizer, Model COG, 1G/L. Ozonation experiments were carried out by continuously feeding an ozone gas stream in a mixed semi-bach bubble reactor (continuous for gas and batch for liquid). The air flow rate was adjusted at 5 L/min. The ozone content of the input air stream was measured as 0.27 g O_3/ h. The cylindrical steel reactor had a 5 cm diameter with a total volume of 1450 ml (with lamp) in which the solution was introduced at the bottom. The irradiation in the photo reactor was obtained by a 15 W UV lamp (Hitachi, emission: 365 nm, constant intensity=60 mW/cm^2) that protected by a quartz jacket, and positioned and immersed in the solution in the center of the reactor. The lighted length of the lamp was 452 mm with a quartz sleeve diameter of 3.5 cm. The reactor was followed by a washing bottles, containing 250 ml of acidified 2 % KI solution for determining of unreacted ozone at influent and effluent in several times.

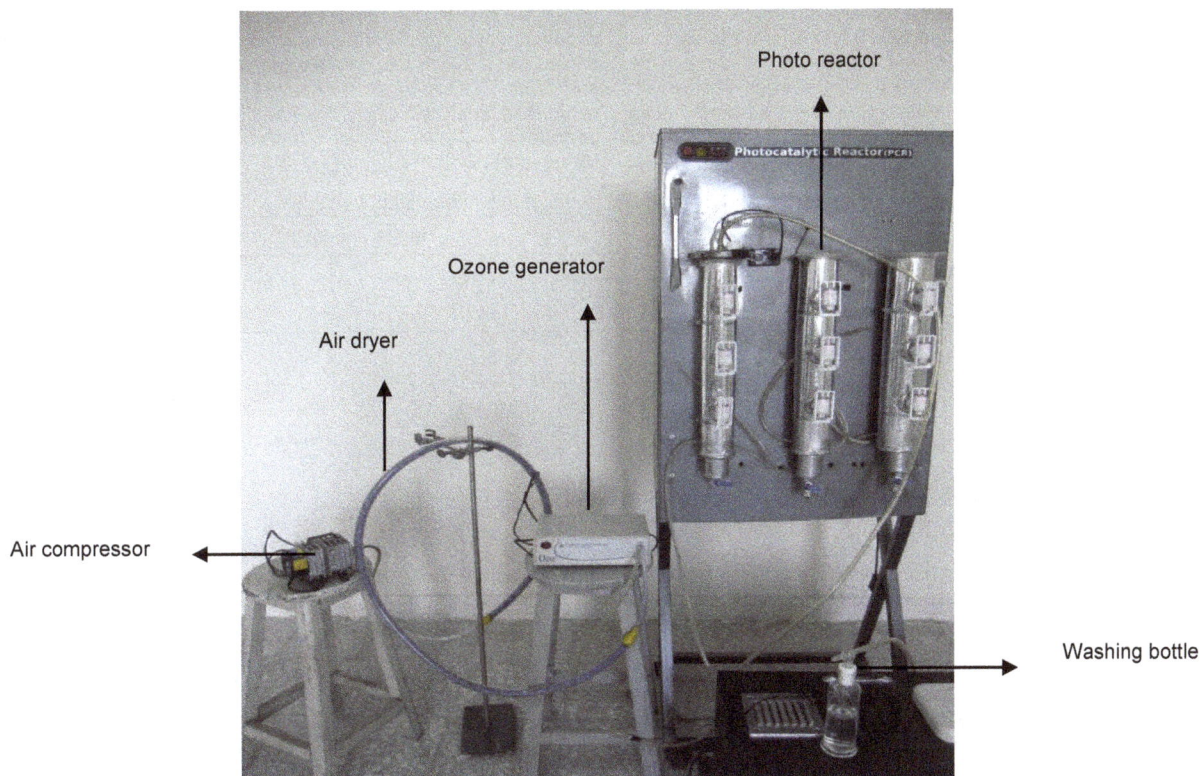

(a)

1. Reservoir
2. Magnetic stirrer
3. Peristaltic pump
4. Power supply
5. UV lamp
6. Qurtz jacket
7. Sampling port
8. Distributor
9. Air compressor
10. Flow meter
11. Air dryer
12. Ozone generator
13. Washing bottle
14. Effluent

(b)

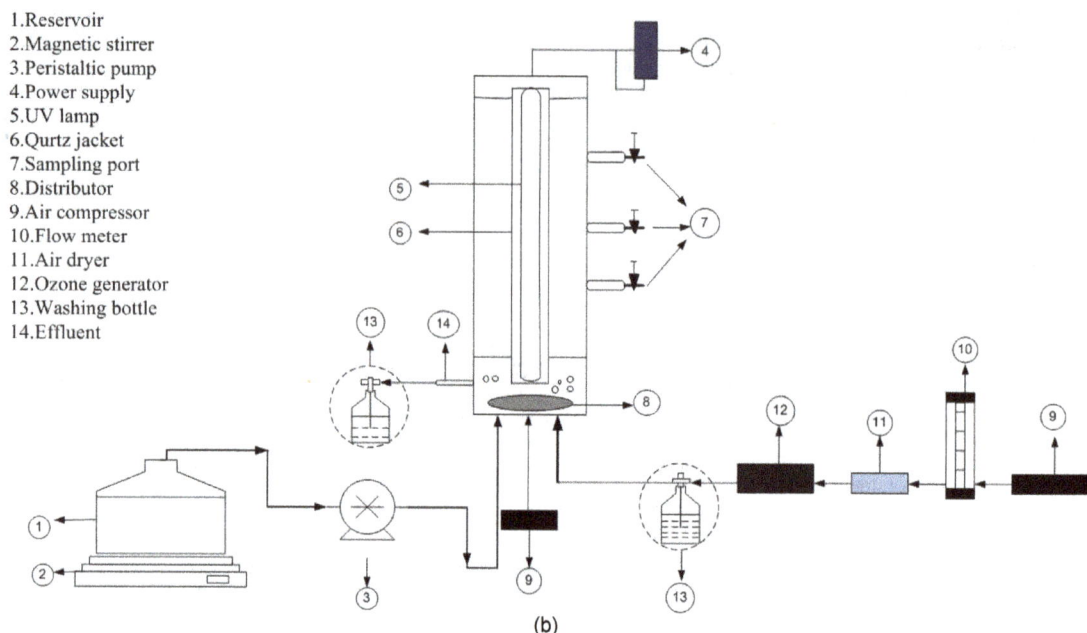

Fig. 1. Laboratory-scale experimental set-up (a) photo reactor, (b) Schematic diagram of the experimental set-up.

2.3. Photo reactor operation

In experimental runs, 1450 mL of TCW with COD concentration 500 mg/L was loaded in the photo reactor. The experiments were performed at selected solution pH which was left uncontrolled during the reaction. The selected solution pHs were 3, 5, 7, 9 and 11. For those runs where the initial pH had to be adjusted, this was done by adding the appropriate amount of 1M NaOH or 1M HCl solutions, as necessary. Concentration of H_2O_2 was selected to be 0, 5, 10, 15 and 20 mM. Categorical variables (ozonation and UV irradiation) were used in several runs by turned on the UV lamp and injection of the ozone in the photo reactor. Air was continuously sparged in the reaction mixture with flow rate of 5 L/min. The ozone content of the input air stream was measured as 0.27 g O_3/h. The ozone of offgas was also measured and the consumed ozone was obtained (3.68 g ozone$_{consumed}$/ gCOD$_{removed}$).

2.4. Experimental design

Effects of four independent factors (initial H_2O_2 concentration and initial pH, ozonation and UV irradiation) on the process performance were investigated. The response surface methodology (RSM) used in the present study was a general factorial design involving two numerical factors, initial H_2O_2 concentration and initial pH, and two categorical factors, ozonation and UV irradiation. The experimental range and levels of the independent variables is shown in Table 2. The experimental conditions are presented in Table 3. All experiments were carried out in batch mode in terms of wastewater input. COD removal and BOD$_5$/COD ratio were dependent responses. The results were completely analyzed using analysis of variance (ANOVA) which was performed by Design Expert Software (version 6.0, State-Ease, Inc., Minneapolis, MN).

Table 2. Experimental range and levels of the independent variables

Type of variables	Variables	Range and levels				
		-1	-0.5	0	+0.5	+1
Numerical	A-Initial pH	3	5	7	9	11
	B-H_2O_2 Conc.	0	5	10	15	20
Categorical	C-Ozonation	Without	-	-	-	With
	D-UV irradiation	Without	-	-	-	With

2.5. Analytical methods

The samples analyzed by GC/MS using an Agilent 6890N (0.25 lm, 30 m) capillary column. Oven conditions: 110 °C (2 min), 200 °C (4 min), 20 °C /min, 250 °C (2 min), 40 °C /min. Injector temperature set up to 180 °C, source temperature 250 °C. Helium (Infra, chromatographic purity) was used as carrier gas at 1 psi of pressure. Main products were identified by comparing their mass spectra with those in NIST library. All the chemicals used in the analysis were analytical grade (Merck, Darmstadt, Germany). Chemical oxygen demand (COD) was measured according to the Standard Methods. A colorimetric method with closed reflux method was developed. Spectrophotometer (DR 5000, Hach, Jenway, USA) at 600 nm was used to measure the absorbance of COD samples. A pH meter (JENWAY 3510) was used for pH measurement. Biodegradability was measured by 5-day biochemical oxygen demand (BOD$_5$) test in a BOD meter (OxiTop IS 6) according to the Standard Methods (American Public Health Association 1999).

In the runs with H_2O_2, MnO$_2$ powder was used for elimination of the interference of residual H_2O_2 in COD test. Then, the sample was centrifuged to remove MnO$_2$ powders; the supernatant was used for COD test (Sousy et al. 2007).

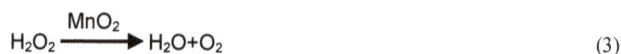

$$H_2O_2 \xrightarrow{MnO_2} H_2O + O_2 \qquad (3)$$

The ozone dosage was determined by an iodometry method using a washing bottle containing 2 wt % KI solution (Nyangiro 2003).

3. Results and discussion
3.1. Process performance
3.1.1. Statistical analysis

In this study, relationship between the two numerical independent variables (initial H_2O_2 concentration and initial pH) and two categorical factors (ozonation and UV irradiation) and two process responses (COD removal efficiency and BOD$_5$/COD ratio) for the TCW photo oxidation process were determined using response surface methodology (RSM). The ANOVA results for all responses have been summarized in Table 4. In order to quantify the curvature effects, the data from the experimental results were fitted to higher degree with cubic model. The model terms in the equations are after elimination of insignificant variables and their interactions. Based on the statistical analysis, the models were highly significant with very low probability values (<0.0001). It was shown that the models terms of independent variables were significant at the 99% confidence level. The square of correlation coefficient for each response was computed as the coefficient of determination (R^2). It showed high significant regression at ≥90% confidence levels. The value of the adjusted determination coefficient (adjusted R^2) was also high to prove the high significance of the model.

Table 3. Experimental conditions for advanced oxidation process

Run No.	Factor1 A: Initial pH	Factor2 B: H_2O_2 concentration, mmol/l	Factor3 C: Ozonation	Factor4 D: UV irradiation	Run No.	Factor1 A: Initial pH	Factor2 B: H_2O_2 concentration, mmol/l	Factor3 C: Ozonation	Factor4 D: UV irradiation
1	3	0	1*	1	27	7	10	1	2
2	3	0	2*	2	28	7	10	1	2
3	3	0	1	2	29	7	10	1	2
4	3	0	2	1	30	7	10	1	2
5	3	20	1	1	31	7	10	1	2
6	3	20	2	2	32	7	10	2	1
7	3	20	1	2	33	7	10	2	1
8	3	20	2	1	34	7	10	2	1
9	5	10	1	1	35	7	10	2	1
10	5	10	2	2	36	7	10	2	1
11	5	10	1	2	37	7	15	1	1
12	5	10	2	1	38	7	15	2	2
13	7	5	1	1	39	7	15	1	2
14	7	5	2	2	40	7	15	2	1
15	7	5	1	2	41	9	10	1	1
16	7	5	2	1	42	9	10	2	2
17	7	10	1	1	43	9	10	1	2
18	7	10	1	1	44	9	10	2	1
19	7	10	1	1	45	11	0	1	1
20	7	10	1	1	46	11	0	2	2
21	7	10	1	1	47	11	0	1	2
22	7	10	2	2	48	11	0	2	1
23	7	10	2	2	49	11	20	1	1
24	7	10	2	2	50	11	20	2	2
25	7	10	2	2	51	11	20	1	2
26	7	10	2	2	52	11	20	2	1

1* without applying the categorical variable   2* with applying the categorical variable

Adequate precision is a measure of the range in predicted response relative to its associated error or, in other words, a signal-to-noise ratio. Its desired value is 4 or more. The value was found more than 30. Simultaneously, low values of the coefficient of variation (CV) (>5%) indicated good precision and reliability of the experiments as suggested by Kuehl (2000), Khuri and Cornell (1996). The predicted versus actual plots for the two responses, COD removal and BOD_5/COD ratio is shown in Figs. 2a and b, respectively. These plots indicate an adequate agreement between real data and the ones obtained from the models.

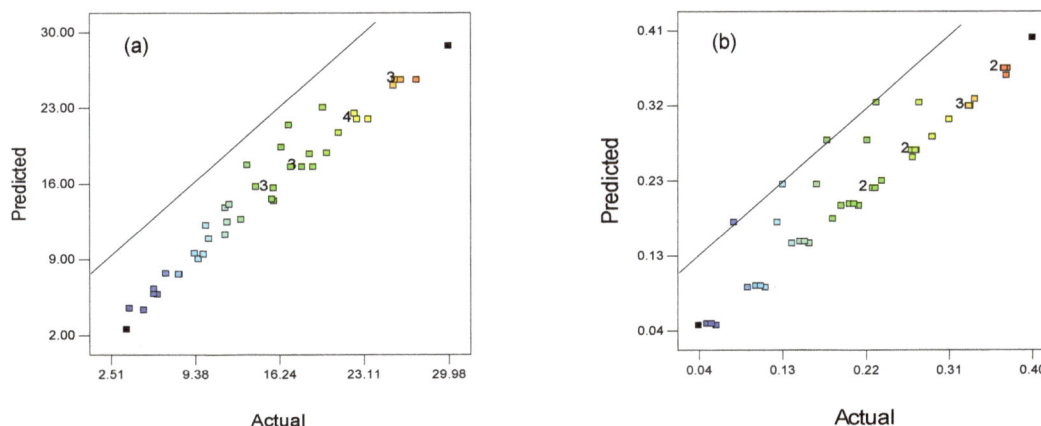

Fig. 2. Predicted vs. actual values plots for (a) COD removal, (b) BOD_5/COD ratio.

3.1.2. COD removal

In this process, COD was measured as a response representing the organic content of the TCW. Figs. 3a and b illustrate the COD removal as a function of initial pH and H_2O_2 concentration at the conditions with and without the categorical variables, respectively. Relationship between the response and the variables is described by the Eq. 4.

$$COD\ removal = 20.15 - 6.27A + 1.63B + 3.60C + 1.40D - 0.917AB + 0.431CD - 12.07B^2 + 6.25AB^2 - 1.525B^2C \quad (4)$$

where, A is initial pH, B is H_2O_2 concentration, C is ozonation and D is UV irradiation. From the Eq. 4, terms B, C and D had a positive effect on the response while term A had a negative effect on the response. As can be seen in the Figs., H_2O_2 concentration showed a reverse impact on the response; an increasing effect at low concentration (0-10 mmol/l) and a decreeing effect at higher concentration (10-20 mmol/l). While for initial pH; a decreasing effect on the process response was found except at the conditions with the lowest and highest levels of H_2O_2 concentration which showed no effect. As seen in the Fig. 3 a (UV/O_3/H_2O_2), maximum COD removal efficiency was found to be about 32% for the H_2O_2 concentration 10 mmol/l and initial pH 3. In other side, minimum COD removal obtained 13% at H_2O_2 concentration 20 mmol/l and initial pH 11. Whereas, under conditions with no ozonation and UV irradiation, the range of changes in the response was 6 to 20 % at H_2O_2 concentration 20 mmol/l and initial pH 11 and H_2O_2 concentration 10 mmol/l and initial pH 3, respectively (Fig. 3 b). In the both systems, minimum efficiency was found at the highest level of H_2O_2 concentration (20 mmol/l). This may be due to auto-decomposition of H_2O_2 to oxygen and water and the recombination of OH· radicals. Since OH· radicals react with H_2O_2, H_2O_2 itself contributes to the OH scavenging capacity. Therefore, H_2O_2 should be added at an optimal concentration to achieve the best degradation (Crittenden et al. 1999; Ghaly et al. 2001; Im et al. 2012). According to the Fig. 3 a, an enhancement of the COD removal was achieved by adding ozone and UV to the hydrogen peroxide in the solution under the $O_3/UV/H_2O_2$ process with a COD/H_2O_2 (w/w) ratio equal to 1.5.

Table 4. ANOVA for response surface models applied.

	Response	Modified Equations with significant terms	Probability	R^2	Adj.R^2	Adeq. precision	S.D.	CV	PRESS
Advanced oxidation process	COD removal	$20.15-6.27A+1.63B+3.60C+1.40D-0.917AB+0.431CD-12.07B^2+6.25AB^2-1.525B^2C$	<0.0001	0.9611	0.9528	40.48	1.48	9.24	121.37
	BOD_5/COD	$0.289-0.076A+0.050C+0.023D-0.168B^2+0.077AB^2$	< 0.0001	0.9030	0.8925	31.77	0.03	14.22	0.06

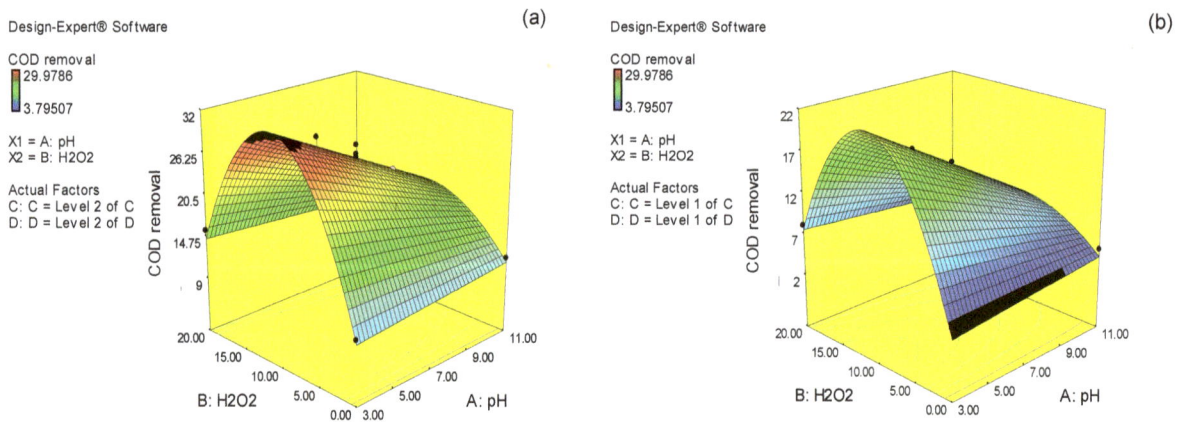

Fig. 3. Response surface plots for COD removal in (a) with and (b) without ozonation and UV irradiation.

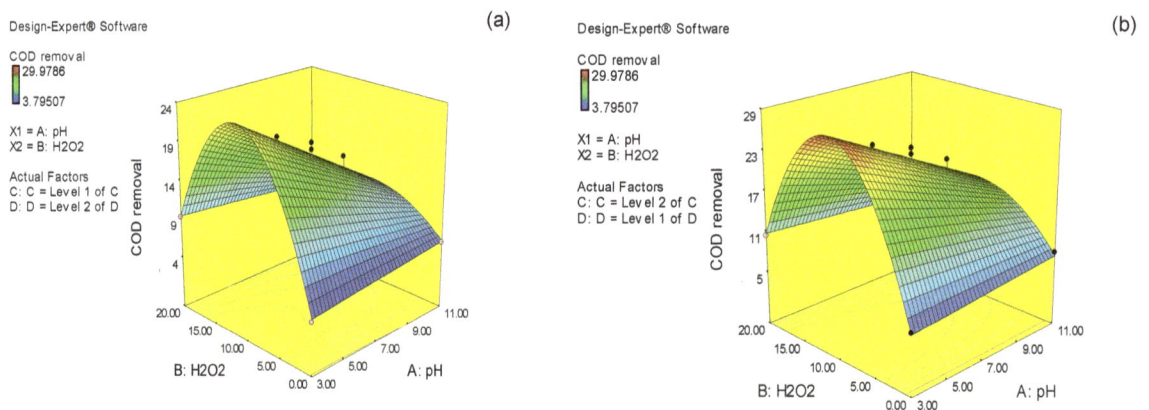

Fig. 4. Response surface plots for COD removal (a) without ozonation and with UV irradiation (b) with ozonation and without UV irradiation.

Figs. 4a and b illustrate the COD removal as a function of initial pH and H_2O_2 concentration at different conditions of ozonation and UV irradiation. As is seen in the Figs., ozonation showed to be a bit more effective in comparison with UV irradiation. In the UV/H_2O_2 system, the COD removal efficiency can be considered to occur mainly through its reaction with highly oxidizing OH· which is produced by the photolysis of H_2O_2 (Safarzadeh-Amiri 2001). More studies were performed with O_3/H_2O_2 and UV/H_2O_2 processes for degradation of different compounds. As a result, O_3/H_2O_2 process showed to be more effective than the UV/H_2O_2 process in terms of COD removal efficiency (Ghaly et al. 2001; Safarzadeh-Amiri 2001; Sanchez-Polo et al. 2007; Guedes Maniero et al. 2008).

In order to clarify the interactive effects of the variables studied, interactive graphs were prepared as shown in Figs. 5a-f and 6a-d. Figs. 5a-f depict interactive effects of DC (UV irradiation-ozonation) on COD removal at different values of pH and H_2O_2 concentration. As illustrated in the Figs., in all conditions, D (UV irradiation) had a slight constant

increasing interaction on C effect. It proves this fact that O_3/H_2O_2 system increases the kinetics of ozone decay and accelerates its transformation into OH·(Katsoyiannis et al. 2011). The principal species reacting with and oxidizing compounds in the O_3/H_2O_2 process are molecular ozone and hydroxyl radicals. The decomposition of ozone catalyzed by hydroperoxide (HO_2^-) generates hydroxyl radicals (Safarzadeh-Amiri 2001). By comparing the Figs., the synergistic effect of the combined system ($O_3/UV/H_2O_2$) showed the maximum COD removal at pH 3 and H_2O_2 concentration 10 mmol/l.

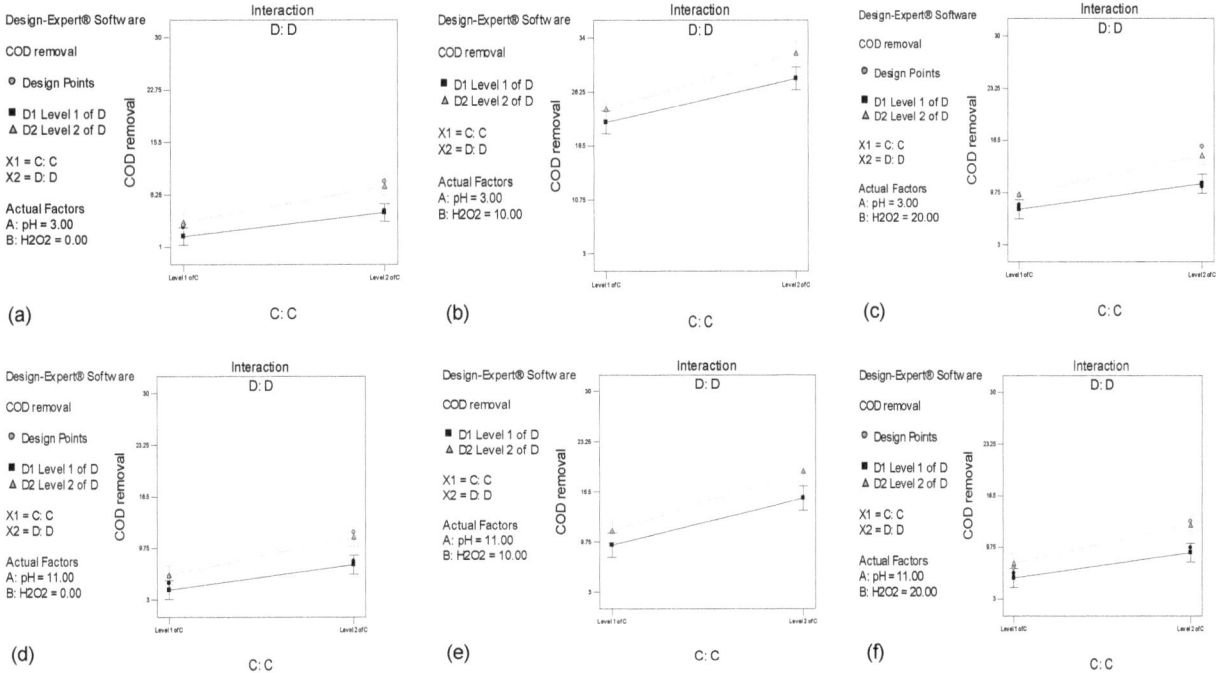

Fig. 5. Interactive effects of DC (UV irradiation-ozonation) on COD removal at different pH and H_2O_2 concentration.

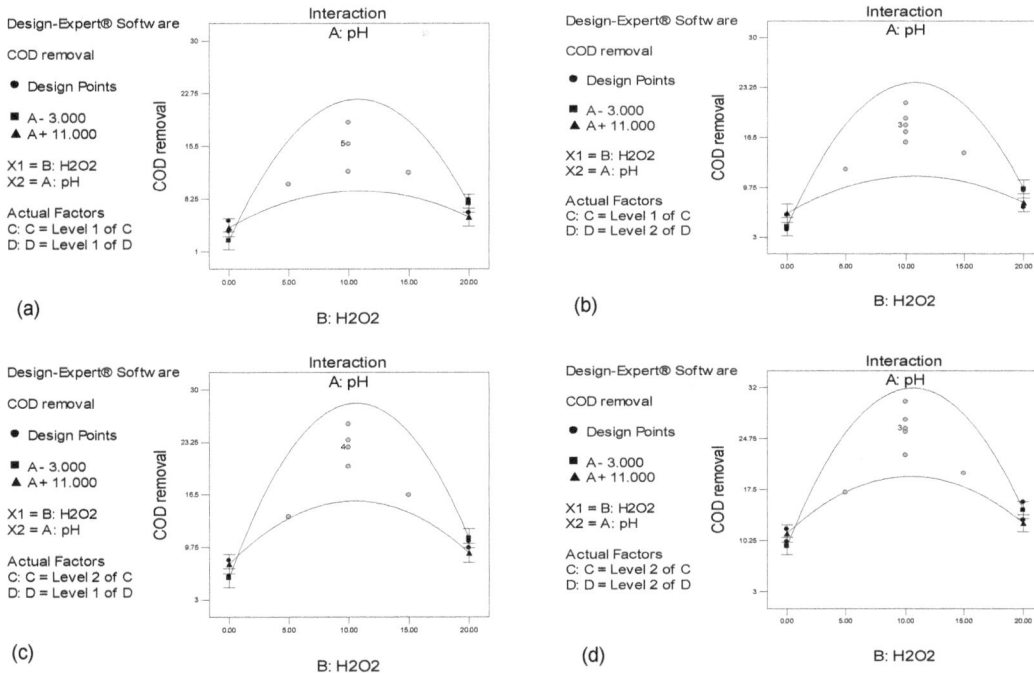

Fig. 6. Interactive effects variable AB (initial pH-H_2O_2 concentration) on COD removal at different conditions of ozonation and UV irradiation.

Figs. 6a-d represent the interactive effect of pH-H_2O_2 on the response in the different conditions studied. As mentioned earlier, pH had a significant effect on the process performance in terms of COD removal which the interactive effect of AB (pH-H_2O_2) proves this matter

as shown in the Figs. The dissociated form of hydrogen peroxide (HO_2^-) in alkaline media reacts with hydroxyl radicals more than two orders of magnitude faster than hydrogen peroxide does. Higher pH values slightly enhance the decomposition of hydrogen peroxide for the same reason. Furthermore, the molar extinction coefficient of HO_2^- is more than ten times greater than that H_2O_2 (with the same quantum yield as H_2O_2), and thus increases the hydrogen peroxide decomposition (Crittenden et al. 1999). Therefore, best degradation efficiency was obtained at acidic condition.

In summary, the effectiveness of the different AOPs examined removing COD content of TCW at neutral pH is shown in Fig. 7. As presented in the Fig., the order of the processes in terms of COD removal efficiency is as follow: $(H_2O_2/O_3/UV) > (H_2O_2/O_3) > (H_2O_2/UV) > (H_2O_2)$. Maximum COD removal efficiencies obtained 26.16, 22.59, 18.11 and 14 %, respectively at 10 mM of H_2O_2. The best COD/H_2O_2 (w/w) ratio was determined 1.5.

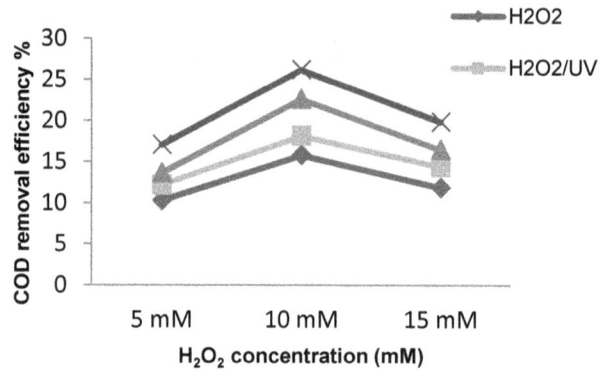

Fig. 7. Performance of different AOPs treating TCW.

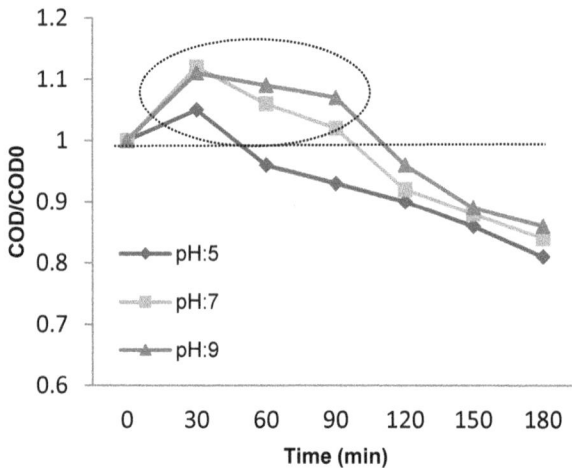

(a)

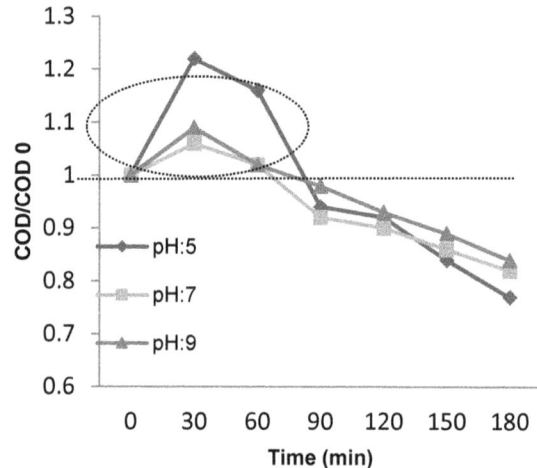

(b)

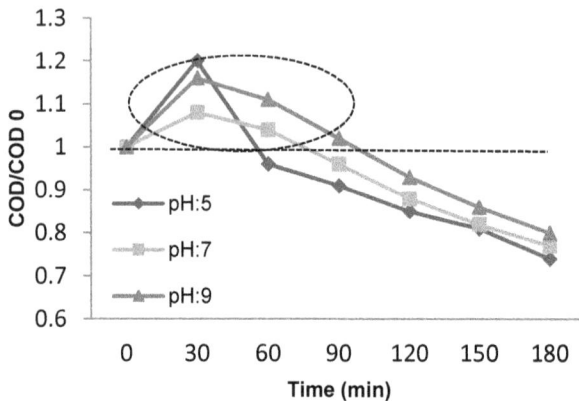

(c)

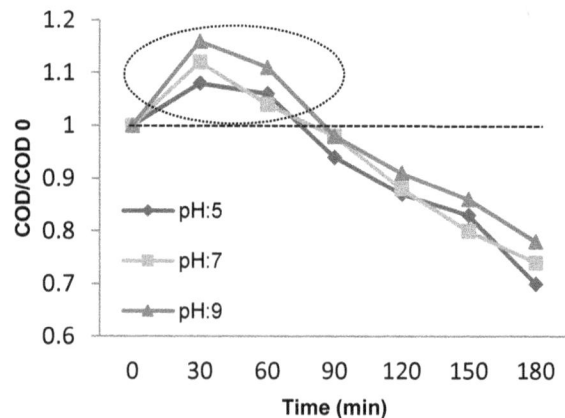

(d)

Fig. 8. Changes in COD/COD_0 versus reaction time at (a) H_2O_2 alone, (b) H_2O_2/UV, (C) H_2O_2/O_3 and (d) $H_2O_2/O_3/UV$

COD concentration was measured every 30 min up to 180 min for all the experiments. In order to investigate the trend of COD reduction during the reaction time, the COD to COD_0 ratio versus reaction time was drawn for few selected conditions (at 10 mM of H_2O_2 and different pH) as shown in Figs. 8a-d. Increase in the ratio at the beginning times indicates the presence of some recalcitrant and refractory organic compounds in TCW which probably are not detected in the COD test (e.g. pyridine 2-ethyl, styrene, pyridine 2-ethenyl and etc.). So that, by progressing the time the compounds are broken to intermediates that observed in the COD test and caused a decrease in the COD removal

efficiency. As observed in the Figs., thereafter degradation of the refractory compounds, the COD/COD_0 was decreased, implying increase in COD removal efficiency. The abovementioned claim was approved by results obtained from GC-MS analysis of the samples.

Fig. 9 represents GC-MS chromatogram of raw TCW sample. The GC-MS analysis confirmed the aforementioned points, so that, the raw wastewater samples contained some recalcitrant and refractory organic compounds which are not detectable in COD test. Table 5 presents these compounds and peak area obtained from the GC-MS analysis.

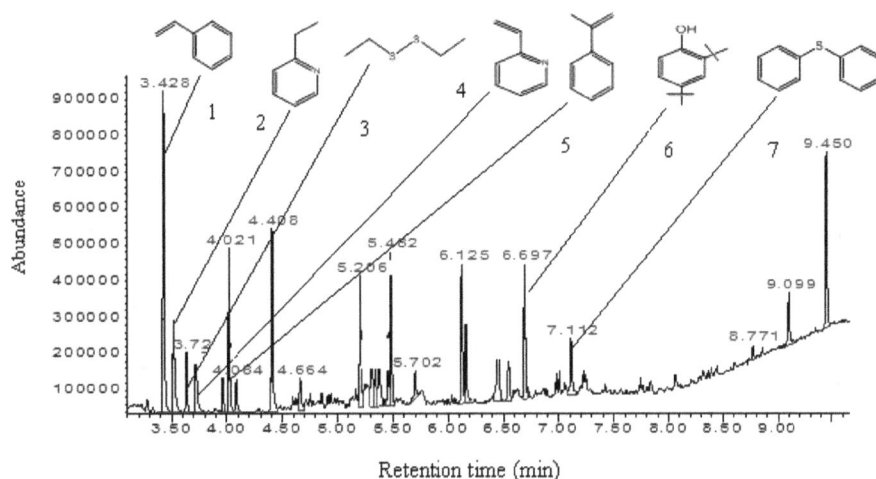

Peak identities are as follows: 1: Styrene, 2: Pyridine 2-ethyl 3: Diethyl disulfide, 4: Pyridine 2- ethenyl, 5: Alpha-methyl styrene, 6: Phenol, 2,4-bis (1,1-dimethylethyl, 7: Diphenyl sulfide

Fig. 9. GC-MS analysis of raw TCW.

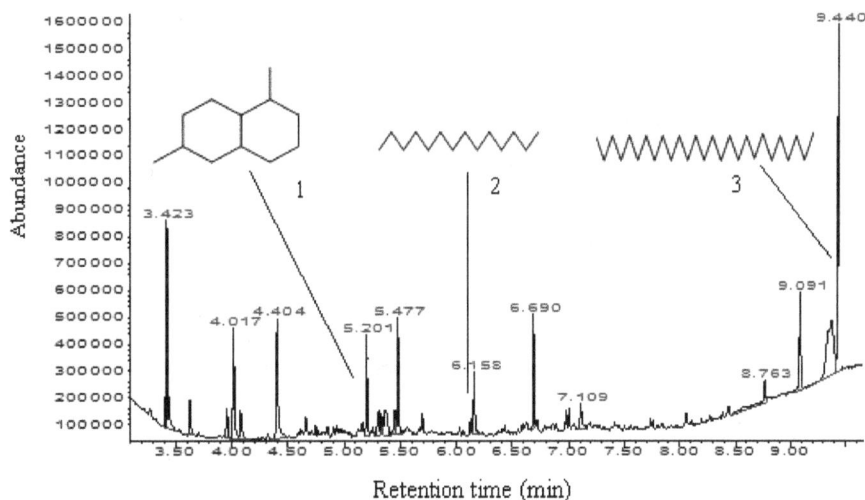

Peak identities are as follows: 1: Naphthalene decahydro-1,6-dimethyl, 2: Tetradecane, 3: Heptacosane.

(a)

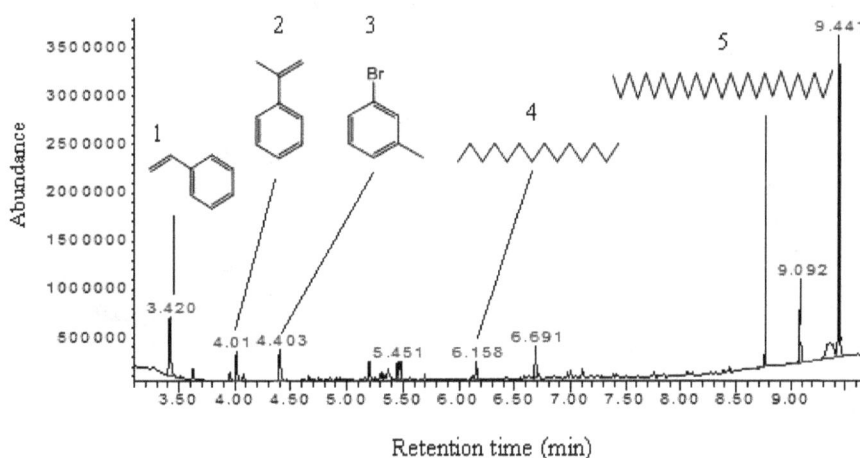

Peak identities are as follows: 1: Styrene, 2: Alpha methyl styrene, 3: Benzene, 1-bromo-3 methyl, 4: Tetradecane, 5: Heptacosane.

(b)

Fig. 10. GC-MS analysis of organic contaminants in the oxidized TCW by (a) H_2O_2 /O_3 at pH 9 and (b) H_2O_2 at pH 5.

Table 5. Details of GC-MS chromatogram of raw TCW.

Type of component	Peak height	Corr. Area
Styrene	902125	9816363
Pyridine 2-ethyl	230586	3142783
Pyridine 2-ethenyl	168635	2336464
Diethyl disulfide	145155	1484417
Alpha methyl styrene	425891	4157105
n-Decane	86184	942783
Benzene, 1-bromo-3 methyl	511086	5637735
Naphthalene,decahydro-1,6-dimethyl	400884	4166187
Naphthalene,decahydro-2,3-dimethyl	140133	2175645
Cyclo undecene,1.methyl	204772	1837840
7-Heptadecene,17-chloro	98631	1062907
Tetradecane	209160	2190089
Phenol,2,4-bis(1,1-dimethylethyl)	378149	3412067
Diphenyl sulfide	147090	1878328
1-monolinoleoylglycerol trimethylsilyl ether	36928	429615
Heptacosane	487483	4861025

Table 6. Details of GC-MS chromatogram of two treatment processes.

Type of treatment	Type of component	Peak height	Corr. Area	Degradation %
H_2O_2 /O_3 at pH 9	Diethyl disulfide	136294	1404791	5.36
	Alpha methyl styrene	383321	4155120	0.048
	Benzene, 1-bromo-3 methyl	430634	5482166	2.76
	Naphthalene,decahydro-1,6-dimethyl	94245	1307860	68.61
	Tetradecane	93926	1281441	41.49
	Heptacosane	321369	3212407	33.92
H_2O_2 at pH 5	Styrene	596855	7460985	23.99
	Alpha methyl styrene	286660	3066820	26.23
	Benzene, 1-bromo-3 methyl	316744	3783160	32.90
	Cyclo undecene,1.methyl	180367	1666486	9.32
	Naphthalene,decahydro-1,6-dimethyl	346351	3470103	16.71
	Tetradecane	185331	1725886	21.20
	Phenol,2,4-bis(1,1-dimethylethyl)	313331	2710295	20.57
	Heptacosane	173604	980132	79.84

In order to trace the fate of the components, two samples after 180 min were analyzed by GC-MS and the results were compared with the raw TCW. Fig. 10a and b shows the GC-MS chromatogram for these samples. Table 6 presents the peak area obtained from the GC-MS analysis. From the Table, the ratio of peak area for a specific compound showed degradation percentage achieved. As observed, some of recalcitrant compounds (e.g. Pyridine 2-ethyl, and Pyridine 2-ethenyl) were disappeared after the photo oxidation process. The other compounds were relatively degraded as shown in the Table.

3.1.3. BOD₅ to COD ratio

Since one of the aims to apply the photo oxidation process is to increase the BOD_5/COD ratio, the ratio was determined to indicate biodegradable fraction of COD at the end of each experiment (after 180 min) as another response. Relationship between the response and the variables is described by the Eq. 5.

$$COD\ removal = 0.289-0.076A+0.050C+0.023D-0.168B^2+0.077AB^2 \qquad (5)$$

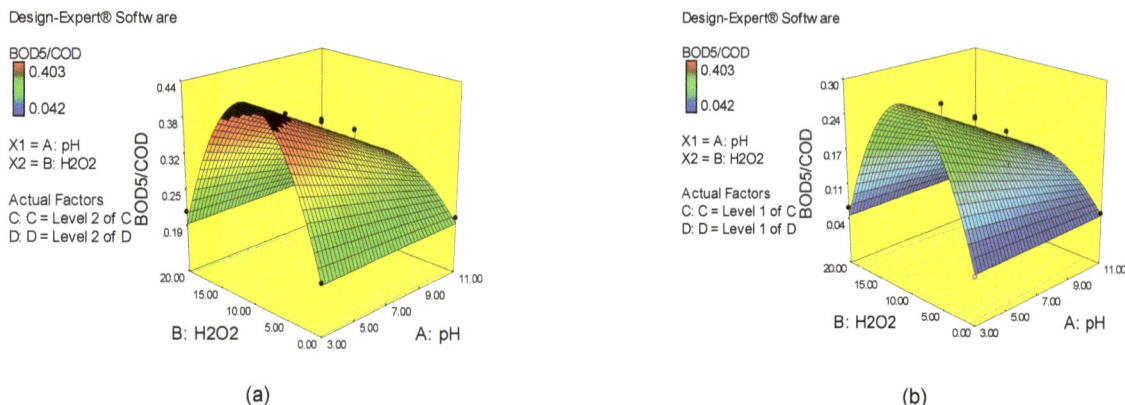

From the Eq. 5, terms C and D had a positive effect on the response while term A had a negative effect on the response. Figs. 11a and b represent the changes in the BOD_5/COD ratio as a function of the initial pH and H_2O_2 concentration at the conditions with and without the categorical variables, respectively. As can be seen in the Figs., similar trends as obtained for COD removal were observed for the BOD_5/COD ratio. As seen in the Fig. 11a, maximum BOD_5/COD ratio was found to be 0.41 at the conditions with H_2O_2 concentration of 10 mmol/l and initial pH of 3. Minimum ratio of BOD_5/COD determined 0.21 for the H_2O_2 concentration 20 mmol/l and initial pH 11. Figs. 12a and b illustrate the BOD_5/COD ratio as a function of initial pH and H_2O_2 concentration at the conditions without ozonation, with UV irradiation and with ozonation, without UV irradiation, respectively. Similar trends but with different amount were found. It showed effect of ozonation was more compared to UV irradiation. As be seen in the Fig. 12 b, maximum BOD_5/COD ratio was found to be 0.37, where ozonation was applied at H_2O_2 concentration of 10 mmol/l and initial pH of 3.

(a)

(b)

Fig. 11. Response surface plots for BOD₅/COD ratio at (a) with (b) without of ozone and UV irradiation.

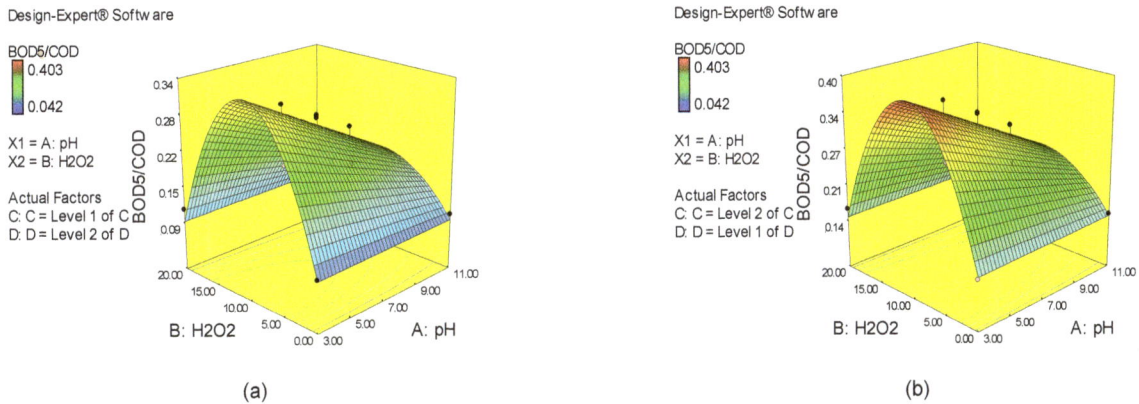

Fig. 12. Response surface plots for BOD5/COD ratio at (a) without ozonation and with UV irradiation (b) with ozonation and without UV irradiation.

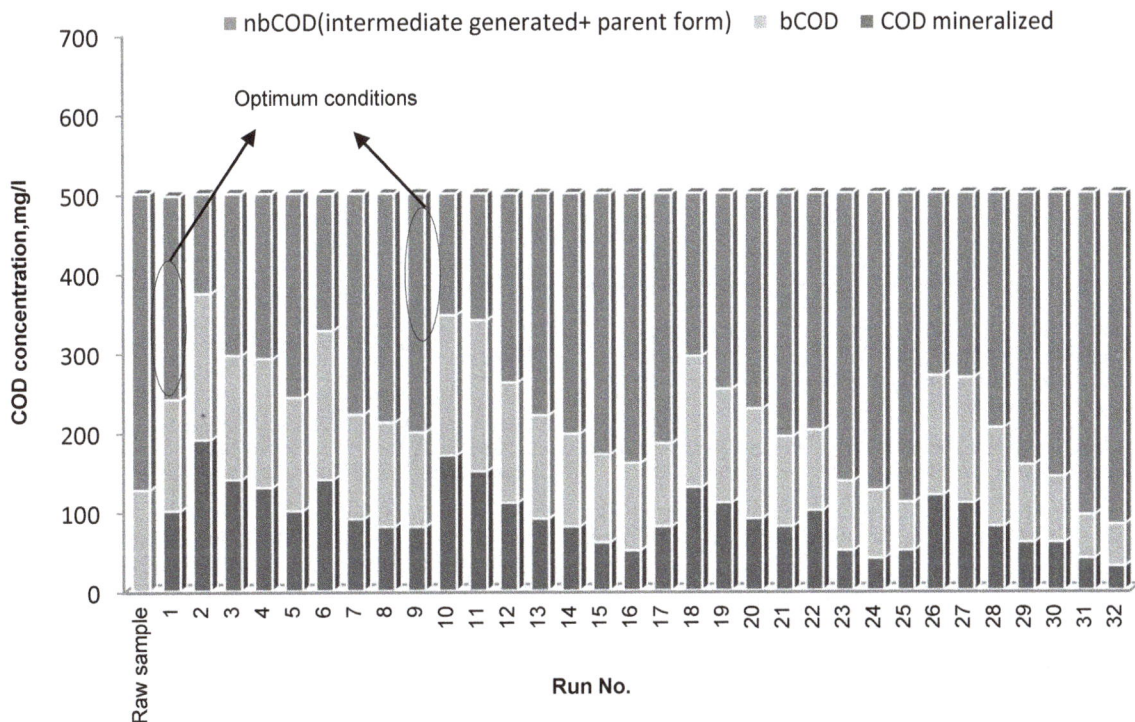

Fig 13. COD fractionation for the samples after the treatment process.

As the ratio of BOD5/COD for the raw TCW was determined to be about 0.1-0.2. In order to monitor the fate of COD contents in the samples after the treatment processes, different fractions of the COD is represented in Fig. 13. Fig. 13 has been drawn according the experiments number as presented in Table 7. As specified in the Fig., maximum mineralization and bCOD fraction were observed at experiments no. 2 and 10 (the operating conditions are presented in the Table 7).

4. Conclusions

This research work showed that the advanced oxidation processes could be applicable for treatment of Tire Cord wastewaters. However, it needs more studies in details to introduce the best practice. The main aims in use of advanced oxidation processes for treatment of refractory wastewaters can be COD mineralization and/or COD to BOD conversion which could be well achieved in this study. It is noted that the when UV irradiation or ozonation were combined with hydrogen peroxide, the process performance in terms of COD removal efficiency increased compared to that of hydrogen peroxide alone treatment. O_3/H_2O_2 process showed to be a bit more effective in comparison with UV/H_2O_2 process. As a result, higher performance was obtained at

H_2O_2 concentration 10 mM and initial pH 3 with 25% COD removal efficiency. The best result was obtained by using $O_3/UV/H_2O_2$ system at same condition with 32%. At low H_2O_2 concentration, $UV/O_3/H_2O_2$ combination improved the oxidation performance slightly, but showed an inhibitory effect at H_2O_2 concentrations higher than 10 mM. The maximum ratio of BOD5/COD was found 0.41 in the $UV/O_3/H_2O_2$ system. The O_3/H_2O_2 found more economical; nevertheless, its performance is slightly low related to others. In the basis of the results obtained, the photo oxidation processes applied could be used as a pretreatment method prior a biological treatment process. And also, other advanced oxidation processes are recommended to be examined.

Acknowledgement

The authors would like to acknowledge Kermanshah Department of Environment for providing analytical equipment to measure quality parameters. The authors would also like to acknowledge the cooperation of Mr. S. Ghanbari for his fantastic and unique job in the fabrication of the immobilized photocatalytic reactor set up. Special thanks also go to Mr. Kaboudi for his assistance in GC-MS analysis.

Table 7. Order of experiments number according to operating conditions.

Run No.	Types of treatment	H_2O_2 concentration (m mol/l)	Initial pH
1		5	7
2		10	3
3		10	5
4	$H_2O_2/ O_3 / UV$	10	7
5		10	9
6		15	7
7		20	3
8		20	11
9		5	7
10		10	3
11		10	5
12	H_2O_2/ O_3	10	7
13		10	9
14		15	7
15		20	3
16		20	11
17		5	7
18		10	3
19		10	5
20	H_2O_2/ UV	10	7
21		10	9
22		15	7
23		20	3
24		20	11
25		5	7
26		10	3
27		10	5
28	H_2O_2	10	7
29		10	9
30		15	7
31		20	3
32		20	11

References

Afzal A., Drzewicz P., Martin JW., El-Din MG., Decomposition of cyclohexanoic acid by the UV/H2O2 process under various conditions, Science of the Total Environment 426 (2012) 387–392.

American Public Health Association (APHA). Standard methods for the examination of water and wastewater. 20th.Washington: 1999.

Crittenden J.C., Hu Sh., Hand D.W., Green S.A., kinetic model for H2O2/UV process in a completely mixed batch reactor, Water Research 33 (1999) 2315-2328.

Di Marzio W.D., Saenz M., Alberdi J., Tortorelli M., Silvana G., Risk assessment of domestic and industrial effluents unloaded into a freshwater environment, Ecotoxicology and Environmental Safety 61 (2005) 380–391.

Garoma T., Gurol M.D., Thotakura L., Osibodu O., Degradation of tert-butyl formate and its intermediates by an ozone/UV process, Chemosphere 73 (2008) 1708–1715.

Ghaly M.Y., Hartel G., Mayer R., Haseneder R., Photochemical oxidation of p-chlorophenol by UV/H2O2 and photo-Fenton process. A comparative study, Waste Management 21 (2001) 41-47.

Ghaly M.Y., Hartel G., Mayer R., Haseneder R., Photochemical oxidation of p-chlorophenol by UV/H2O2 and photo-Fenton process. A comparative study, Waste Management 21 (2001) 41-47.

Guedes Maniero M., Maia Bila D., Dezotti M., Degradation and estrogenic activity removal of 17β-estradioland 17α-ethinylestradiol by ozonation and O3/H2O2, Science of the Total Environment 407 (2008) 105-115.

Haji Sh, Benstaali B, Al-Bastaki N. Degradation of methyl orange by UV/H2O2 advanced oxidation process, Chemical Engineering Journal 168 (2011) 134–139.

Horsing M., Kosjek T., Andersen H.R., Heath E., Ledin A., Fate of citalopram during water treatment with O3, ClO2, UV and fenton oxidation, Chemosphere 89 (2012) 129–135.

Im J.K., Cho I.I.H., Kim S.K., Zoh K.D., Optimization of carbamazepine removal in O3/UV/H2O2 system using a response surface methodology with central composite design, Desalination 285 (2012) 306–314.

Katsoyiannis I.A., Canonica S., Von Gunten U., Efficiency and energy requirements for the transformation of organic micropollutants by ozone, O3/H2O2 and UV/H2O2, Water Research 45 (2011) 3811-3822.

Katsoyiannis I.A., Canonica S., Von Gunten U., Efficiency and energy requirements for the transformation of organic micropollutants by ozone, O3/H2O2 and UV/H2O2, Water Research 45 (2011) 3811-1822.

Keller V., Bernhardt P., Garin F., Photocatalytic oxidation of butyl acetate in vapor phase on TiO2, Pt/TiO2 and WO3/TiO2 catalysts, Journal of catalysis 215 (2003) 129–138.

Khuri AI, Cornell JA. Response surfaces: design and analyses. 2nd edition., Marcel Dekker, New York, 1996.

Kosaka K, Yamada H., Shishida K, Echigo Sh, Minear RA, Tsuno H, Matsui S. Evaluation of the treatment performance of a multistage ozone/ hydrogen peroxide process by decomposition by-products, Water Research 35 (2001) 3587–3594.

Kuehl R.O., Design of Experiments: Statistical Principles of Research Design and Analysis. 2nd edition., Duxbury Press, Pacific Grove, (2000) 2–225.

Kusic H., Koprivanac N., Srsan L., Azo dye degradation using Fenton type processes assisted by UV irradiation: A kinetic study, Journal of Photochemistry and photobiology A: Chemistry 181 (2006) 195–202.

Latifoglu A., Gurol M.D., The effect of humic acids on nitrobenzene oxidation by ozonation and O3/UV processes, Water Research 37 (2003) 1879–1889.

Lin F., Zhang Y., Wang L., Zhang Y., Wang D., Yang M., Yang J., Zhang B., Jiang Z., Li C., Highly efficient photocatalytic oxidation of sulfur-containing organic compounds and dyes on TiO2 with dual cocatalysts Pt and RuO2, Applied Catalysis B: Environmental 127 (2012) 363– 370.

Lucas M.S., Peres J.A., Li Puma G., Treatment of winery wastewater by ozone-based advanced oxidation processes (O3, O3/UV and

$O_3/UV/H_2O_2$) in a pilot-scale bubble column reactor and process economics, Separation and Purification Technology 72 (2010) 235–241.

Lucas M.S., Peres J.A., Li Puma G., Treatment of winery wastewater by ozone-based advanced oxidation processes (O_3, O_3/UV and $O_3/UV/H_2O_2$) in a pilot-scale bubble column reactor and process economics, Separation and Purification Technology 72 (2010) 235–241.

Nyangiro D., New application of ozone in the treatment of chemical pulps. Ph.D. Thesis, EFPG/INPG. France: 2003.

Poyatos J.M., Munio M.M., Almecija M.C., Torres J.C., Hontoria E., Osorio F., Advanced Oxidation Processes for Wastewater Treatment: State of the Art, Water, Air, and Soil Pollution 2010;205:187–204.

Qiang Z, Liu Ch, Dong B, Zhang Y. Degradation mechanism of alachlor during direct ozonation and O_3/H_2O_2 advanced oxidation process. Chemosphere 78 (2010) 517–526.

Rosenfeldt E.J., Linden K.G., Canonica S., Von Gunten U., Comparison of the efficiency of OH radical formation during ozonation and the advanced oxidation processes O_3/H_2O_2 and UV/H_2O_2, Water Research 40 (2006) 3695 – 3704.

Safarzadeh-Amiri A., O_3/H_2O_2 treatment of methyl-tert-butyl ether (MTBE) in contaminated waters, Water Research 35 (2001) 3706–3714.

Sanchez-Polo M, Rivera-Utrilla J, Mendez-Diaz JD, Canonica S, Von-Gunten U. Photooxidation of naphthalenesulfonic acids: Comparison between processes based on O_3, O_3/activated carbon and UV/H_2O_2, Chemosphere 68 (2007) 1814–1820.

Shen Y.Sh., Ku Y., Treatment of Gas-phase Volatile Organic Compounds (VOCs) by the UV/O_3 process, Chemosphere 38 (1999) 1855-1866.

Shi H., Industrial wastewater-types, amounts and effects. Point sources of pollution: local effects and its control (2003) 1:442.

Sousy E., Hussen K., Hartani A., Elimination of Organic Pollutants using Supported Catalysts with Hydrogen Peroxide, Jordan Journal of Chemistry 2 (2007) 97-103.

21

Optimization of adsorption removal of ethylene glycol from wastewater using granular activated carbon by response surface methodology

Behnaz Jalili[1], Seyed Mehdi Borghei[1], Vahid Vatanpour[2,*], Christopher Sarkizi[3]

[1]Department of Chemical and Petroleum Engineering, Sharif University of Technology, Tehran, Iran.
[2]Department of Applied Chemistry, Faculty of Chemistry, Kharazmi University, Tehran, Iran.
[3]Chemical Engineering Department, Tarbiat Modares University, Tehran, Iran.

ARTICLE INFO	ABSTRACT
Keywords: Adsorption Granular activated carbon RSM Wastewater treatment Environmental pollution	Wastewater reuse has been attracted a lot of attention in recent years especially in places with low water availability. The effluents that were considered to be discharged are now could be used as potential sources of reusable water. In this study, variables affecting the removal of ethylene glycol (EG) by adsorption on granular activated carbon (GAC) from the synthetic wastewater solutions were optimized by response surface methodology (RSM) using a central composite design. The investigated factors were temperature, EG concentration, contact time, activated carbon amount and granular size. Adsorption kinetic was also studied and an acceptable correlation between Langmuir model and experimental data was observed. As a result, a modified third degree equation was proposed and used to find the optimized condition. The maximum adsorption was achieved at 27.7 °C with 0.8 g of 20-30 mesh activated carbons for an EG feed concentration of 135 mg/L at 210 minutes.

---

1. Introduction

The water management in dry areas of the world has been a critical challenge for the centuries, which is caused by limitations on water availability. Critical point refers to the condition when the balance between the usable water and the demand cannot be maintained. In most areas affected by drought or low water accessibility, all-out industrialization, urbanization and high growth rate of population, increase the need to address the serious pressure on available water resources. Furthermore, climatic changes are envisaged to have more negative effects on natural water resources and their water quality, emphasizing the urgency of this severe problem (Meyer et al. 2009; Sowers et al. 2011). Reusing of wastewater is one of the most popular strategies of water management employed to handle these situations (Pereira et al. 2002). Generally, the industries with large share of water consumption are more likely to have a wide choice when it comes to water reuse strategies. Depending on the usage, different practices could be employed. In the most cases, after physical and chemical treatments, the wastewater achieves required standards for general maintenance applications i.e. washing floors or cooling purposes (Rebhun and Engel, 1988; Mohsen and Jaber, 2003; Farahani et al. 2016).

In recent years, with realization of fresh water resources limitation and depletion, water recycling and reuse has become one of the main priorities of the industrial and urban communities (Petrinic et al. 2015). Researchers have concluded that ecological footprint of water and wastewater system can be dropped for about 25% just by benefiting from recycled wastewater in water management (Anderson 2003). In order to use the wastewater as a valuable water resource, some typical treatments are required to improve the water quality. On the other hand, treatment costs rise drastically for higher qualities (Feng and Chu, 2004).

Mono-ethylene glycol (MEG) is an odorless and colorless clear liquid, which is miscible with water and it is known for its low volatility

(Eisenreich et al., 1981). In 2004, worldwide MEG production was roughly about 18500 kilotons. Although it might be known for its use as a common coolant, it's also being used in a variety of industries from deicing fluids for airstrips to being a component of beauty products. With these amounts of consumption, there is a considerable volume of wastewater generated in these industries contaminated with MEG (Staples et al., 2001; Devlin and Schwartz, 2014).

Ethylene glycol (EG) in its effluent form typically after using as a runway deicing agent or as an industrial wastewater has high mobility and therefore, it has high potential to contaminate the soil and any water bodies that it comes into contact with. Also, it undergoes an approximately 2-days photo-chemical degradation process in atmosphere. Its degradation in soil and water occurs under both aerobic and anaerobic conditions, which range from a day to about a few weeks depending on the environmental conditions. The EG as a component of wastewater stream increases the effluents biological oxygen demand; therefore contaminated aqua-ecosystems will have higher chance to be disrupted (Carnegie and Ramsay, 2009). The EG as a sole pollutant has low level of toxicity. But once it's metabolized by the microorganisms present in the environment, it breaks down to different types of toxic components. It is considered toxic to central nervous system. However, since its adsorption through human skin happens at a very slow rate, it is very unlikely to reach toxic dosage (Dye 2001). Although aerobic digestion of EG has been the subject of numerous research articles (Gonzalez et al. 1972; Staples et al. 2001; Revitt and Worrall, 2003), some suggest that acidogenesis in anaerobic digestion in soil is able to significantly degrade ethylene glycol and is more effective method comparing to aerobic techniques (McVicker et al. 1998).

The anaerobic digestion of ethylene glycol has been subjected to several researches but only fermentation (Straß and Schink, 1986; Elreedy and Tawfik, 2015) and methanogenesis stages have been discussed in details (Dwyer and Tiedje, 1983). Dwyer and Tiedje (1983) proposed the anaerobic pathway, which methanogens utilize ethanol,

---

*Corresponding author Email: vahidvatanpoor@yahoo.com

acetic acid and acetaldehyde for better treatments of ethylene glycol. McGahey and Bouwer (1992) have studied subsurface cases and reported degradation of EG by native consortium in soil and water bodies. They pointed out that efficiency of the process was highly dependent on oxygen availability, therefore ethylene glycol concentrations near surface drops more quickly.

Another technique for removal of EG is membrane processes. Polyamide and polydopamine thin film composite membranes was used for dehydration of ethylene glycol by pervaporation process. The results showed that temperature has a positive effect on permeation flux. The higher temperatures resulted in better separation when NaCl was present in the feed (Wu et al. 2015). There have been some approaches to the ethylene glycol oxidation in fuel cells. In one the recent studies, it was concluded that an effective promoter for Pd electro-catalysis of EG oxidation was a possible candidate for industrial ethylene glycol fuel cells (Yang et al. 2015).

Polyethylene glycol (PEG) is another abundant pollutant present in industrial wastewater which made its biodegradation interesting for many researchers (Haines J., Alexander, 1975; Huang et al. 2005; Cadar et al. 2012). The investigations include biological aerobic and anaerobic treatment of PEGs with different molecular weights (Huang et al. 2005), comparison of biodegradation of polyethylene glycols and polypropylene glycols (Zgoła-Grześkowiak et al. 2006), and anaerobic digestion of PEGs with different molecular weights (Bernhard et al. 2008).

One of the most commonly used material for separating a wide spectrum of unwanted chemicals in liquid and gas phases is granular activated carbon (GAC) which can be produced using wood-like carbon rich materials (McQuillan et al. 2018; Jaria et al. 2019). Generally, the term of activated carbon is used for a large group of carbonic material with high levels of porosity and surface area. In recent years, the activated carbon which has been used in chemical industries mostly in adsorption processes is used for environmental purposes (Bansal and Goyal, 2005). It is used to treat industrial wastewaters in order to meet the environmental standards that allow discharge of the effluent to receiving waters.

Recently, response surface methodology (RSM) has been used for modeling adsorption of different pollutants on activated carbon adsorbents (Hameed et al. 2009; Sahu et al. 2009; Arulkumar et al. 2011; Esfandiar et al. 2014; Hajati et al. 2015). The RSM based statistical analysis, which is extensively utilized to multivariable optimization studies (Wongkaew et al. 2016), is a collection of mathematical and statistical techniques. This statistical technique decreases the number of essential tests considerably without overlooking the interactions among the experiment variables (Vatanpour et al. 2017). Since it is important to understand the interaction effects between several factors, the RSM provides a better understanding of the process than the standard methods of experimentation. It can predict in a complex process how the inputs influence the outputs where different factors can interact among themselves.

In this study, the removal of ethylene glycol from wastewater was examined by adsorption on granular activated carbon. The investigated parameters were granular activated carbon amount and different meshes of it, retention time, temperature, and feed concentration of ethylene glycol. A statistical modeling was used to estimate the remained concentration of EG. A series of central-composite-design based experiments were conducted using response surface methodology and conditions of the adsorption process were optimized. We used RSM for experimental design and investigated 5 effective parameters in three levels, which this report is comprehensive investigation of theses parameters in removal of ethylene glycol by the GAC.

Table 1. Investigated factors and their levels.

Independent variable	Level				
	-α	-1	0	+1	+α
(A) Temperature (°C)	20.0	23.8	27.5	31.2	35.0
(B) Feed concentration (mg/L)	90	135	180	225	270
(C) Granular activated carbon (g)	0.2	0.4	0.6	0.8	1
(D) Time (min)	120	150	180	210	240

2. Materials and methods
2.1. Materials and instruments

Ethylene glycol was obtained from Maroon petrochemical complex (Iran). Granular activated carbon processed from Coconut by Jacobi®. In order to take particle size into account GAC was screened into two mesh size categories 10-20 (841-2000 μm) and 20-30 (595-841 μm). NaOH and HNO₃ from Merck were used for pH adjustment. HPLC-grade water was prepared by reverse osmosis de-ionizing apparatus (water purification system ultraclear direct, SG waters, Germany). A pH meter (Meterohm, Switzerland) and a balance scale (AND-HR200, Japan) were used to determine the solution pH and weigh the adsorbent, respectively. The samples were incubated for the duration of retention time in a shaken incubator (Labron, South Africa).

2.2. Measurement of EG concentration by high performance liquid chromatography (HPLC)

In order to measure the concentrations of EG duration of the adsorption process, HPLC analysis was applied. The HPLC (Waters, USA) was equipped with C18 column (250×4.60 mm, 10 micron) from Waters, μbondapak™, Ireland, which attached to IR absorbance detector set at 254 nm. The mobile phase was pure water at a flow rate of 2 mL/min. The samples were injected in 20 μL duplicates. The peak areas for each compound were averaged and percent concentration was calculated by comparison to the peak areas.

2.3. Adsorption process

Batch adsorption process was carried out in 250 mL Erlenmeyer flasks. Different concentrations of EG were obtained by dilution of a 1 g per liter stock solution of the ethylene glycol. Afterwards, certain amounts of GAC adsorbent were added to the solutions. The pH was adjusted in 7.0 and the samples were incubated at different temperatures. All samples were stirred at 170 RPMs.

2.4. Experimental design

Design of experiments is a procedure which allows the experiments to be carried out in a way that sufficient data for optimization is obtained from the minimum numbers of experiments. Therefore, defining a definite number of runs in a specific order according to number of variables is possible. There are different methods to use designing experiments. Among them, RSM is one of the most frequently used. Central composite design (CCD) is a robust and accurate design which falls into the RSM category. It comprises of 5-level design for each factor included in the model.

In this study, 52 experiments were carried out according to the CCD design. Table 1 illustrates the investigated factors affecting outlet concentration of ethylene glycol and their modified ranges. Since two different mesh sizes (10-20 and 20-30) of GAC were used and it was difficult to obtain specific values for it, therefore the GAC mesh sizes effect was tested as a categorical factor. In order to estimate remained concentration of ethylene glycol, final equation was derived from a raw cubic model which was a third degree polynomial and it was generally described as following:

$$R = a_0 + \sum_{i=1}^{k} a_i x_i + \sum_{i=1}^{k} a_{ii} x_i^2 + \sum_{i=1}^{k} a_{iii} x_i^3 + \sum_i^k \sum_j^k a_{ij} x_i x_j + \sum_i^k \sum_z^k a_{iz} x_i x_z$$

$$+ \sum_j^k \sum_z^k a_{jz} x_j x_z + \sum_i^k \sum_j^k \sum_z^k a_{ijz} x_i x_j x_z + \sum_i^k \sum_j^k a_{ij} x_i^2 x_j$$

$$+ \sum_i^k \sum_z^k a_{iz} x_i^2 x_z + \sum_j^k \sum_i^k a_{ji} x_j^2 x_i + \sum_j^k \sum_z^k a_{jz} x_j^2 x_z + \sum_z^k \sum_i^k a_{zi} x_z^2 x_i +$$

$$\sum_z^k \sum_j^k a_{zj} x_z^2 x_j + \varepsilon \qquad (1)$$

In which a_0 is constant, a_i, a_{ii} and a_{iii} are respectively coefficients for first, second and third degree terms of a factor. a_{ij}, a_{iz}, a_{zj} represent interaction coefficients and x_i is an independent parameter, which is called a factor, k is the number of factors and ε is the associated model with the model (Mason et al. 2003).

Table 3. ANOVA parameters for the proposed model.

Source	Sum of squares	DF	Mean square	F value	p-value
Model	22377.05	14	1598.36	34.33	< 0.0001
Residual	1722.64	37	46.56		
Lack of fit	1707.50	35	48.79	6.45	0.1431
Pure error	15.13	2	7.57		
Core total	24099.69	51			
$R^2 = 0.928$	adequate precision = 28.765				

3. Results and discussion
3.1. Analysis of variance (ANOVA)

Investigation of operation parameters which affect the adsorption process was conducted and the experiments were carried out according to the CCD design matrix. Table 2 shows design points and the relevant remained concentrations of ethylene glycol after adsorption process. Design of experiments was done in such a way that effect of uncontrolled factors was reduced and the experiments were generated in a random order by the Design Expert (Ver. 8.0.1). After analysis, a cubic equation which its coefficients were calculated by least-squares regression was found to be fit for representing the concentration of ethylene glycol in the remained solution. Third-order equations of ethylene glycol's outlet concentration considering only the significant terms are shown in Eqs. 2 & 3.

EG' outlet (mesh of AC= 10-20) = +1129.9 − 44.62 A − 12.96 B − 41.63C +0.947 D +0.537 AB +0.0334 B^2 -3.84 D^2 − 0.001 A^{2}B + 0.0022 A^{2}C + 0.0037 A^{2}D -0.0012 B^{2}A $\qquad (2)$

EG' outlet (mesh of AC= 20-30) = +1132.49 − 44.63 A − 12.98 B − 41.63C +0.947 D +0.537 AB +0.0334 B^2 -3.84 D^2 − 0.0013 A^{2}B + 0.0023 A^{2}C + 0.0037 A^{2}D -0.0013 B^2 $\qquad (3)$

In constructing the model, a series of lack-of-fit tests should be performed in order to determine significance of the proposed model. The mentioned model has F-value of 34.33 shown in Table 3 accompanying with other analysis of variance (ANOVA) parameters. The p-value is an index that shows the significance of a term or a model and in this case terms with p-values less than 0.1 were considered significant and those with p-values more than 0.1 were considered insignificant and removed from the model manually. The p-value for the final model was less than 0.0001, which indicated that the model was significant. The lack-of-fit term for the remained concentration of ethylene glycol had a p-value of 0.1431.

Another parameter which indicates robustness and statistical importance of the proposed model is the adequate precision value which is a measure of "signal-to-noise ratio". In this case, the adequate precision was 28.765. The models with an adequate precision higher than 4 are more likely to make acceptable predictions in the central composite design defined space. The normal probability plot of residuals, which is shown in Fig. 1, indicated that error distribution was normal throughout the model. Another important plot that can illustrate the quality of the model is the predicted vs. actual diagram. As it is shown in Fig. 2, data points were located close to the diagonal line. A relatively high R2 confirms the model integrity.

3.2. Investigation of factors and their interactions

To observe how the factors affect the response, they were studied in pairs in three different levels of initial feed concentration of ethylene

glycol (160, 185 and 210 mg/L). As shown in Fig. 3 (a & b), in low concentrations of EG in high levels of granular activated carbon and retention time, response surface becomes horizontal. In early stages of the adsorption (low retention times), adsorption occurs mainly on the outer surface of the adsorbent. The observed concentration gradient in Fig. 3 (a & b) is due to this happening. The flattening of concentration surface in higher retention times shows that in order to exploit inner porous areas of the adsorbent, a higher level of driving force is required for happening of the diffusion. Driving force can be altered by changes in concentration levels of ethylene glycol which is affected by the adsorption; hence, decrease in ethylene glycol concentration lowers the driving force needed to utilize inner porous areas of the adsorbent needed to further decrease the concentration of the EG. In Fig. 3 (c, d, e, f), as it is expected, increment in levels of activated carbon and retention time have positive effects on the adsorption process. It is obvious that by increasing of GAC amount and adsorption time, the removal efficiency improves.

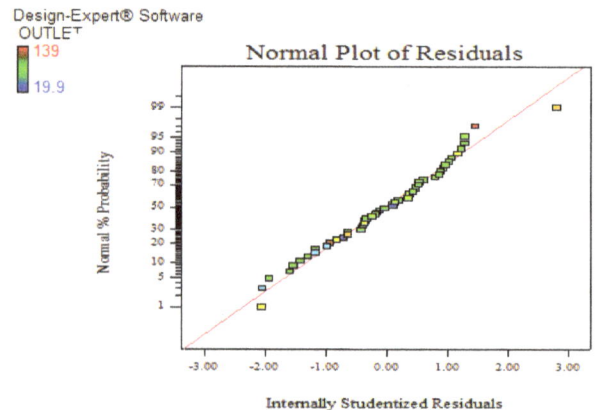

Fig. 1. Normal plot of residual for remained concentration of EG.

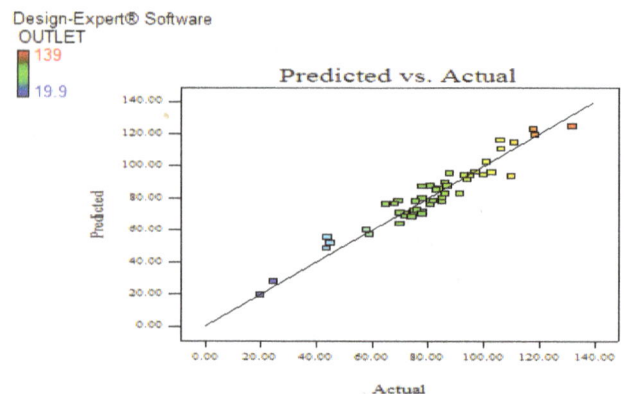

Fig. 2. Plot of predicted response vs. actual value for remained concentration of EG.

Table 2. Response surface methodology central composite design of experiments.

Run	(A) Temperature (°C)	(B) Feed Conc. (mg/L)	(C) GAC (g)	(D) Time (min)	(E) Mesh of AC	Remained EG
1	35.0	180	0.60	180.0	10-20	118.5
2	31.2	135	0.80	150.0	20-30	71.9
3	23.8	225	0.80	150.0	10-20	81.0
4	27.5	180	0.60	180.0	10-20	94.1
5	27.5	180	1.00	180.0	20-30	45.1
6	27.5	180	0.20	180.0	10-20	118.0
7	23.8	225	0.40	210.0	20-30	81.0
8	31.2	135	0.80	210.0	10-20	82.0
9	27.5	180	0.60	240.0	10-20	59.0
10	23.8	135	0.40	210.0	20-30	75.2
11	27.5	180	0.60	120.0	10-20	101.0
12	20.0	180	0.60	180.0	20-30	44.0
13	27.5	90	0.60	180.0	10-20	24.5
14	31.2	225	0.40	210.0	10-20	110.0
15	27.5	270	0.60	180.0	10-20	132.0
16	27.5	180	0.60	180.0	20-30	86.0
17	23.8	225	0.80	150.0	20-30	78.0
18	27.5	270	0.60	180.0	20-30	106.1
19	20.0	180	0.60	180.0	10-20	70.2
20	23.8	225	0.80	210.0	20-30	74.3
21	31.2	135	0.40	210.0	10-20	83.7
22	31.2	225	0.80	210.0	10-20	100.4
23	31.2	225	0.40	150.0	20-30	102.9
24	27.5	180	0.60	120.0	20-30	93.1
25	27.5	180	0.60	240.0	20-30	43.7
26	31.2	135	0.80	150.0	10-20	85.2
27	31.2	225	0.80	150.0	20-30	69.5
28	31.2	135	0.40	150.0	20-30	78.2
29	23.8	225	0.40	210.0	10-20	83.1
30	31.2	225	0.40	210.0	20-30	95.1
31	31.2	135	0.40	210.0	20-30	75.6
32	23.8	135	0.80	210.0	20-30	70.0
33	27.5	180	0.20	180.0	20-30	111.2
34	27.5	180	0.60	180.0	20-30	91.5
35	27.5	180	0.60	180.0	10-20	94.2
36	23.8	225	0.40	150.0	10-20	78.1
37	35.0	180	0.60	180.0	20-30	106.2
38	31.2	225	0.40	150.0	10-20	87.7
39	31.2	135	0.40	150.0	10-20	87.1
40	31.2	135	0.80	210.0	20-30	72.0
41	23.8	225	0.80	210.0	10-20	65.1
42	23.8	225	0.40	150.0	20-30	85.2
43	31.2	225	0.80	150.0	10-20	97.1
44	23.8	135	0.40	210.0	10-20	81.0
45	27.5	90	0.60	180.0	20-30	19.9
46	23.8	135	0.80	210.0	10-20	70.2
47	23.8	135	0.80	150.0	10-20	75.1
48	23.8	135	0.40	150.0	20-30	77.9
49	31.2	225	0.80	210.0	20-30	68.3
50	23.8	135	0.80	150.0	20-30	76.2
51	27.5	180	1.00	180.0	10-20	58.1
52	23.8	135	0.40	150.0	10-20	86.1

However, by considering of economical and feasibility aspect, it is not reasonable to use higher concentration, higher temperature and higher retention time. The used area in RSM design is usual concentration, temperature and time of adsorbent in the industrial application. In all cases, the efficiency of activated carbon with 20-30 mesh was higher than the GAC with 10-20 mesh. By decreasing of the adsorbent size, the surface area of the material improves and the adsorption sites reachability increases. The falling slopes of the diagrams shown in Fig. 3 indicate saturation of the adsorbent and reach to equilibrium.

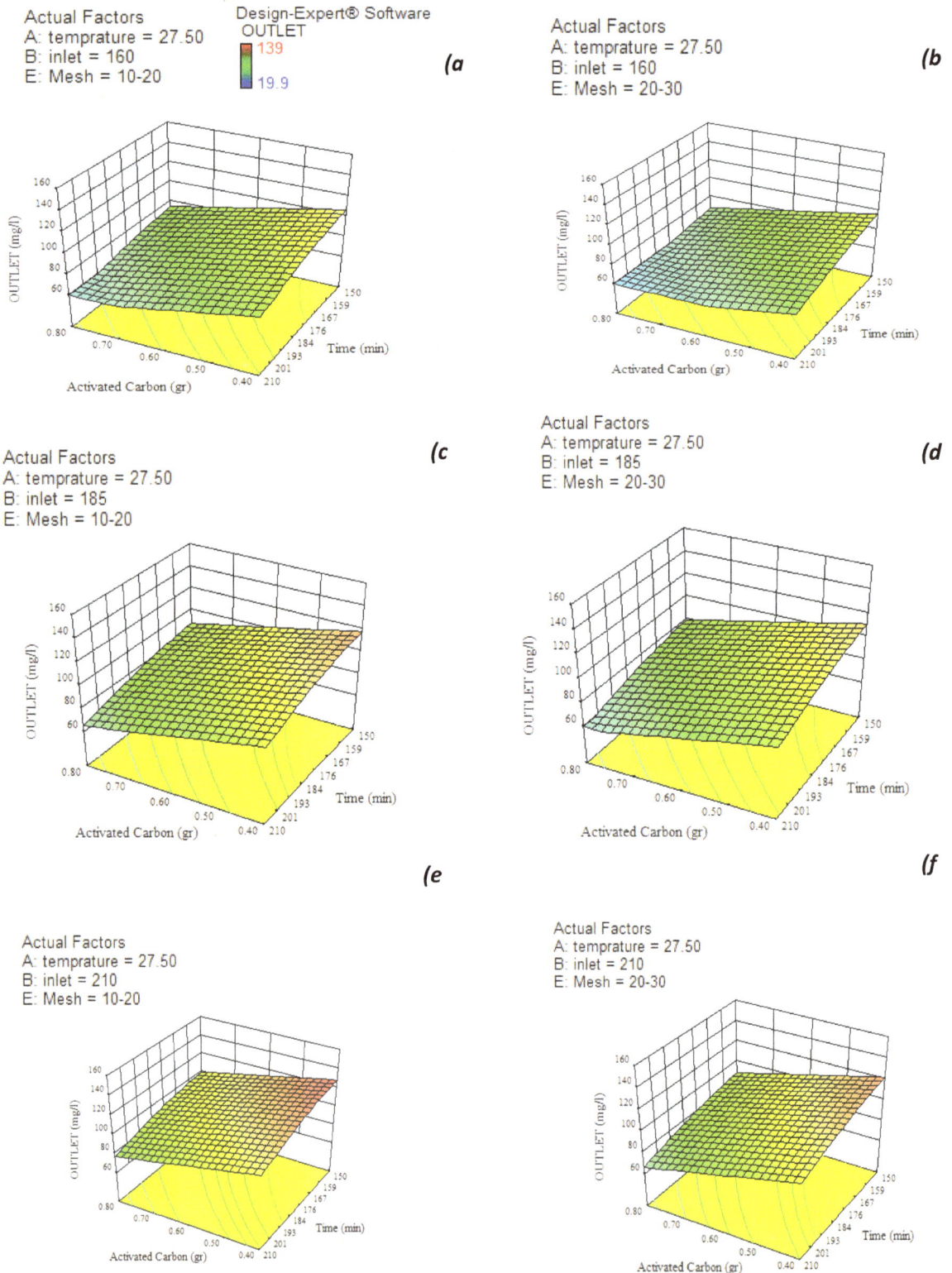

Fig. 3. Effects of activated carbon amount and retention time on the remained concentration of ethylene glycol in different levels of feed concentration and two different meshes.

Interactions of granular activated carbon and temperature were investigated for all three concentrations (160, 185 and 210 mg/L) and the results are presented in Fig. 4 (a-f). In all three concentrations, increment of activated carbon's amount and decrease in temperature result in higher ethylene glycol removal. It is probably due to the exothermic nature of the adsorption process (Vatanpour et al. 2018).

As it is expected, longer retention times and low temperatures improve the adsorption. In this case, the lower temperatures elevate the driving force needed to render inner porous areas available by improving diffusion phenomenon. Fig. 5 (a-f) shows how the outlet concentration responded to variations in retention time and temperature.

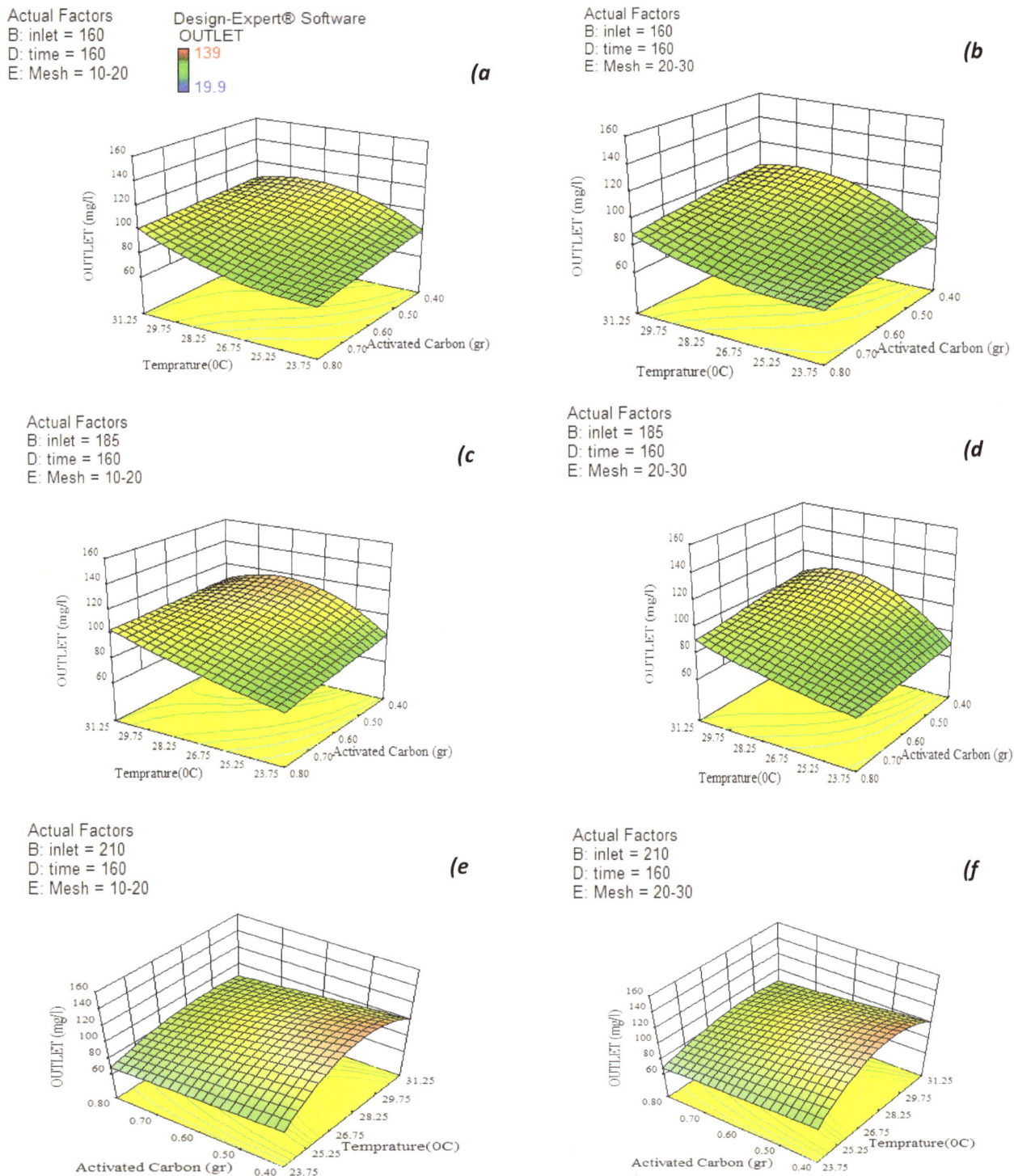

Fig. 4. Effects of activated carbon amount and temperature on the outlet concentration of ethylene glycol in different levels of feed concentration and two different meshes.

Interactions of granular activated carbon and temperature were investigated for all three concentrations (160, 185 and 210 mg/L) and the results are presented in Fig. 4 (a-f). In all three concentrations, increment of activated carbon's amount and decrease in temperature result in higher ethylene glycol removal. It is probably due to the exothermic nature of the adsorption process (Vatanpour et al. 2018).

As it is expected, longer retention times and low temperatures improve the adsorption. In this case, the lower temperatures elevate the driving force needed to render inner porous areas available by improving diffusion phenomenon. Fig. 5 (a-f) shows how the outlet concentration responded to variations in retention time and temperature.

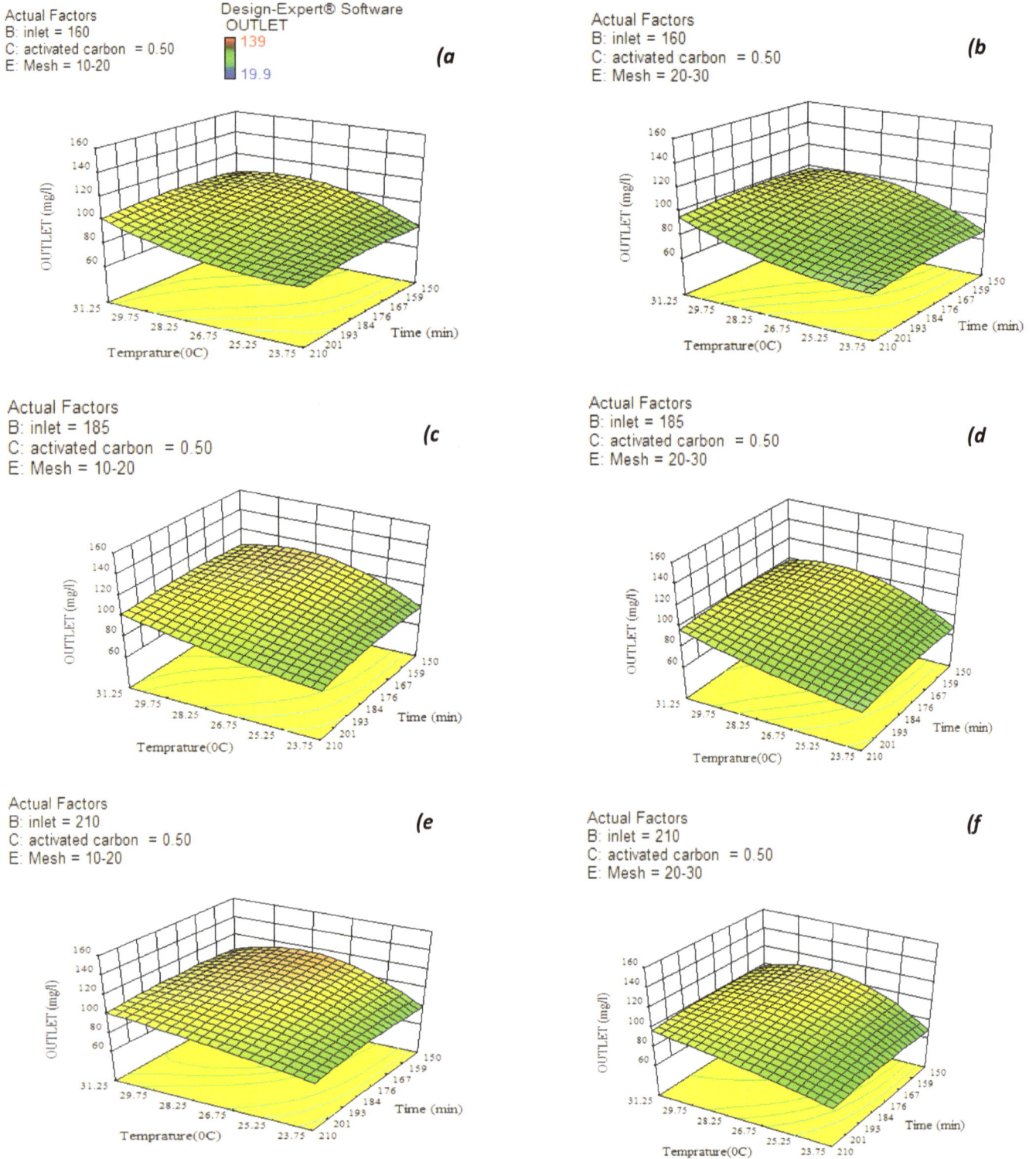

Fig. 5. Effects of retention time and temperature on the outlet concentration of ethylene glycol in different levels of feed concentration and two different meshes.

3.3. Optimization and validation
3.3.1. Optimization

Design-Expert 8.0.1 is equipped with an optimization toolbox was employed for optimizing the adsorption process of ethylene glycol. Each possible solution was evaluated and marked by a "Desirability" criteria ranging between 0 and 1. If a solution satisfies all conditions its "Desirability" would be 1 and 0 for the least desired situation when none

of the conditions, any other case would be between these two values. In this investigation, the main goal was to minimize ethylene glycol outlet concentration and limiting factors were not needed. Therefore, the optimization criteria for all factors were defined as "within range" and "minimize" was selected for remained concentration of ethylene glycol.

3.3.2. Validation

After optimization, the most desirable solution was selected which had a "Desirability" value of 0.995. At these conditions, the predicted response had a value of 31.20. In order to verify the cubic model assigned to the adsorption process, a validation experiment was carried out. Table 4 shows the acceptable agreement between the experimental data and the predicted values for remained concentration of ethylene glycol.

Table 4. Predicted optimized conditions and corresponding experimental validation.

Optimum condition								
Temperature (°C)	Feed conc. (mg/L)	GAC (g)	Time (min)	Mesh of GAC	predict	experiment	Low level	High level
27.7	135	0.80	210.0	20-30	27.1	31.2	20.1	34.1

Table 5. Freundlich and Langmuir isotherm parameters for adsorption of EG.

Isoterm	Parameters	
Langmuir	Q_m	25.38
	K_1	0.006
	R^2	0.9902
Freundlich	N	2.25
	K_F	1.44
	R^2	0.9671

3.4. Equilibrium isotherms

Conventional batch method was used to measure the equilibrium adsorption of ethylene glycol onto the GAC. Two adsorption isotherms, Langmuir and Freundlich models, were employed to explain the EG adsorption equilibrium. In this section, the investigation of ethylene glycol removal was carried out using 0.4 g of the adsorbent in the concentration range of 50-300 mg/L. Samples were kept in 25 °C for 2 h. To further investigate the case, equilibrium data obtained from the experiments were compared with Freundlich and Langmuir models.

The Langmuir isotherm is a theoretical model derived base on the assumptions that adsorption is monolayer, homogeneous and without lateral interactions between adsorbing species (Salehi et al. 2012). Considering linear form of Langmuir model and data at hand, correlation coefficients can be determined using these equations:

$$\frac{C_e}{q_e} = \frac{1}{K_1 Q_m} + \frac{C_e}{Q_m} \quad (4)$$

where Ce (mg/L) is equilibrium EG concentration and qe is the equilibrium adsorption capacity of the adsorbent. By determining the slope and the intercept of Langmuir isotherm (Qm and K1), the constants were calculated. In order to fit the experimental data, linear Freundlich equation was used:

$$\log(q_e) = \log(K_F) + \frac{1}{n_F} \log(C_e) \quad (5)$$

where n_F and KF are Freundlich constants, which depend on adsorbate and adsorbent and on the temperature which adsorption is taking place. Plotting log (qe) versus log (Ce) will result in a linear equation. The slope can be used to calculate n_F and the intercept reveals KF. Both Freundlich and Langmuir isotherm models exhibited acceptable correlation with the experimental data. Figs. 6 and 7 show linear approximations used to estimate contains for Langmuir and Freundlich isotherms. The calculated isotherm parameters for adsorption of EG were depicted in Table 5. The Langmuir isotherm provided much better fit to the equilibrium data compared with the Freundlich isotherm based on the higher R^2 values.

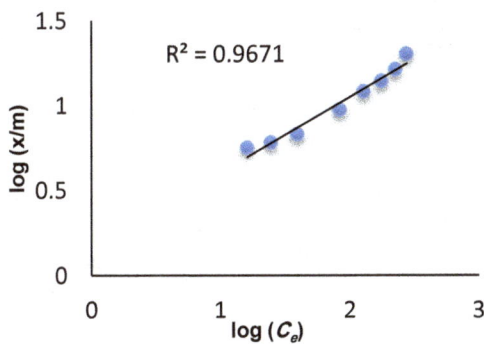

Fig. 6. Freundlich adsorption isotherm for EG on granular activated carbon.

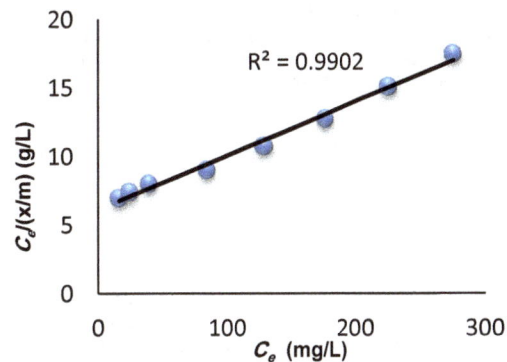

Fig. 7. Langmuir adsorption isotherm for EG on granular activated carbon.

4. Conclusions

In this research, efficiency of granular activated carbon for removal of EG was investigated, which it presented outstanding characteristics as a cheap adsorbent. Amount of activated carbon, temperature, retention time and feed concentration of ethylene glycol along side of different meshes of granular activated carbon were chosen as the factors influencing the adsorption process. Using these factors, a statistical modeling approach was taken to estimate the remained concentration of ethylene glycol. A series of central composite design based experiments were conducted using RSM. Observations showed that increment of retention time and amount of the adsorbent had positive effects on adsorption phenomenon while an increase in temperature resulted in remaining higher levels of ethylene glycol in the feed solution, i.e. decreasing EG removal. Maximum adsorption was achieved at 27.7 °C with 0.8 g of 20-30 mesh activated carbons for a feed concentration of 135 mg/L ethylene glycol at 210 minutes. Also, the kinetics of the adsorption process was studied and experimental data had a good agreement with Langmuir isotherm.

References

Anderson J., Walking like dinosaurs: Water, reuse and urban jungle footprints, in: Water Recycling Australia, AWA 2nd National Conference, Brisbane, Australia, 2003.

Arulkumar M., Sathishkumar P., Palvannan T., Optimization of Orange G dye adsorption by activated carbon of Thespesia populnea pods using response surface methodology, Journal of Hazardous Materials 186 (2011) 827-834.

Bansal R.C., Goyal M., Activated carbon adsorption, CRC press, 2005.

Bernhard M., Eubeler J.P., Zok S., Knepper T.P., Aerobic biodegradation of polyethylene glycols of different molecular weights in wastewater and seawater, Water Research 42 (2008) 4791-4801

Cadar O., Paul M., Roman C., Miclean M., Majdik C., Biodegradation behaviour of poly (lactic acid) and (lactic acid-ethylene glycol-malonic or succinic acid) copolymers under controlled composting conditions in a laboratory test system, Polymer Degradation and Stability 97 (2012) 354-35.

Carnegie D., Ramsay J., Anaerobic ethylene glycol degradation by microorganisms in poplar and willow rhizospheres, Biodegradation 20 (2009) 551-558.

Dwyer D.F., Tiedje J.M., Degradation of ethylene glycol and polyethylene glycols by methanogenic consortia, Appl. Environmental Microbiology 46 (1983) 185-190.

Dye R.F., Ethylene glycols technology, Korean Journal of Chemical Engineering18 (2001) 571-579.

Devlin J., Schwartz M., Ethylene glycol, Encyclopedia of toxicology, third ed. Elsevier Inc, (2014) 525-527.

Eisenreich S.J., Looney B.B., Thornton J.D., Airborne organic contaminants in the Great Lakes ecosystem, Environmental Science & Technology 15 (1981) 30-38.

Elreedy A., Tawfik A., Effect of Hydraulic Retention Time on Hydrogen Production from the Dark Fermentation of Petrochemical Effluents Contaminated with Ethylene Glycol, Energy Procedia 74 (2015) 1071-1078.

Esfandiar N., Nasernejad B., Ebadi T., Removal of Mn (II) from groundwater by sugarcane bagasse and activated carbon (a comparative study): Application of response surface methodology (RSM), Journal of Industrial and Engineering Chemistry 20 (2014) 3726-3736.

Farahani M.H.D.A., Borghei S.M., Vatanpour V., Recovery of cooling tower blowdown water for reuse: Theinvestigation of different types of pretreatment prior nanofiltrationand reverse osmosis, Journal of Water Process Engineering 10 (2016) 188–199.

Feng X., Chu K., Cost optimization of industrial wastewater reuse systems, Proc. Process Safety and Environmental Protection 82 (2004) 249-255.

Gonzalez C.F., Taber W.A., Zeitoun M., Biodegradation of ethylene glycol by a salt-requiring bacterium, Applied microbiology, 24 (1972) 911-919.

Hajati S., Ghaedi M., Yaghoubi S., Local, cheep and nontoxic activated carbon as efficient adsorbent for the simultaneous removal of cadmium ions and malachite green: Optimization by surface response methodology, Journal of Industrial and Engineering Chemistry 21 (2015) 760-767.

Haines J., Alexander M., Microbial degradation of polyethylene glycols, Journal of Applied Microbiology 29 (1975) 621-625.

Hameed B., Tan I., Ahmad A., Preparation of oil palm empty fruit bunch-based activated carbon for removal of 2, 4, 6-trichlorophenol: Optimization using response surface methodology, Journal of Hazardous Materials 164 (2009) 1316-1324.

Huang Y.-L., Li Q.-B., Deng X., Lu Y.-H., Liao X.-K., Hong M.-Y., Wang Y., Aerobic and anaerobic biodegradation of polyethylene glycols using sludge microbes, Process Biochemistry 40 (2005) 207-211.

Jaria G., Calisto V., Silva C.P., Gil M.V., Otero M., Esteves V.I., Obtaining granular activated carbon from papermill sludge– A

challenge for application in the removal of pharmaceuticals from wastewater, Science of the Total Environment 653 (2019) 393–400.

Mason R.L., Gunst R.F., Hess J.L., Statistical design and analysis of experiments: with applications to engineering and science, John Wiley & Sons, 2003.

McGahey C., Bouwer E., Biodegradation of ethylene glycol in simulated subsurface environments, Water Science & Technology 26 (1992) 41-49.

McQuillan R.V., Stevens G.W., Mumford K.A., The electrochemical regeneration of granular activated carbons: A review, Journal of Hazardous Materials 355 (2018) 34–49.

McVicker L., Duffy D., Stout V., Microbial growth in a steady-state model of ethylene glycol-contaminated soil, Current Microbiology 36 (1998) 136-147.

Meyer J.L., Sale M.J., Mulholland P.J., LeRoy Poff N., Impacts of climate change on aquatic ecosystem functioning and health, JAWRA Journal of the American Water Resources Association 35 (1999) 1373-1386.

Mohsen M.S., Jaber J.O., Potential of industrial wastewater reuse, Desalination 152 (2003) 281-289.

Pereira L.S., Oweis T., Zairi A., Irrigation management under water scarcity, Agricultural Water Management 57 (2002) 175-206.

Petrinic I., Korenak J., Povodnik D., Hélix-Nielsen C., A feasibility study of ultrafiltration/reverse osmosis (UF/RO)-based wastewater treatment and reuse in the metal finishing industry, Journal of Cleaner Production 101(2015) 292-300.

Rebhun M., Engel G., Reuse of wastewater for industrial cooling systems, Journal of the Water Pollution Control Federation 60 (1988) 237-241.

Revitt D., Worrall P., Low temperature biodegradation of airport de-icing fluids, Water Science & Technology 48 (2003) 103-111.

Sahu J., Acharya J., Meikap B., Response surface modeling and optimization of chromium (VI) removal from aqueous solution using Tamarind wood activated carbon in batch process, Journal of Hazardous Materials 172 (2009) 818-825.

Salehi E., Madaeni S., Rajabi L., Vatanpour V., Derakhshan A., Zinadini S., Ghorabi S., Monfared H.A., Novel chitosan/poly (vinyl) alcohol thin adsorptive membranes modified with amino functionalized multi-walled carbon nanotubes for Cu (II) removal from water: preparation, characterization, adsorption kinetics and thermodynamics, Separation and Purification Technology 89 (2012) 309-319.

Sowers J., Vengosh A., Weinthal E., Climate change, water resources, and the politics of adaptation in the Middle East and North Africa, Climatic Change 104 (2011) 599-627.

Staples C.A., Williams J.B., Craig G.R., Roberts K.M., Fate, effects and potential environmental risks of ethylene glycol: a review, Chemosphere 43 (2001) 377-383.

Straß A., Schink B., Fermentation of polyethylene glycol via acetaldehyde in Pelobacter venetianus, Applied Microbiology and Biotechnology 25 (1986) 37-42.

Vatanpour V., Salehi E., Sahebjamee N., Ashrafi M., Novel chitosan/polyvinyl alcohol thin membrane adsorbents modified with detonation nanodiamonds: Preparation, characterization, and adsorption performance, Arabian Journal of Chemistry (2018), In press.

Vatanpour V., Sheydaei M., Esmaeili M., Box-Behnken design as a systematic approach to inspect correlation between synthesis conditions and desalination performance of TFC RO membranes, Desalination 420 (2017) 1–11.

Wongkaew K., Wannachod T., Mohdee V., Pancharoen U., Arpornwichanop A., Lothongkum A.W., Mass transfer resistance and response surface methodolog y for separation of platinum(IV) across hollow fiber supported liquid membrane, Journal of Industrial and Engineering Chemistry 42 (2016) 23–35.

Wu D., Martin J., Du J., Zhang Y., Lawless D., Feng X., Thin film composite membranes comprising of polyamide and polydopamine for dehydration of ethylene glycol by pervaporation, Journal of Membrane Science 493 (2015) 622-635.

Yang Y., Wang W., Wang F., Liu Y., Chai D., Lei Z., Partially oxidized NiFe alloy: An effective promoter to enhance Pd electrocatalytic performance for ethylene glycol oxidation, International Journal of Hydrogen Energy 40 (2015) 12262-12267.

Zgoła-Grześkowiak A., Grześkowiak T., Zembrzuska J., Łukaszewski Z., Comparison of biodegradation of poly (ethylene glycol)s and poly (propylene glycol)s, Chemosphere 64 (2006) 803-809.

22

Kinetics of photocatalytic degradation of reactive black B using core-shell TiO$_2$-coated magnetic nanoparticle, Fe$_3$O$_4$@SiO$_2$@TiO$_2$

Saeed Aghel, Nader Bahramifar*, Habibollah Younesi

Department of Environmental pollution, Faculty of Natural Resources & Marine Science, Tarbiat Modares University, Tehran, Iran.

ARTICLE INFO

Keywords:
Photocatalytic Degradation
kinetics
Reactive Black B
magnetic nanoparticle

ABSTRACT

In particular, the inappropriate/inevitably discharge of dye-containing effluents is undesirable because of their color, resistant to the biological treatment systems, toxic, and their carcinogenic or mutagenic nature to life forms. About fifty percent of the washing dye liquor is discharged into the water environment. Fe$_3$O$_4$@SiO$_2$@TiO$_2$ nanoparticles, were prepared, characterized and tested as photocatalyst in the removal of Reactive Black B (RBB) dye by a photocatalytic process. The effect of photocatalyst concentration, pH and temperature in the photodegradation kinetics is discussed in terms of the Langmuir–Hinshelwood (L–H) model. SEM and TEM characterizations confirmed the Fe$_3$O$_4$@SiO$_2$@TiO$_2$ nanoparticel, revealed that the obtained particles a spherical morphology with sizes about 100 nm. The DRS pattern of Fe$_3$O$_4$@SiO$_2$@TiO$_2$ shows the energy band gap value of photocatalyst is 2.75 eV. The presence of Fe$_3$O$_4$, SiO$_2$ and anatase TiO$_2$ in the as-synthesis magnetic nanoparticle were confirmed by FTIR and XRD analysis. The Fe$_3$O$_4$@SiO$_2$@TiO$_2$ photocatalyst in combination with ultraviolet irradiation and under optimal conditions can destroy 100% of RBB after 120 min. Furthermore, the magnetic photocatalyst was efficiently separated from the solution with the help of a magnet and shown the capable of reusability up to 10 times without reducing their efficiency.

---

1. Introduction

Textile industries widely use complex synthetic organic colorants as coloring agents which produce colored wastewater, since the dye is not absolutely adsorbed by the textile. The release of this wastewater to the environment is Dangerous to aquatic life (King-Thom. 1993) and mutagenic to humans (Lucas and Peres. 2006). These textile wastewater, usually with dye contents about 10–200 mg/L (F.P. van der Zee. 2002) create a bulky problem to wastewater treatment plants in the whole world. The customary treatment methods usually applied to textile wastewaters, such as flocculation/coagulation, adsorption on the surface of activated carbon or membrane separation, only lead to the phase transference of the contaminant from liquid phase to the solid phase (Lucas et al. 2007). Biological treatments are also not a good solution because of the biological resistance of most pigment. Therefore, the removal of these dyes before release into the environment is a precedence. Advanced oxidation processes (AOPs) arose as a suitable method for the degradation of organic pollutants in wastewater (Litter. 1999).

AOPs inclusive Fenton reaction, Ozonation, photolysis, wet air oxidation, ultrasounds and photocatalysts. Photocatalysts are a wide application in the total degradation of organic pollutants into CO$_2$ and H$_2$O (Xu. 2007; Beydoun et al. 2000) Among all semiconductor photocatalysts, TiO$_2$ is the widely applied as photocatalyst because it is non-toxic, chemically stable, inexpensive and its photogenerated electrons and holes are highly reducing and oxidizing, respectively (Litter. 1999). A Usual method to separation the catalyst in slurry-type reactors is by sedimentation of TiO$_2$ particles after pH adjustment followed by a flocculation–coagulation process; however, these processes are expensive in terms of time, manpower and reagents. One Technique to overcome the separation problem is through the making of titanium dioxide with magnetic properties, Magnetic separation constitutes a more sustainable process since it prevents catalyst mass losses and the use of additional solvents. It has been established that the direct contact between the TiO$_2$ shell and the iron oxide magnetic cores may lead to low photoactivity because Fe$_3$O$_4$ act as recombination centers for electrons and positive holes, especially in the case of Fe$_3$O$_4$ (Beydoun et al. 2000) The use of a buffer SiO$_2$ layer between the magnetic core and the TiO$_2$ shell, i.e. Fe$_3$O$_4$@SiO$_2$@TiO$_2$, improves the photocatalytic activity of the nanomaterial by preventing the injection of charges from TiO$_2$ shell to the magnetic cores (Chen et al. 2001) In this study, the kinetic and equilibrium data were modeled using Langmuir–Hinshelwood (L–H) model.

2. Materials and methods
2.1. Materials and reagents

Hydrochloric acid (HCl), Iron (III) chloride hexahydrate, sodium acetate anhydrous (NaAc), tetrabutyl orto titanate (TBOT, 97 %), n-hexane, ethylene glycol, Ethylene Diamine Tetra Acetic acid disodium salt (EDTA-2Na) and sodium silicate were purchased from Merck Company. during all steps when water was needed, deionized water was used throughout.

2.2. Synthesis of Fe$_3$O$_4$

Fe$_3$O$_4$ magnetic Nano particles were synthesized according to the method reported by Lin et al. 2013. first (2.5 mmol, 0.68 g) FeCl$_3$·6H$_2$O, and (0.015 mol, 1.2 g) NaAc were dissolved in (20 ml) ethylene glycol under magnetic stirring until the solution became clear, then (0.01 mmol, 0.034 g) EDTA-2Na was add to this solution and sonicate at 37 Hz for 30 min, the resulting was transferred to a Teflon-lined stainless-steel autoclave. The autoclave was sealed and heated at 180 −200 °C for 8–10 h and naturally cooled to room temperature. After that, the black particles were washed with deionized water and ethanol three times and then dried under vacuum at 50 °C (Lin et al. 2013).

---

*Corresponding author Email: n.bahramifar@modares.ac.ir

2.3. Synthesis of Fe_3O_4@SiO_2

For Synthesis of Fe_3O_4@SiO_2 Briefly 1.3 g of sodium silicate was dissolved in 100 ml deionized water (which was heated to 80 °C) under magnetic stirring to form a clear solution. then 0.3 g of Fe_3O_4 put into the solution. The pH value of the mixture was adjusted to 6.5 with 2 mol/L HCl solution. The mixture was further stirred at 80 °C for 180 min. The resulting silica-coated Fe_3O_4 nanoparticles were collected with the help of a magnet and washed with deionized water and, followed by drying in Freeze Dryer for 24 h (Wang et al. 2010).

2.4. Synthesis of Fe_3O_4@SiO_2@TiO_2

The preparation procedure of Fe_3O_4@SiO_2@TiO_2 described by Lirong et al. 2014. Briefly 0.2 g as-prepared Fe_3O_4@SiO_2 particles were dispersed in a mixture of hexane (70 ml) and deionized water (0.2 ml), followed by the addition of TBOT (0.5ml) under ultrasonication treatment at 50 Hz for 90 min. The mixture was transferred to Teflon-lined autoclave. It was heated at 100 °C for 180 min. The precipitates were collected by magnet and washed with hexane three times, then dried at room temperature. the obtained Fe_3O_4@SiO_2@TiO_2 particles were calcined at 500 °C for 180 min (Lirong et al. 2014).

2.5. Synthesis of Fe_3O_4@SiO_2@TiO_2

For assessment of the photocatalytic activity of the Fe_3O_4@SiO_2@TiO_2 nanoparticel, we used a solution of RBB. The photocatalytic experiments were done in beaker 400ml including 100 ml of 50 mg/L dye solution. For unlimbering favorite Ultraviolet photon, we implemented UV lamps of 8 W (Philips, Holland) in Quartz chamber in beaker Content of 50 mg/L RBB. In order to make the solution homogenous used the magnetic stirrer during the irradiation. At the beginning of each experiment, we turned off the UV lamps for 15 minutes and then we turned them on. Samples of the solution were obtained after, 15, 30, 45,60,75,90,105,120 min.

2.6. Parameter effects on the removal of RBB

The effects of the Fe_3O_4@SiO_2@TiO_2 dose (50, 100, 150 and 200 mg/l) on the removal of RBB were Examined. In the photocatalytic process, pH is one of the most important operating parameters that affect the charge on the catalyst. The effect of different initial pH on the dye removal was investigated by mixing 150 mg of Fe_3O_4@SiO_2@TiO_2 photocatalyst dose at 25 °C for 120 min. The pH was adjusted to values 3-9 (from acidic to basic) using 1 M HCl and 1 M NaOH solution.

2.7. Kinetics of photocatalytic degradation of RBB

For investigation photodegradation reaction, we use the Langmuir–Hinshelwood model to description the initial rates of photodegradation of dye.

$$r_0 = \frac{-dC}{dt} = \frac{k_r KC}{1+KC} \quad k_r KC = k_{app}C \tag{1}$$

where r_0 is the initial rate of the removal of the RBB. t the reaction time and, C, the equilibrium bulk-solute dose. K indicant the equilibrium constant for absorption of dye on the surface of photocatalyst and kr represent the limiting rate constant of the reaction at maximum coverage under the given experimental conditions. In the case of highly diluted solution and the term KC becomes less than 1, when the denominator of Eq. (1) neglected and the rate data can be modeled by the apparent first-order kinetics as in the following equation:

$$r = \frac{-dC}{dt} = k_r KC = k_{app}C \tag{2}$$

where C is the dye concentration at time t, and kapp the apparent first-order rate constant. Integrating Eq. (2) And using boundary condition C= C_0 at t = 0 gives:

$$\ln\left(\frac{c_t}{c_0}\right) = -k_{app}t \tag{3}$$

where C_0 is the dye concentration at the initial time.

2.8. Physico-chemical characterization

X-ray diffraction (XRD) of the products were obtained on Philips Xpert MPD diffractometer. The morphology and microstructure of products were characterized by scanning electron microscopy (SEM, LEO,1455VP, Cambridge, U.K) and CM120 transmission electron microscopy (TEM). Fourier Transform Infrared (FT-IR) spectra were recorded on (Shimadzo, FT_IR1650 spectrophotometer, Japan) using KBr pellets for samples. The magnetic properties of the photocatalyst were quantified using a Vibrating Sample Magnetometer (VSM) at a temperature range of 1.8 to 310 K with a Meghnatis Daghigh Kavir Co. UV-Vis absorption spectra of products were recorded by a Diffuse Reflectance Spectroscopy (DRS) V/650 spectrophotometer (Jasco Inc., Japan) with a wavelength range from 220 to 1020 nm. XRD patterns of Fe_3O_4, Fe_3O_4@SiO_2 and Fe_3O_4@SiO_2@TiO_2 are shown in Fig1. As shown, in general, all the powders are well-crystalline materials. For Fe_3O_4 core, seven diffraction peaks were observed in 2θ =18.5, 30.5, 35.5, 43.2, 53.5, 57.1 and 62.6 (Fig. 1a). This pattern is corresponding to crystalline Fe_3O_4 magnetic nanostructures. (Wang et al. 2012) The XRD pattern of the Fe_3O_4@SiO_2 sample shows almost the same feature as pure Fe_3O_4, except that a broad peak centered at 15-25 of 2θ corresponding to SiO_2 was observed, indicating that the prepared SiO_2 is amorphous (Jian et al. 2012) (Fig. 1b). In the XRD results for Fe_3O_4@SiO_2@TiO_2 some peaks were emerged that is related to TiO_2 semiconductor. The emerged peak at 2θ = 25.37, 38.11, 48.07, 54.14 is shown the anatase phase for TiO_2 (Jian et al. 2012) (Fig. 1C). In addition, no other peak is observed belonging to any adsorbed impurities or phase in the sample structure. Moreover, the sharp diffraction peaks show that the obtained nanoparticles have high crystallinity.

FT-IR measurements were performed for Fe_3O_4, Fe_3O_4@SiO_2 and Fe_3O_4@SiO_2@TiO_2 samples as shown in Fig. 2 all spectra present absorption peak at 578 cm^{-1}, corresponding to the Fe–O vibration from the magnetite phase (Yamaura et al. 2004). Spectrum of Fe_3O_4@SiO_2 present the typical Si-O-Si bands of the inorganic symmetric vibration modes around 786 cm^{-1}, asymmetric stretching vibration around 1033 – 1100 cm^{-1} and the band at 964 cm^{-1} is assigned to the Si-O stretch that indicates the silica layer around the Fe_3O_4 (Innocenzi. 2003; Pillay et al. 2013). The Fe_3O_4@SiO_2@TiO_2 spectra are shown in Fig. 2C. The peak at 500-850 cm^{-1}, which corresponds to the Ti-O band and new absorption band covering a range from 859 to 1087 cm-1, which corresponds to the stretching vibration of Ti-O-Si, was appeared in the FT-IR spectra of Fe_3O_4@SiO_2@TiO_2 as compared to Fe_3O_4@SiO_2 (Ghasemi et al. 2016).

Therefore, the photocatalyst is not a simple mechanical mixture of TiO_2 anatase and Fe_3O_4@SiO_2 but it is a nanocomposite regarding the formation of mentioned new bonds. The morphologies of Fe_3O_4@SiO_2@TiO_2 powder were studied by scanning electron microscopy illustrated in Fig 3. In SEM, a focused electron beam scans conductive sample surface reveals information about the sample including external morphology (texture) and topography. This analysis reveals that for Fe_3O_4 synthesized by Solvothermal method, particles almost are nano-size with mediocre size about 80-90 nm (Fig. 3A). When the surface of Fe_3O_4@SiO_2 particles was coated with layers of TiO_2 via sol–gel procedures, the roughness and the size of particles have been increased to mediocre size about 100 nm for Fe_3O_4@SiO_2@TiO_2 nanostructures (Fig. 3B).

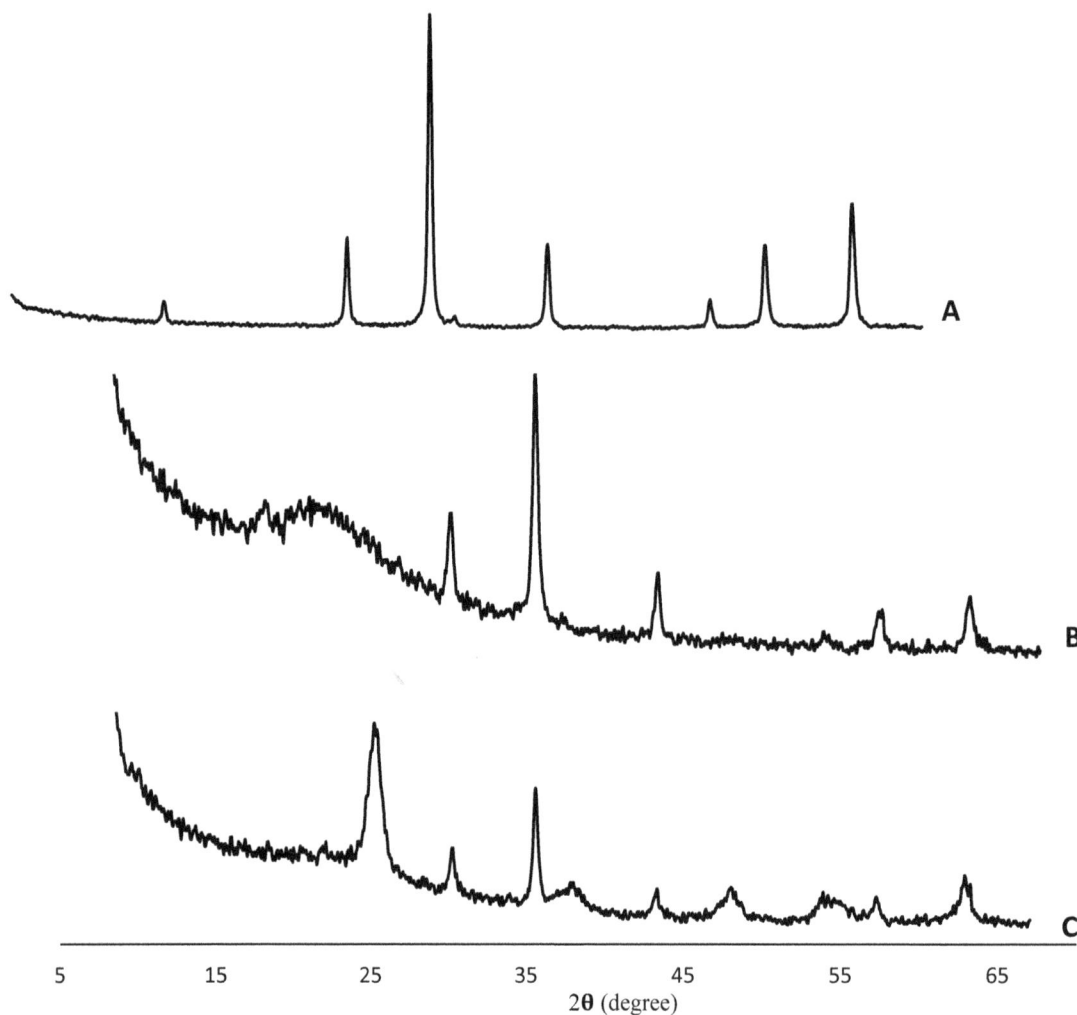

Fig 1. XRD pattern of Fe_3O_4 (A), $Fe_3O_4@SiO_2$ (B) and $Fe_3O_4@SiO_2@TiO_2$ composites (C).

The TEM images of the $Fe_3O_4@SiO_2@TiO_2$ nanoparticles are shown in Fig. 4, The core-shell structure can be clearly distinguished because of the different color contrast between the cores and shells. It can be seen that the shape of nanoparticles is almost spherical with an average diameter of 100 nm.

For studying magnetization treatment of the as-prepared nanoparticles, the magnetization curve of Fe_3O_4, silica coated Fe_3O_4 and $Fe_3O_4@SiO_2@TiO_2$ samples was measured at room temperature, as shown in Fig3. Results indicate that the magnetic hysteresis loops are S-like curves. Also, the magnetic remanence of the all samples was nearly zero, suggesting that samples exhibit a superparamagnetic behavior. The specific saturation magnetizations (Ms) are 76, 23 and 12 emu/mg for Fe_3O_4, $Fe_3O_4@SiO_2$ and $Fe_3O_4@SiO_2@TiO_2$ samples, respectively. The reduction in the value of Ms could be attributed to the rather smaller size of the Fe_3O_4 nanoparticles and the added mass of some layers which were nonmagnetic on them. Due to this magnetic property, $Fe_3O_4@SiO_2@TiO_2$ could move regularly under the action of an external magnet after they congregated.

The absorption coefficient and optical band gap of a material are two important parameters which controlling a photocatalytic activity and this feature relevant to the electronic structure of the material. It can be seen from the Fig4 that the absorption is around 450 nm for $Fe_3O_4@SiO_2@TiO_2$, this absorption extends into the visible region. The band gap energy value of corresponding spectrum was calculated using the equation $E_{bg} = 1239.8/\lambda$ nm (Velmurugan et al.2011) The band gap energy value of for $Fe_3O_4@SiO_2@TiO_2$ is 2.75 eV.

3. Results and discussion
3.1. Effect of photocatalyst concentration

The effect of $Fe_3O_4@SiO_2@TiO_2$ dosage on the rate of dye removal was Investigated (Fig. 7A). The first-order rate constant Was obtained for photocatalyst dosage of 50, 100, 150 and 200 mg/l (Fig. 7B). The design of $\ln(C_t/C_o)$ against reaction time gives a good linear relevance (with $R^2 > 0.96$), which is witness of the good agreement of fitting the reaction data in first-order reaction. According the dosage of the $Fe_3O_4@SiO_2@TiO_2$ and their rate constants in the dye removal reaction (Fig. 7B), it can be said the concentration of the $Fe_3O_4@SiO_2@TiO_2$ Positive correlation with photocatalytic operation, it shows that removal of RBB is dependent on the number of available electron hole pair. The photocatalytic removal of dye in numerous cases Revealed this behavior (Chen and Liu. 2007). The maximum photocatalytic degradation efficiency was obtained at a photocatalyst dosage of 200 mg/l. It has been found that the rate of photocatalytic degradation increases with increasing for photocatalyst dosage from 50 to 200 mg/l; it is clear that k_{app} increased with an increase in $Fe_3O_4@SiO_2@TiO_2$ dosage.

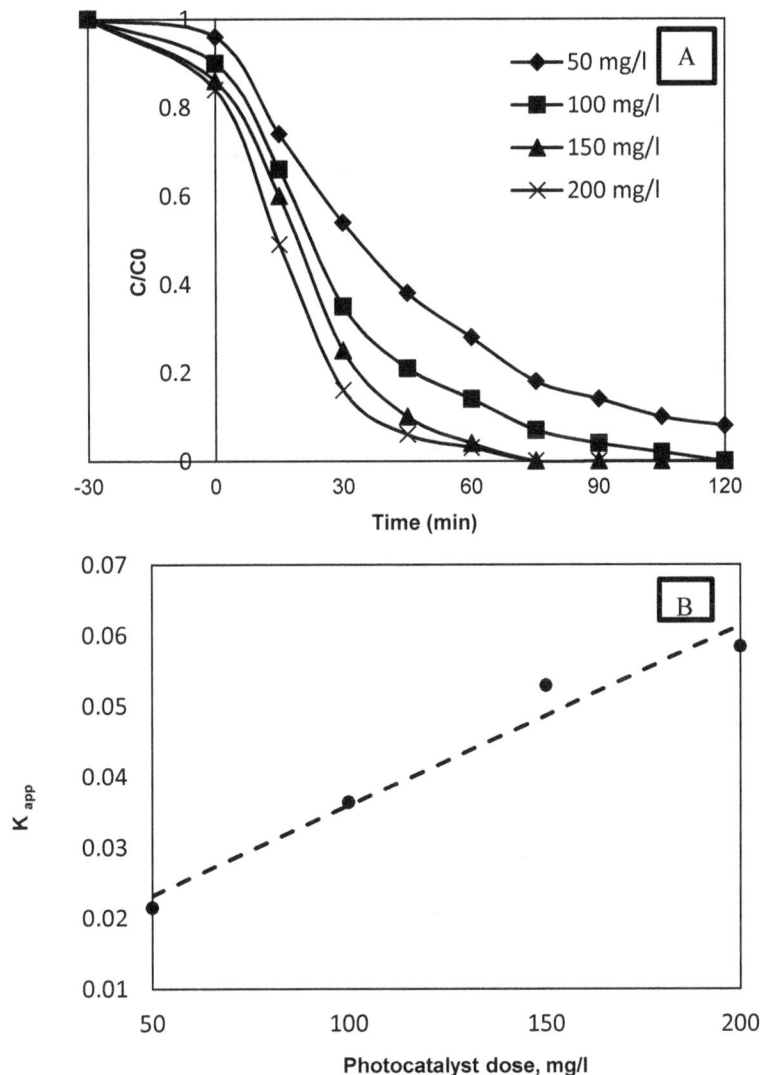

Fig 7. Effect of photocatalyst concentration on dye removal rate (A) and effect of different amounts of photocatalyst concentration on the kinetic rate constant (B).

3.2. Effect of pH

The pH value of the Solution is an essential operational factor on the surface charge properties of the TiO_2, the absorption behavior of dye and removal of dye taking place on the surface of $Fe_3O_4@SiO_2@TiO_2$ Nanoparticle (Khodadoust et al. 2012). Therefore, it is important to study the role of pH on Photocatalytic process and specify the optimal pH for dye removal (Diya'uddeen et al. 2011). Fig. 8A shows the efficiency of $Fe_3O_4@SiO_2@TiO_2$ Nanoparticle for dye removal as a function of pH under using 150 mg/l photocatalyst dosage at a temperature of 25 °C for 120 min. The apparent first-order rate constant for the reaction of $Fe_3O_4@SiO_2@TiO_2$ with dye decreased linearly with pH increasing (Fig. 8B). The maximum efficiency of removal of RBB 100 % was observed at pH 3. The influence of pH on the photocatalytic processes can be described on the basis of the point of zero charge (PZC) of TiO_2 and the absorption of the Organic materials on the TiO_2 in different pH values (Evgenidou et al. 2005). According to the PZC of TiO_2 catalyst, its surface charge is positive in acidic solution and negative in basic solution, respectively (Malato et al. 2009). The RBB is an anionic species of dye, the adsorption of RBB on the surface of TiO_2 is better in acidic pH (Lucas et al. 2013).

3.3. Effect of temperature

The effect of temperature on the dye removal rate was significant (Fig. 9A). The results showed that the dye removal rate Decrease as the temperature rose. The photocatalytic activity of $Fe_3O_4@SiO_2@TiO_2$ was tested at 150 mg/l photocatalyst concentration and pH under varying temperatures from 15 to 65 °C. The maximum removal efficiency was achieved at a temperature of 15 °C. The apparent first-order rate constant for the reaction of $Fe_3O_4@SiO_2@TiO_2$ with dye Decrease linearly with increasing temperature (Fig. 9B), which further confirms that the dye removal is not an Endothermic reaction.

3.4. Studies upon photocatalyst recycling

The Durability and Reusability of the magnetic nanoparticle upon several times is a key issue, since one of the basic Bugs of non-magnetic photocatalyst process is the catalyst separating and reuse. The Greatest advantage of the magnetic photocatalyst is the easily separation after the photocatalytic Process. The $Fe_3O_4@SiO_2@TiO_2$ was separated and reused in further 10 times. Fig. 10 summarizes the efficiency of dye removal (in percentage).

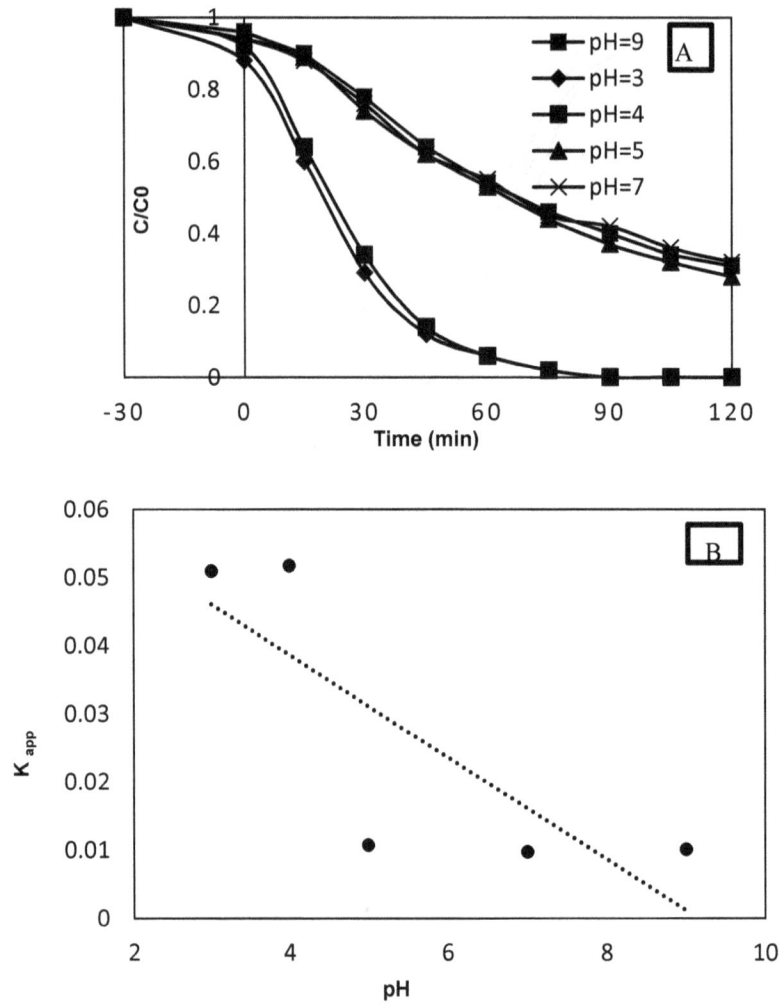

Fig 8. Effect of pH on dye removal rate(A) and photocatalytic rate constant as a function of pH(B).

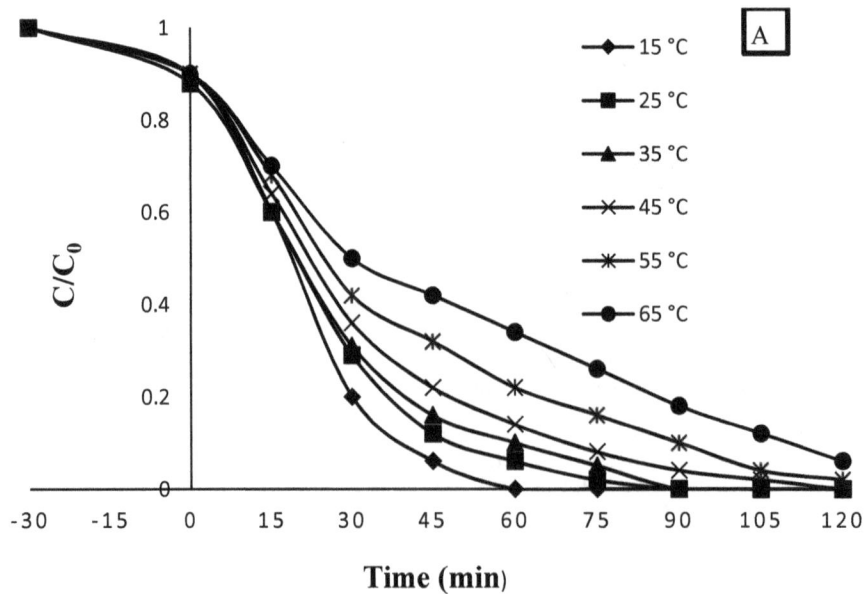

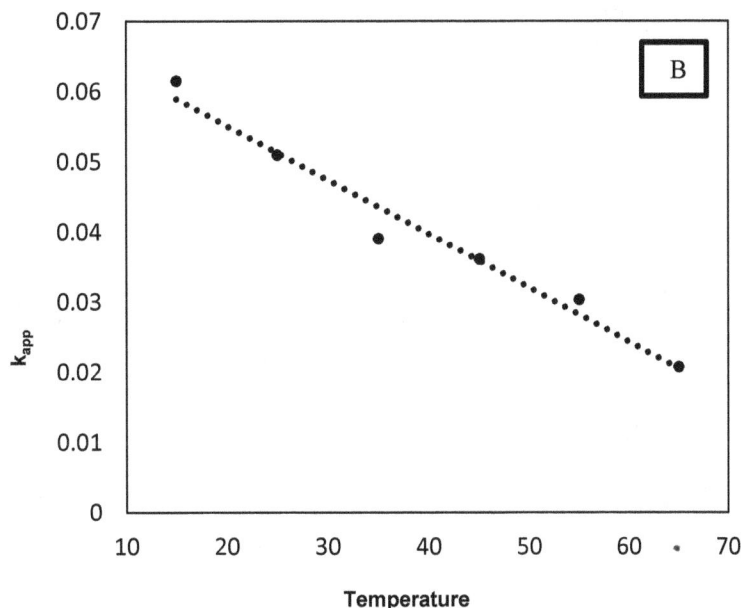

Fig 9. (A) Effect of temperature on dye removal rate and (B) photocatalytic rate constant as a function of temperature.

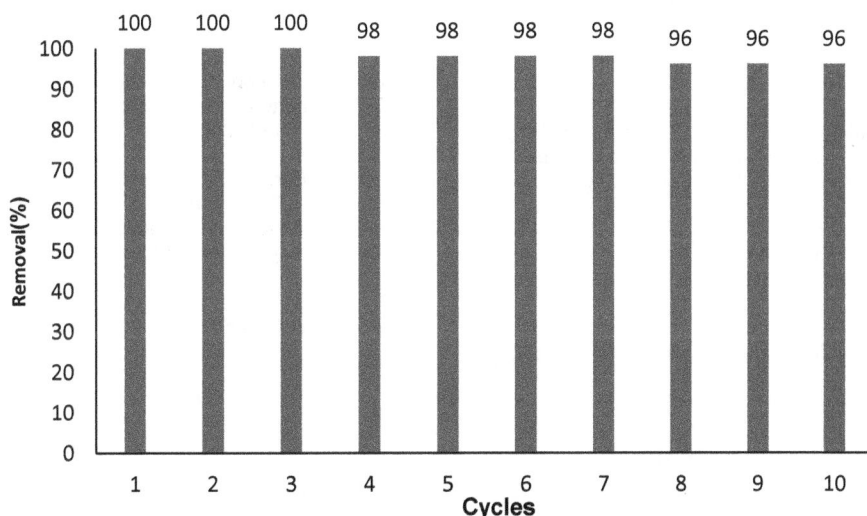

Fig 10. Reusability of photocatalyst for dye removal.

4. Conclusions

A magnetic nanoparticel Fe_3O_4@SiO_2@TiO_2 was successfully synthesized and tested in the RBB removal. The fabricated materials were used in the photodegradation of RBB under UV light irradiation, and it was shown that nanospheres had high photodegradation efficiency. In addition, the current study reveals that the fabricated nanospheres displayed good magnetic properties at room temperature, which can create a fast separation photocatalyst after reaction. The results indicate that after ten cycles of use, the high photocatalytic activity of the Fe_3O_4@SiO_2@TiO_2 nanocatalyst did not decrease. Therefore, the nanocatalyst are promising agents and highly beneficial for various potential applications for the treatment of textile effluents containing harmful organic dyes. The dye removal rate increased with the increasing of Fe_3O_4@SiO_2@TiO_2 dosage to 200 mg/l. pH value of 3 was determined to be Optimize for dye removal and removal rate was much faster at pH less than 4 compared with higher.

References

Beydoun D., Amal R., Low G.K.C., McEvoy S., Novel photocatalyst: titania-coated magnetite. Activity and photodissolution, The Journal of Physical Chemistry B 104 (2000) 4387-4396.

Chen F., Xie Y., Zhao J., Lu G., Photocatalytic degradation of dyes on a magnetically separated photocatalyst under visible and UV irradiation, Chemosphere 44 (2001) 1159-1168.

Chen S., Liu Y. Study on the photocatalytic degradation of glyphosate by TiO$_2$ photocatalyst, Chemosphere 67 (2007) 1010-1017.

Evgenidou E., Fytianos K., Poulios I., Photocatalytic oxidation of dimethoate in aqueous solutions, Journal of Photochemistry and Photobiology A: Chemistry 175 (2005) 29-38.

Ghasemi Z., Younesi H., Zinatizadeh A.A., Preparation, characterization and photocatalytic application of TiO$_2$/Fe-ZSM-5 nanocomposite for the treatment of petroleum refinery wastewater: Optimization of process parameters by response surface methodology, Chemosphere 159 (2016) 552-564.

Innocenzi P., Infrared spectroscopy of sol–gel derived silica-based films: a spectra-microstructure overview, Journal of Non-Crystalline Solids 316 (2003) 309-319.

Jian G., Liu Y., He X., Chen L., Zhang Y., Click chemistry: a new facile and efficient strategy for the preparation of Fe$_3$O$_4$ nanoparticles covalently functionalized with IDA-Cu and their application in the depletion of abundant protein in blood samples, Nanoscale 4 (2012) 6336-6342.

Khodadoust S., Sheini A., Armand N. Photocatalytic degradation of monoethanolamine in wastewater using nanosized TiO$_2$ loaded on clinoptilolite, Spectrochimica Acta Part A: Molecular and Biomolecular Spectroscopy 92 (2012) 91-95.

King-Thom C., Degradation of azo dyes by envi ronmental microorganisms and helminthes, Environmental Toxicology and Chemistry 13 (1993) 2121-2132.

Lin M., Huang H., Liu Z., Liu Y., Ge J., Fang Y., Growth–dissolution–regrowth transitions of Fe$_3$O$_4$ nanoparticles as building blocks for 3D magnetic nanoparticle clusters under hydrothermal conditions, Langmuir 29 (2013) 15433-15441.

Lirong M., Jianjun S., Ming Z., Jie H., Synthesis of Magnetic Sonophotocatalyst and its Enhanced Biodegradability of Organophosphate Pesticide, Bulletin of the Korean Chemical Society 35 (2014) 21-35.

Litter M.I., Heterogeneous photocatalysis: transition metal ions in photocatalytic systems. Applied Catalysis B: Environmental, 23 (1999) 89-114.

Lucas M.S., Peres J.A., Decolorization of the azo dye Reactive Black 5 by Fenton and Photo-Fenton oxidation, Dyes and Pigments 71 (2006) 236-244.

Lucas M.S., Dias A.A., Sampaio A., Amaral C., Peres J.A., Degradation of a textile reactive Azo dye by a combined chemical–biological process: Fenton's reagent-yeast, Water research 41 (2007) 1103-1109.

Lucas M.S., Tavares P.B., Peres J.A., Faria J.L., Rocha, M., Pereira C., Freire C., Photocatalytic degradation of Reactive Black 5 with TiO$_2$-coated magnetic nanoparticles, Catalysis today 209 (2013) 116-121.

Malato S., Fernández-Ibáñez P., Maldonado M.I., Blanco J., Gernjak W., Decontamination and disinfection of water by solar photocatalysis: recent overview and trends, Catalysis Today 147 (2009) 1-59.

Pillay K., Cukrowska E.M., Coville N.J., Improved uptake of mercury by sulphur-containing carbon nanotubes, Microchemical Journal 108 (2013) 124-130.

Velmurugan, R., Sreedhar, B., Swaminathan, M. Nanostructured AgBr loaded TiO$_2$: an efficient sunlight active photocatalyst for degradation of reactive Red 120. Chemistry Central Journal, 5 (2011). 46.

Wang J., Zheng S., Shao Y., Liu J., Xu Z., Zhu D., Amino-functionalized Fe$_3$O$_4$@SiO$_2$ core–shell magnetic nanomaterial as a novel adsorbent for aqueous heavy metals removal, Journal of Colloid and Interface Science 349 (2010) 293-299.

Wang R., Wang X., Xi X., Hu R., Jiang G., Preparation and photocatalytic activity of magnetic Fe$_3$O$_4$/SiO$_2$/TiO$_2$ composites, Advances in Materials Science and Engineering (2012) 1-8.

Xu S., Shangguan W., Yuan J., Chen M., Shi J., Preparations and photocatalytic properties of magnetically separable nitrogen-doped TiO$_2$ supported on nickel ferrite, Applied Catalysis B: Environmental 71 (2007) 177-184.

Yamaura M., Camilo R.L., Sampaio L.C., Macedo M.A., Nakamura M., Toma H.E., Preparation and characterization of (3-aminopropyl) triethoxysilane-coated magnetite nanoparticles, Journal of Magnetism and Magnetic Materials 279 (2004) 210-217.

Zee F.P., Anaerobic azo dye reduction, Ph.D. Thesis, Wageningen University, Wageningen, The Netherlands, (2002), p. 142.

23

Performance of advanced oxidation process (UV/O₃/H₂O₂) degrading amoxicillin wastewater

Zahra Shaykhi Mehrabadi

Young Researches Club, Kermanshah Islamic Azad University (IAUKSH).

ARTICLE INFO	ABSTRACT
Keywords: Amoxicillin wastewater Advanced oxidation Ozonation H_2O_2/UV BOD_5/COD	The degradation of synthetic amoxicillin wastewater (SAW) treated with advanced oxidation process (AOP) including $UV/O_3/H_2O_2$ was investigated in the present study. In order to investigate the impacts of effective factors on the process performance, four variables involving two numerical factors, initial H_2O_2 concentration, and initial pH, and two categorical factors, ozonation and UV irradiation, were selected. Enhancement of ozonation processes by the addition of H_2O_2 and different initial pH was also evaluated. The process was modeled and analyzed using response surface methodology (RSM). The region of exploration for the process was taken as the area enclosed by initial H_2O_2 concentration (0-20 mM) and initial pH (3-11) boundaries. For two categorical factors (ozonation and UV irradiation), the experiments were performed at two levels (with and without application of each factor). Ozone was the most effective factor with a direct effect on the response in this research. The variables had a synergistic impact on the response. Maximum chemical oxygen demand (COD) removal efficiency was obtained at H_2O_2 concentration 20 mM at initial pH 11. As a result, O_3/H_2O_2 system at pH 5 showed better performance in terms of BOD_5/COD ratio (0.40). From the HPLC chromatograms, complete degradation of amoxicillin (AMX) was achieved. The O_3/H_2O_2 process showed to be more effective in comparison with UV/H_2O_2 system.

1. Introduction

The antibiotics have been known to be presented in the ecosystem for almost 30 years. However, it was only in the mid-1990s. when the use of these compounds was widespread and new analytical technologies were developed, that their presence became an emerging concern (Lissemore et al. 2006; Hernando et al. 2006; Bound and Voulvoulis. 2006). Antibiotics enter the environment from different sources including industrial and domestic wastewaters which may cause resistance in bacterial populations and make them ineffective in the treatment processes (Schwartz et al. 2003; Schwartz et al. 2006; Baquero et al. 2008; Rosenblatt-Farrel. 2009; Martinez. 2009). Most of the WWTPs are not designed to remove highly polar micro pollutants like antibiotics (Xu et al. 2007). Therefore, they can be transported to rivers and reach groundwater after leaching. Therefore, in order to reduce the antibiotics discharged into the environment, practical and economical solutions must be achieved. A wide range of chemical and physical techniques for antibiotics removal can be used, for example, chemical oxidation and biodegradation (destructive methods), adsorption, liquid extraction and membrane techniques (nondestructive processes) (Adams et al. 2002; Putra et al. 2009; Acero et al. 2010; Radjenovic et al. 2008; Jacobsen et al. 2004; Thompson and Doraiswamy. 2000).

The non-biodegradable nature of the effluents containing antibiotics residues interferes with the removal of these compounds by conventional biological treatments. In these cases, one solution is to apply advanced oxidation processes (AOPs). In antibiotic wastewater treatment, advanced processes can be applied to increase the BOD/COD ratio, degrading non-biodegradable wastewater. Several laboratory studies have been performed to investigate the ability of different wastewater treatment methods in removing antibiotics and other pharmaceuticals from wastewater with increasing biodegradability (Elmolla and Chaudhuri. 2010; Elmolla and Chaudhuri. 2011; Elmolla and Chaudhuri. 2012; Halling-Sorensen et al. 2000; Ternes et al. 2002; Perez et al. 2005). One of the antibiotic classes is β-Lactams that are narrow spectrum antibiotics, which are highly

effective against the gram-positive genera viz. Streptococcus, Gonococcus, and Staphylococcus. These antibiotics act as bacteriostatics by inhibiting the synthesis of the bacterial peptidoglycan cell wall (Marzo and Dal Bo. 1998). The degradation of β-lactam antibiotics such as amoxicillin (AMX), takes place under acidic and alkaline conditions or by reactions with weak nucleophiles, such as water or metal ions (Aksu and Tunc. 2005). And they are not very stable due to hydrolysis of the β-lactam ring (Langin et al. 2009).

Advanced oxidation processes (AOPs) are quite efficient novel methods and effective in the degradation of most pollutants in wastewater (Pera-Titus et al. 2004; Ahmed et al. 2010). Ozonation like the other AOPs is based on the generation of intermediate radicals, the hydroxyl radicals (OH·), which are extremely reactive and less selective than other oxidants (e.g. chlorine, molecular ozone...). Chemical mineralization can occur by direct reaction with the applied oxidant and/or via the generation of highly oxidative secondary species, most commonly, hydroxyl radicals (OH·), which is one of the most powerful oxidants known (Metcalf. 2007). Processes that promote the enhanced formation of OH· are generally referred to as AOPs. UV radiation is commonly used to enhance the formation of OH·. This can be achieved by a number of methods such as O3/H2O2 and UV/O3 are also considered AOPs since they promote the formation of hydroxyl radicals. One of the most widely used antibiotics is amoxicillin. In this study, an advanced oxidation process including UV radiation, ozonation at different dosage of H_2O_2 was examined for the treatment of SAW in a batch experiment. The process performance was evaluated in terms of COD removal efficiency and final BOD_5/COD.

2. Materials and methods
2.1. Wastewater preparation

The synthetic antibiotic wastewater (SAW) was prepared by dissolving two capsules of amoxicillin (AMX 500 mg) in tap water in the laboratory scale. The stock SAW was prepared in CODin 2000 mg/L. Other solutions were prepared by dilution the main solution. Furthermore, the actual COD values have been verified each time

Corresponding author Email: sheikhizahra@yahoo.com

before initiation of experimental work. pH and BOD5/COD ratio were about 7.5 0.1 and 0.18- 0.2, respectively.

2.2. Experimental set-up

The experimental rig is shown in Fig. 1. The reactor was of stainless steel with 70 mm in diameter, 456 mm in height, and 1 L in volume. The influent air to the ozone generator (COG, 1G/H) supplied by an air pump ($Q = 0.075$ $m^3.min^{-1}$) and air dryer. The air dryer consisted of a column which was filled with a high adsorptive molecular sieve (silica). Ozonation experiments were carried out by continuously feeding an ozone gas stream in a mixed semi-batch bubble reactor (continuous for gas and batch for liquid). Influent ozone concentration in the gas phase

was adjusted to 112.5 mg/L at a flow rate of 5 L/min for all the experiments. The irradiation in the photoreactor was obtained by a 15 W UV/A lamp (HITACHI, emission: 365 nm, constant intensity 60 mW/cm^2) that protected by a quartz jacket, and positioned and immersed in the solution in the center of the reactor. The lighted length of the lamp was 452 mm with a quartz sleeve diameter of 3 cm. The reactor was followed by a washing bottle, containing 250 mL of acidified 2 % wt. KI solution for determining of inject and unreacted ozone at influent and effluent in several times. The air flow rate was adjusted 5 L/min. The ozone content of the input air stream was measured as 0.27 g O_3/h. The ozone in off-gas was also measured and the consumed ozone was obtained 3.68 as g ozone$_{consumed}$/g COD$_{removed}$.

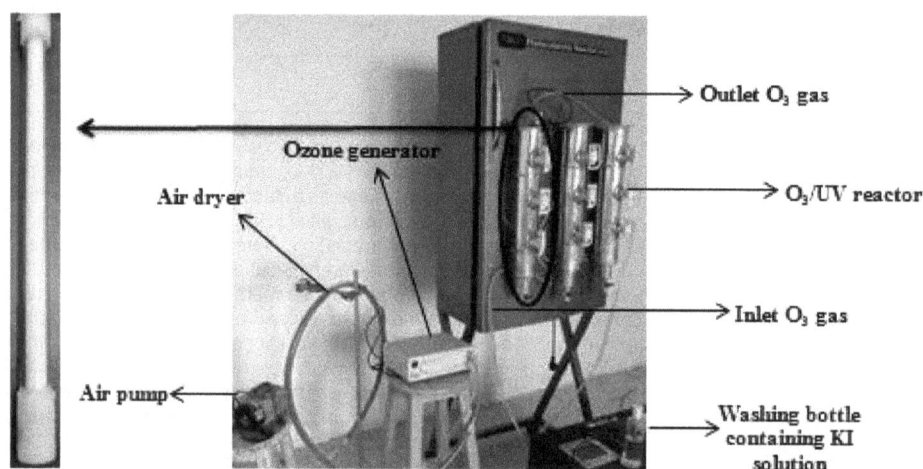

Fig. 1. Laboratory-scale experimental set-up for advanced oxidation reactor (O_3/UV).

2.3. Experimental Design

In order to investigate the effects of influential variables on the process performance, four independent factors (initial H_2O_2 concentration and initial pH, ozonation and UV irradiation) were considered to design the experiments. The experiments were designed using RSM. The RSM used in the present study was a general factorial design involving two different numerical factors, initial H_2O_2 concentration and initial pH, and two categorical factors, ozonation and

UV irradiation. The experimental range and levels of the independent variables are shown in Table 1. All experiments were carried out in batch mode in terms of wastewater input. The results were completely analyzed using analysis of variance (ANOVA) automatically performed by Design Expert software (Stat-Ease Inc., version 6.0). The results can be obtained as 3D presentations for visualization and also as contours to study the effect of system variables on responses. From these three-dimensional plots, the simultaneous interaction of the two factors on the responses was studied.

Table 1. Experimental range and levels of the independent variables.

Variables	-1	-α	0	+α	+1
		Numerical Variables			
pH	3	5	7	9	11
H_2O_2 mM	0	5	10	15	20
		Categorical Variables			
Variables	Level 1			Level 2	
Ozonation	Without			With	
UV Irradiation	Without			With	

2.4. Analytical methods

Antibiotic concentration was determined by a high performance liquid chromatograph (HPLC) equipped with a micro-vacuum degasser (m), quaternary pump, diode array and multiple wavelength detector (DAD) (m) at wavelength 254 nm. The column was ECLIPSE XDD-C18 (4.6 mm×150 mm, 5 μm) and its temperature was 60 °C. The content of the mobile phase was potassium dihydrogen phosphate and acetonitrile in the ratio (98:02 v/v). The flow rate of the mobile phase was maintained at 1.0 mL/min. Chemical oxygen demand (COD) was measured according to the Standard Methods (APHA. 1992). The entire chemicals used in the analysis were analytical grade (Merck, Darmstadt, Germany). A pH meter (JENWAY 3510) and a pH electrode

were used for pH measurement. Biodegradability was measured by 5-day biochemical oxygen demand (BOD$_5$) (OxiTop IS 6) test according to the Standard Methods [28].

3. Result and discussion
3.1. Statistical analysis of the model developed

A mathematical equation was developed to describe this relationship based on the results obtained. The results obtained were then analyzed by ANOVA to assess the "goodness of fit". Equation from the first ANOVA analysis was modified by eliminating the terms found statistically insignificant. Table 2 illustrates the reduced cubic model in terms of coded factors and also shows other statistical parameters.

Data given in this Table demonstrates that all the model terms were significant at the 5 % confidence level since P values were less than 0.05 (Table 2). The R^2 coefficient gives the proportion of the total variation in the response predicted by the model, indicating the ratio of sum of squares due to regression (SSR) to the sum of squares (SST). A high R^2 coefficient ensures a satisfactory adjustment of the model to the experimental data. A value of 0.86 for R^2 indicates that some other variables might have interaction on the response and have not considered in this study. Adequate precision (AP) compares the range of the predicted values at the design points to the average prediction error. Ratios greater than 4 indicate adequate model discrimination (Mason et al. 2003). Diagnostic plots such as the predicted versus actual values (Fig. 2) help us judge the model satisfactoriness. The predicted versus actual values plot of COD removal is presented in Fig. 2.

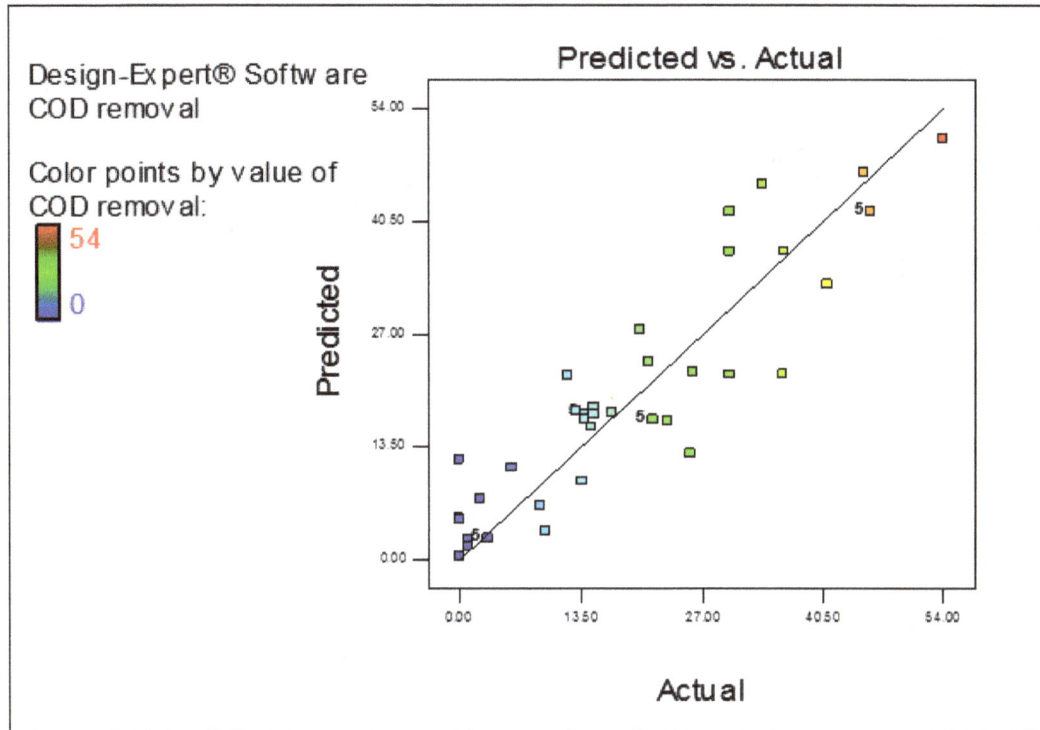

Fig. 2. Actual vs. predicted values of COD removal.

Table 2. ANOVA for response surface model applied.

Response	Model type						
		ANOVA					
		source	Sum of Squares	Degrees of freedom (DF)	Mean square	F value	P value
COD removal (%)	Reduced cubic model	Model	9914.2	8	1239.3	32.4	<0.0001
		A	408.9	1	408.9	10.7	0.0021
		B	557.78	1	557.8	14.6	0.0004
		C	5121.1	1	5121.1	134.0	<0.0001
		D	2969.4	1	2969.4	77.7	<0.0001
		AB	250.6	1	250.6	6.5	0.0140
		BC	262.1	1	262.1	6.8	0.0121
		CD	325.2	1	325.2	8.5	0.0056
		A^2D	902.6	1	902.6	23.6	<0.0001
		Residual	1643.1	43	38.2	-	-

(R^2 = 0.8578, Adj. R^2 = 0.8314, Adeq. Precision = 19.41, Std. Dev. = 6.12, C.V % = 31.62, PRESS = 2637.23)

3.2. Process performance of AOP (UV/O₃/H₂O₂) treating SAW

In this study, AMX content is expressed as COD. As this investigation deals with the effect of advanced oxidation processes removing the 16-C organic molecule-AMX, the compound may be degraded to various intermediates. So, in order to assess the performance of the processes studied, different possible intermediates are classified into four categories of COD as (1) COD degraded into bCOD or BOD, (2) COD mineralized (in the form of $CO_2 + H_2O$), (3) COD broken down to intermediate compounds but in non-biodegradable form (nbCOD) and (4) COD as parent form.

3.2.1. COD removal

TCOD concentration was monitored throughout the experiments representing the organic content of SAW at initial and final samples. The contribution of aeration in COD removal efficiency was determined in an experiment with aeration alone and no effect was observed. The process performance was monitored in terms of COD removal every 30 min up to 180 min. The data obtained after 180 min were selected for the process analysis and modeling. TCOD removal variation as a function of the variables was described by a reduced cubic model in the coded form as below:

TCOD removal, % = 19.55 + 4.77 A + 5.57 B + 9.92C
+ 9.57D + 3.96 AB + 3.82 BC + 2.50 CD − 9.4 A^{2}D   (1)

From Table 2, A, B, C, D, AB, BC, CD and A^{2}D are selected as the effective terms with P value less than 0.05. Where, A is initial pH, B is H$_2$O$_2$ concentration, C and D are respectively, ozonation and UV irradiation. AB, BC, CD and A^{2}D also indicate the interactive terms. All the model terms in the Eq. except A^{2}D, had a positive effect on the COD removal efficiency. From the Eq., term C (O$_3$) was the most effective factor with direct effect on the response with the largest coefficient.
Figs. 3a-d represent three dimensional graphs for COD removal as a function of pH and H$_2$O$_2$ concentration at different combination of O$_3$ and UV. In all of the Figs pH showed an inverse effect on the response. in the condition with no UV irradiation, by increasing pH from 3 to7, the COD removal was decreased. But by increasing pH up to 11, it had a positive effect on the response. On the other hand, in the experiment with UV irradiation, by increasing pH from 3 to about the neutral value, the COD removal was increased. But by increasing pH from about 7 to 11, the response was decreased.
Fig. 3a shows a three dimensional graph for COD removal as a function of pH and H$_2$O$_2$ with no ozonation and UV irradiation. As can be seen in the Fig., pH showed almost no effect at its low levels (3-7) and a remarkable increasing effect on the response at the high values (7-11). H$_2$O$_2$ concentration also showed no impact at the low levels of pH while by increasing pH from 7 to 11 an increasing effect on the response was observed (Arslan-Alaton et al. 2011). Maximum COD removal efficiency was obtained to be about 20 % at the highest value of the variables.
The effect of the numerical variables at the condition with UV irradiation and without ozonation is illustrated in the Fig. 3b. From the Fig., The effect of H$_2$O$_2$ concentration was slight in the range studied for the variables (a slight decreasing effect at low pH and an increasing effect at high values of pH). pH had an inverse effect on the response. Comparing the results presented in the Fig. 3b with Fig. 3a showed that UV irradiation caused an inverse curvature, so that, the maximum response was found at neutral pH (about 21 %). However, it did not have a significant change in the response.

Fig. 3c illustrates the effect of pH and H$_2$O$_2$ under ozonation on the COD removal. By comparing the results with those presented in the Fig. 4a, the results (Fig. 3c) showed an improvement in the COD removal efficiency with almost the same trend when O$_3$ is applied. The addition of high concentrations of H$_2$O$_2$ will not improve the process efficiency, it might be due to H$_2$O$_2$ deforming to hydroxyl radicals under ozonation conditions or it may act as a free radical scavenger (Homem and Santos. 2011). As can be seen in the Fig. 3c, H$_2$O$_2$ concentration showed a linear increasing impact on the response in the range of pH examined. But this effect was more in alkaline condition. Minimum to maximum of the efficiency obtained was varied between 15 to 45 %.
Fig. 3d depicts three dimensional graphs for COD removal as a function of pH and H$_2$O$_2$ under ozonation and UV irradiation. The maximum efficiency of COD removal was obtained to be 54 % at the high value of pH and H$_2$O$_2$ (11 and 20 mM, respectively), where both categorical variables (O$_3$ and UV) are applied. It showed a synergistic effect of the factors studied. Under this condition, the initiation of ozone decomposition is accelerated by increasing the pH value (more changes are observed in the range of pH from 3 to 7). It should be mentioned that according to the actual data, the maximum efficiency was found at pH 11. Based on the modeled data, the optimum condition was determined to be pH 7 and H$_2$O$_2$ 20 mM under ozonation and UV irradiation (Esplugas et al. 2007). Comparing Figs 3b and d, showed that when ozonation be applied, the COD removal efficiency was improved about 30 %. It should be due to O$_3$ was a strong oxidizing relative to UV irradiation.
In general, From the Fig., the oxidation rate is H$_2$O$_2$< (H$_2$O$_2$/UV) < O$_3$ < (O$_3$/UV/ H$_2$O$_2$).
In a research work performed by Andreozzi et al. (2003), removal of AMX (C$_0$= 401.5 mg/L) in an ozonation process was investigated and low mineralization was reported. In the purpose of comparison with the present work, 36 % COD removal was achieved at the initial concentration of 1100 mg/L. The authors also examined UV/H$_2$O$_2$ process for treatment of AMX in their work and not clear finding on the process performance is reported. Another study carried out on the effect of ozonation (2.96 g/L. h) on human antibiotic removal (450 mg/L COD) which a 74 % COD removal was reported (Balcıoglu and Otker. 2003).

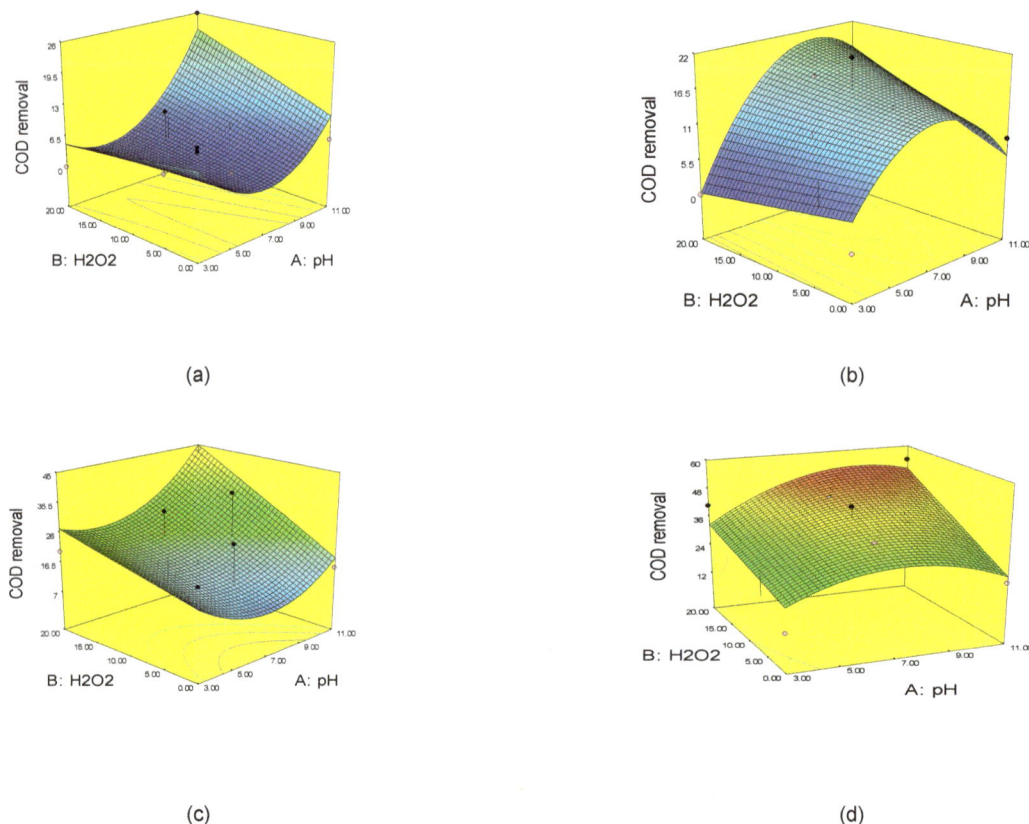

(a)

(b)

(c)

(d)

Fig. 3. Three dimensional graphs for COD removal as a function of pH and H$_2$O$_2$ at (a) without O$_3$ and UV, (b) with UV and without O$_3$, (c) with O$_3$ and without UV, and (d) with O$_3$ and UV.

3.3. Interactive effects of the variables studied
3.3.1. Effect of H_2O_2

Figs. 4a-c show the interactive effects of H_2O_2 concentration-ozone on the COD removal efficiency at different levels of initial pH without UV irradiation. As shown in the Figs., at the condition with ozonation, asimilar trend is observed when ozone is applied which is increasing. The Figs. 4a-c illustrate that the effect of ozone is more significant than initial pH and H_2O_2 concentration. Synergistic interactive effects of the three variables are seen in the Figs., so that by increasing the numerical variables (initial pH and H_2O_2 concentration) under ozonation, the response was increased.

The Figs. show that H_2O_2 had a slight effect on the response at the condition without ozone while showed a positive impact in the presence of ozone. Fig. 4 showed that perozonation (O_3/H_2O_2) enhances the ozonation performance. Similar finding about the effect of H_2O_2 is reported in the literature (Glaze and Kang. 1989; Witte et al. 2009). Effect of hydrogen peroxide combined by ozone to enhance oxidizing ability has been extensively researched recently and is considered to be a promising alternative for refractory organics removal from aqueous solutions (Masten et al. 1997; Lin et al. 2009). It is important to recognize that $OH^{\cdot}$ and related free-radical degradation pathways occur readily even in treatment with O_3 alone. Where a wastewater containing contaminants that warrant enhanced treatment via degradation by $OH^{\cdot}$, H_2O_2 should be applied sparingly as to promote the generation of free radicals but not so high as to consume and deplete aqueous O_3 or even scavenge the resultant $OH^{\cdot}$ rapidly that would end up with no benefits in treatment (Lin et al. 2009). The maximum degradation rate occurred at H_2O_2 concentration of 20 mM. Increasing hydrogen peroxide concentration up to an optimum concentration enhanced about 15 % increase in the oxidation rate in the case of the antibiotic wastewater (Esplugas et al. 2007).

From the result, the effect of H_2O_2 on AMX degradation was obtained that by increasing H_2O_2, COD removal was increased (Fig. 5). These results confirmed that the oxidation of AMX proceeded mainly by hydroxyl radicals in accordance with the results obtained at initial pH 3, 7 and 11.

The trend and interactive effects of H_2O_2 Concentration-Ozone on of the COD removal are the same under UV irradiation at different initial pH but with higher values when UV irradiation is applied.

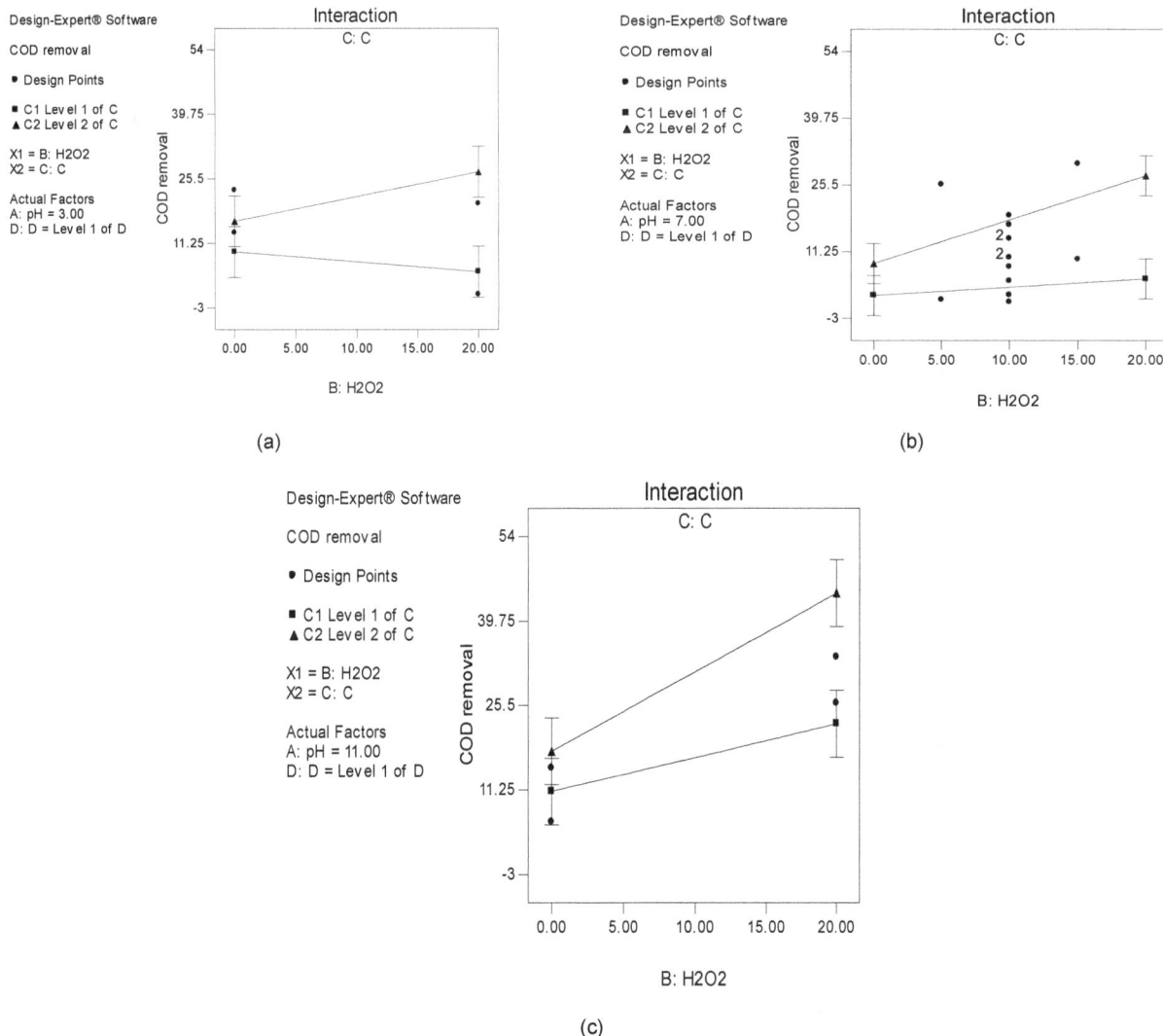

(a)

(b)

(c)

Fig. 4. Interactive effect of H_2O_2 concentration-Ozone on the COD removal without UV irradiation at (a) pH 3, (b) pH 7, and (c) pH 11.

3.3.2. Effect of pH

Initial pH is an important parameter that can affect advanced oxidation reactions (Andreozzi et al. 2005). In overall, when $OH^{\cdot}$ is scarce (i.e., low pH), the depletion reaction is slow allowing for accumulation of dissolved O_3 to a high level, whereas when $OH^{\cdot}$ is abundant (i.e., high pH), the depletion reaction is rapid thus prohibiting dissolved O_3 to accumulate (e.g., dissolved O_3 rarely exceeds 1 mg/L at pH 11 or higher). With the increment of the solution pH from 3 to alkaline pH, overall COD abatement is enhanced for most wastewaters as expected. In general, ozone reacts with organic compounds found in water and wastewater via two different pathways namely direct molecular and indirect radical chain type reaction depending upon pH and composition of water (Andreozzi et al. 2005). It is expected that molecular ozone is the major oxidant at acidic pH, whereas less selective and faster radical oxidation (mainly hydroxyl radical) becomes dominant at pH > 7 as a consequence of OH· accelerated ozone decomposition (Beltran et al. 2001). Since the oxidation potential of hydroxyl radicals is much higher than that of ozone molecule, direct oxidation is slower than radical oxidation and furthermore, causes incomplete oxidation of organic compounds as observed in this study. However, low pH is known to suppress the formation of hydroxyl radicals from ozone and ozone is reacting directly by an electrophilic attack that led to 13 % lower COD removal than that obtained in the highest initial pH. In other words, at optimum values of the variables with the lowest initial pH, the COD removal achieved 41 %.

Figs. 5a and b also show the effect of different levels of initial pH applied on the degradation of AMX relative to reaction time at different oxidation conditions. It was found that by increasing in initial pH, the response (C/C_0) was decreased, corresponding to the improvement in COD removal efficiency.

3.3.3. BOD_5/COD ratio

Further investigations to determine the bioactivity of the observed unknown by-products produced in the AOPs processes would be necessary after application of the advanced oxidation processes. The BOD_5/COD ratio indicates the biodegradable fraction of the remaining COD content in the treated wastewater. However, BOD_5 measurements do not mirror the actual situation in biological treatment units and hence is only used as an approximate estimation of biodegradability improvement. Fig. 6 demonstrates the BOD_5 and different form of COD in selected runs carried out by $O_3/UV/H_2O_2$. Fig. 6 has been drawn according to the selected experiments number as presented in Table 3. From the Fig., the BOD_5 fractions in the most oxidized samples were increased in comparison with the value in the raw SAW. It is noted that the ratio of BOD_5/COD for the raw SAW was determined to be less than 0.2. As observed in the Fig., the maximum ratio of BOD_5/COD was found 0.40 under run no. 1 (ozonation, H_2O_2 10 mM and initial pH of 5). From the Table 3, various advanced oxidation processes have increased the ratio from 0 to 0.4 for different antibiotics. Balcıoglu and Otker (2003) reported an increase in the ratio from 0 to 0.27 for the treatment of human antibiotic containing penicillin (Balcıoglu and Otker. 2003).

It is figured out from the Fig. 6 that 54 % of COD could be removed and 25 % of the COD contents in the treated sample could be converted into the biodegradable form (BOD_5) under optimum conditions (run no. 1, 5). However, about 42 % of total COD has still remained as the non-biodegradable form that may have been changed during the oxidation process, but this is unknown so far and need more investigation. In order to clarify this point, the compounds content of the selected samples was analyzed by HPLC. Figs. 7 show the HPLC chromatogram for the sample and the HPLC analysis results are presented in Table 3. As specified in the illustrations (the Figures. and Table), in all the samples except run no. 3 and 4, AMX could not be detected. It implies that the residual COD (e.g. 42 % at optimum condition) has been degraded into secondary non-biodegradable intermediates.

Table 3. Order of experiments number according to operating conditions.

Run No.	Type of process	H_2O_2 concentration(mM)	Initial pH	BOD_5/COD ratio	Area peak	Degradation percent (%)
1	O_3/H_2O_2	10	5	0.40	0	100
2	UV/ H_2O_2	10	5	0.32	10.5	96
3		20	11	0.23	0	100
4	$O_3/UV/H_2O_2$	5	7	0.20	0	100
5		20	11	0.25	---	---
-	Raw SAW	---	7.5	0.2	266.5	---

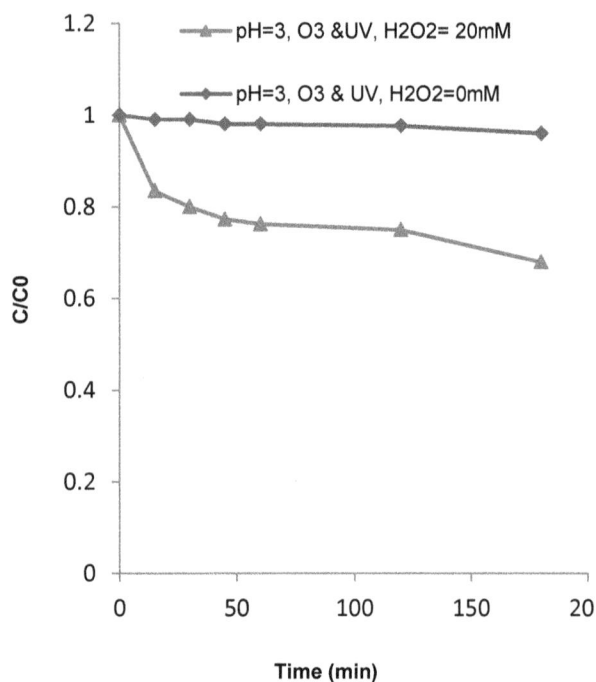

(a)

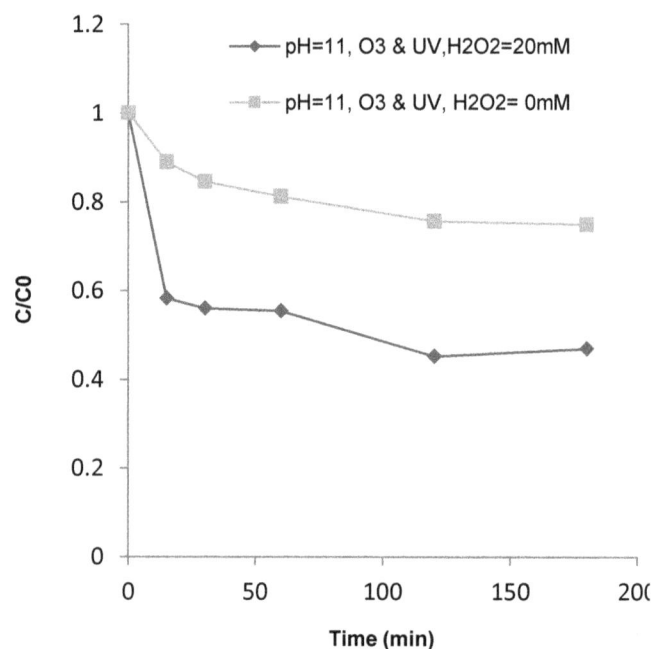

(b)

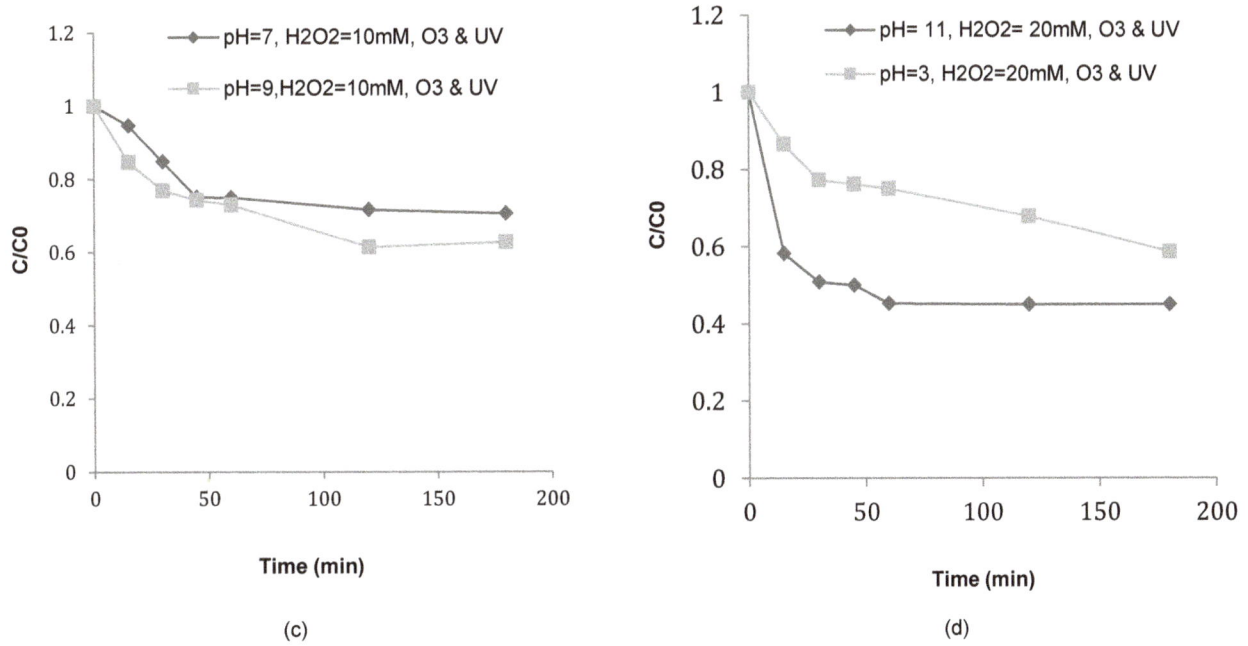

Fig. 5. (a and b) Effect of various H2O2 concentrations on degradation of AMX as a function of reaction time, (c and d) Effect of various initial pH on COD removal as a function of reaction time at different oxidation conditions.

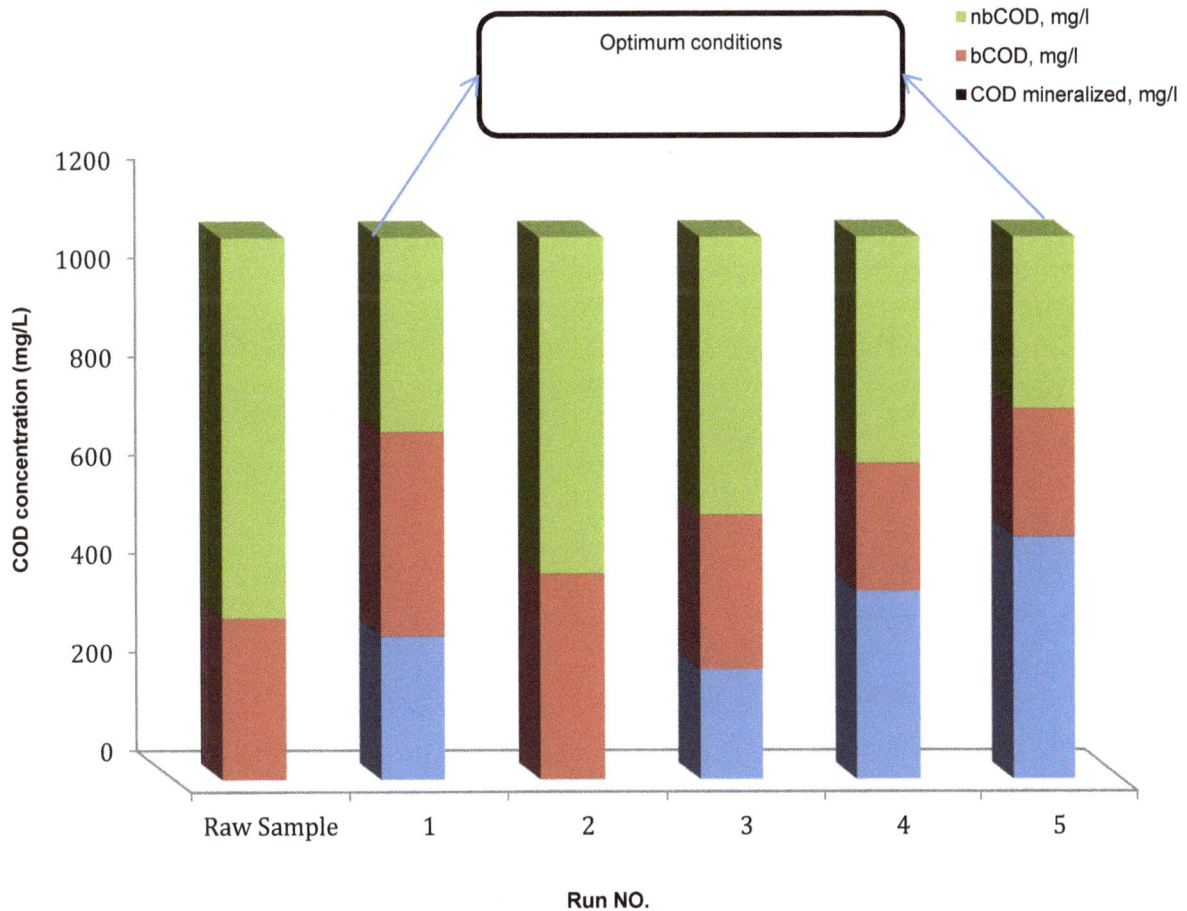

Fig. 6. COD fractionation for the raw and treated samples and run no. (1-5), 1) pH= 5, H$_2$O$_2$= 10, O$_3$, 2) pH= 5, H$_2$O$_2$= 10, UV, 3) pH= 11, H$_2$O$_2$= 20, UV, 4) pH= 7, H$_2$O$_2$= 5, O$_3$ & UV, 5) pH= 11, H$_2$O$_2$= 20, O$_3$ & UV is present in Table 3.

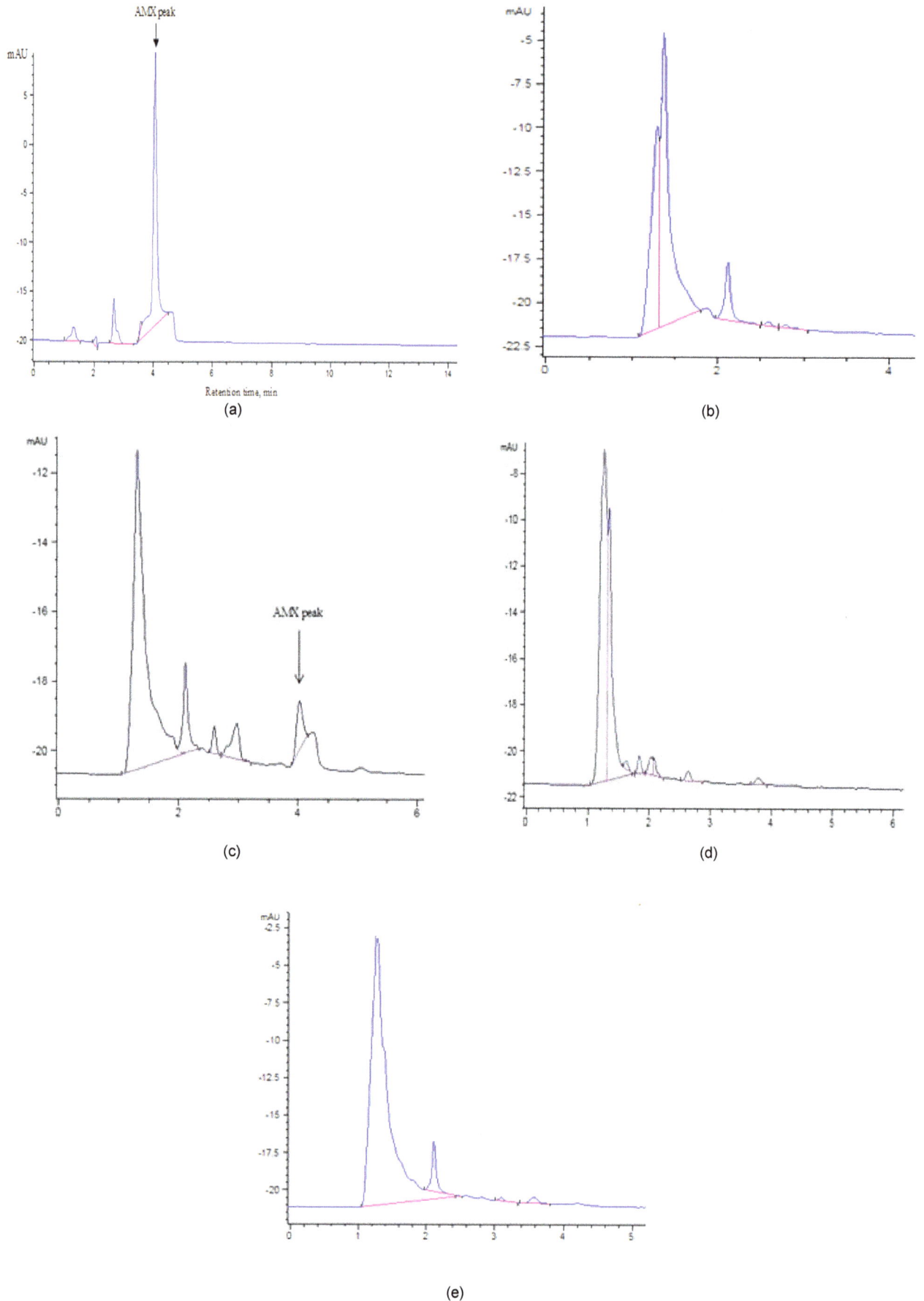

Fig. 7. HPLC chromatograms for a) stock SAW, b) pH= 5, H$_2$O$_2$= 10, O$_3$, c) pH= 5, H$_2$O$_2$= 10, UV, d) pH= 11, H$_2$O$_2$= 20, UV, e) pH= 7, H$_2$O$_2$= 5, O$_3$ & UV.

4. Conclusions

The degradation of AMX was investigated via ozone, direct UV photolysis advanced oxidation processes with adding H_2O_2 in different initial pH. The results of the present study have clearly delineated that using of ozone and UV with adding H_2O_2 at alkalinity pH values provides a promising technique for the treatment of AMX wastewater. (Ozonation was demonstrated to be an effective method in increasing the biodegradability of SAW). From the results, the oxidation rate is $H_2O_2 <$ $(H_2O_2/UV) < O_3 < (O_3/UV/ H_2O_2)$ and effect of H_2O_2 on AMX degradation was obtained that by increasing H_2O_2, COD removal was increased. When UV irradiation was combined with hydrogen peroxide and ozonation, the process performance in terms of COD removal efficiency increased significantly compared to that of single oxidants. The experiment with UV irradiation in neutral pH and experiment without UV irradiation had a significant COD removal in alkaline pH. The best result was obtained at pH 11 with 54 % COD removal efficiency. The maximum ratio of BOD_5/COD was found 0.40 in the O_3/H_2O_2 system. From the HPLC results, the run with only UV irradiation and low H_2O_2 concentration, residual AMX was found in the solution while complete degradation was achieved in the others. In the basis of the results obtained, the AOP applied could be used as a pretreatment method prior a biological treatment process. And also, other advanced oxidation processes are recommended to be examined.

Acknowledgements

The authors would like to acknowledge Kermanshah Department of Environment for providing analytical equipment to measure. The authors would also like to acknowledge the cooperation of Mr. S. Ghanbari for his fantastic and unique job in the fabrication of the immobilized photocatalytic reactor set up. Special thanks also go to Dr. R. Heydari for his assistance in HPLC analysis.

References

Acero J., Benitez F., Roldan G., Fernandez L., Biochemical Engineering Journal 2010, 160, 72-78.

Adams C., Asce M., Wang Y., Loftin K., Meyer M., Removal of antibiotics from surface and distilled water in conventional water treatment processes, Journal of Environmental Engineering 128 (2002) 253-260.

Ahmed S., Rasul M.G., Martens W.N., Brown R., Hashib M.A., Heterogeneous photocatalytic degradation of phenols in wastewater: a review on current status and developments, Desalination 261 (2010) 3–18.

Aksu Z., Tunc O., Application of Biosorption for Penicillin G Removal: Comparison with Activated Carbon, Process Biochemistry 40 (2005) 831-847.

Andreozzi R., Canterino M., Marotta R., Paxeus N., Antibiotic removal from wastewaters: The ozonation of amoxicillin, Journal of Hazardous Materials 122 (2005) 243–250.

APHA, AWWA, WPCF, Standard methods for the examination of water and wastewater, American Public Health Association, American Water Works Association, 18th edition, Water Pollution Control Federation, Washington, DC, USA, 1992.

Arslan-Alaton I., Dogruel S., Baykal E., Gerone G., Combined chemical and biological oxidation of penicillin formulation effluent, Journal of Environmental Management 73 (2004) 155-163.

Balcıoglu I.A., Otker M., Treatment of pharmaceutical wastewater containing antibiotics by O_3 and O_3/H_2O_2 processes, Chemosphere 50 (2003) 85–95.

Baquero F., Martinez J.L., Canton R., Antibiotics and antibiotic resistance in water environments, Biotechnology 19 (2008) 260-265.

Bound J.P., Voulvoulis N., Predicted and measured concentrations for selected pharmaceuticals in UK rivers: Implications for risk assessment, Water Research 40 (2006) 2885 – 2892.

De Witte B., Dewulf J., Demeestere K., Van Langenhove H., Ozonation and advanced oxidation by the peroxone process of ciprofloxacin in water, Journal of Hazardous Materials 161 (2009) 701–708.

Elmolla E.S., Chaudhuri M., Comparison of different advanced oxidation processes for treatment of antibiotic aqueous solution, Desalination 256 (2010) 43–47.

Elmolla E.S., Chaudhuri M., The feasibility of using combined TiO_2 photocatalysis-SBR process for antibioticwastewater treatment, Desalination 272 (2011) 218–224.

Elmolla E.S., Chaudhuri M., The feasibility of using combined Fenton-SBR for antibiotic wastewater treatment, Desalination 285 (2012) 14–21.

Esplugas S., Bil D.M., Gustavo L., Kraus T., Dezotti M., Ozonation and advanced oxidation technologies to remove endocrine disrupting chemicals (EDCs) and pharmaceuticals and personal care products (PPCPs) in water effluents, Journal of Hazardous Materials 149 (2007) 631–642.

Glaze W.H., Kang J.H., Advanced oxidation processes: Description of a kinetic model for the oxidation of hazardous materials in aqueous media with ozone and hydrogen peroxide in a semibatch reactor, Industrial Engineering and Chemistry Research 28 (1989) 1573–1580.

Halling-Sorensen B., Lutzhoft H.C.H., Andersen H.R., Ingerslev F., Section of Environmental Chemistry, Department of Analytical and Pharmaceutical Chemistry, Journal of Antimicrobial Chemotherapy 46 (2000) (Suppl. 1) 53-58.

Hernando M.D., Mezcua M., Fernandez-Alba A.R., Barcelo D., Environmental risk assessment of pharmaceutical residues in wastewater effluents, surface waters and sediments, Talanta 69 (2006) 334–342.

Homem V., Santos L., Degradation and removal methods of antibiotics from aqueousmatrices - A review, Journal of Environmental Management 92 (2011) 2304-2347.

Jacobsen A.M., Halling-Sorensen B., Ingerslev F., Hansen S.H., Simultaneous extraction of tetracycline, macrolide and sulfonamide antibiotics from agricultural soils using pressurised liquid extraction, followed by solid-phase extraction and liquid chromatography–tandem mass spectrometry, Journal of Chromatography A, 1038 (2004) 157–170.

Langin A., Alexy R., Konig A., Kummerer K., Deactivation and transformation products in biodegradability testing of ß-lactams amoxicillin and piperacillin, Chemosphere 75 (2009) 347–354.

Lissemore L., Hao C., Yang P., Sibley P.K., Mabury S., Solomon K.R., An exposure assessment for selected pharmaceuticals within a watershed in Southern Ontario, Chemosphere 64 (2006) 717–729.

Marzo A., Dal Bo L., Chromatography as an analytical tool for selected antibiotic classes: a reappraisal addressed to pharmacokinetic applications, Journal of Chromatography A, 812 (1998) 17–34.

Martinez J.L., Environmental pollution by antibiotics and by antibiotic resistance determinants, Environmental Pollution 157 (2009) 2893-2902.

Mason R.L., Gunst R.F., Hess J.L., Statistical Design and Analysis of Experiments, Eighth Applications to Engineering and Science, second edition, Wiley, New York, 2003.

Masten S.J., Galbraith MJ., Davies SHR., Oxidation of trichlorobenzene using advanced oxidation process, Ozone Science Engineering 18 (1997) 535–548.

Metcalf, Eddy, Water Reuse: Issues, Technologies, and Applications, McGraw Hill, New York, 2007.

Pera-Titus M., Garcıa-Molina V., Banos M.A., Gimıenez J., Esplugas S., Degradation of Chlorophenols by Means of Advanced Oxidation Processes: A General Review, Applied Catalysis B Environmental 47 (2004) 219–256.

Perez S., Eichhorn P., Aga D.S., Evaluating the biodegradability of sulfamethazine, sulfamethoxazole, sulfathiazole, and trimethoprim at different stages of sewage treatment, Environmental Toxicol Chemistry 24 (2005) 1361–1367.

Putra E.K., Pranowo R., Sunarso J., Indraswati N., Ismadji S., Performance of activated carbon and bentonite for adsorption of amoxicillin from wastewater: Mechanisms, isotherms and kinetics, Water Research 43 (2009) 2419-2430.

Radjenovic J., Petrovic M., Ventura F., Barcelo D., Rejection of Pharmaceuticals in Nanofiltration and Reverse Osmosis Membrane Drinking Water Treatment, Water Research 42 (2008) 3601-3610.

Rosenblatt-Farrel N., The Landscape of Antibiotic Resistance, Environmental Health Perspect 117 (2009) 245-250.

Schwartz T., Kohnen W., Jansen B., Obst U., Detection of antibiotic-resistant bacteria and their resistance genes in wastewater, surface water, and drinking water biofilms, FEMS Microbiology Ecology 43 (2003) 325-335.

Schwartz T., Volkmann H., Kirchen S., Kohnen W., Schön-Hölz K., Jansen B., Obst U.,Real-time PCR detection of Pseudomonas aeruginosa in clinical and municipal wastewater and genotyping of the ciprofloxacin-resistant isolates, FEMS Microbiology Ecology57 (2006) 158-167.

Ternes T.A., Meisenheimer M., McDowell D., Sacher F., Brauch H.J., Haist-Gulde B., Preuss G., Wilme U., Zulei-Seibert N., Removal of pharmaceuticals during drinking water treatment, Environmental Science Technology 36 (2002) 3855–3863.

Thompson L.H., Doraiswamy L.K., The rate enhancing of ultrasound by inducing supersaturation in a solid-liquid system, Chemical Engineering Science 55 (2000) 3085-3090.

Xu W.H., Zhang G., Zou S.C., Li X.D., Liu Y.C., Determination of selected antibiotics in the Victoria Harbour and the Pearl River, South China using high-performance liquid chromatography-electrospray ionization tandem mass spectrometry, Environmental Pollution 145 (2007) 672-679.

24

Treatment of landfill leachate by electrochemicals using aluminum electrodes

Soraya Mohajeri[1], Hamidi Abdul Aziz[1], Mohamed Hasnain Isa[2], Mohammad Ali Zahed[*,3]

[1]School of Civil Engineering, Universiti Sains Malaysia, Nibong Tebal, Pinang, Malaysia.
[2]Civil Engineering Department, Universiti Teknologi PETRONAS, Perak, Malaysia.
[3]Faculty of biological sciences, Kharazmi University, Tehran, Iran.

ARTICLE INFO

Keywords:
COD
Color removals
Electrochemical oxidation
Landfill leachate
Aluminum electrodes

ABSTRACT

Electrochemical oxidation process has been shown to be a favourable choice for Chemical oxygen demand (COD) and color removals from various types of wastewaters. The technique was employed for mineralization of semi-aerobic landfill leachate. Leachate sampling were carried out from Pulau Burung Landfill Site (PBLS), Penang, Malaysia. The main objective was to determine the effectiveness of electrochemical oxidation in leachate treatment using aluminum electrodes which are relatively nontoxic and cost-effective. The influence of pH, reaction time, current density, electrolyte concentration, agitation rate and dilution on COD and color removals was investigated. The highest COD and color removal were obtained as 57.1% and 72.0% respectively at pH 8, current density 60 mA/cm2, electrolyte concentration 2000 mg/L, agitation rate 400 rpm, dilution 50% and reaction time 4 h. The energy consumption was determined as 128 kWh/m^3 for this type of landfill leachate. The study shows that electrochemical oxidation can be used as a step of shared treatment.

---

1. Introduction

Municipal landfill leachate is complex wastewater that often comprises of organic substances, heavy metals, chloride and many other soluble compounds. Factors such as different type of waste and landfill age affect the composition and concentration of contaminants (Wang et al. 2018; Yusoff et al. 2018). Inadequate management of landfill, may develop it to a source of pollution due to leachate infiltration into soil and underlying water (Mohajeri et al., 2019; Vlyssides et al. 2003).

Given the biological refractory character of old landfill leachates, utilization of techniques rather than biodegradation necessitates to efficiently lessen the contaminant load of these sewage discharge (Rivas et al. 2005). Using of sole technology is often inadequate to achieve appropriate levels of pollution reduction and integration of biological, physical and chemical processes need to be employed (Fan et al. 2010).

In the event of landfill leachate, electrochemical oxidation process using various electrodes has been revealed to be capable of reducing high molecular weight organic compounds from leachates in comparison to other physiochemical technologies which only raise a question about phase transfer of the contaminants and do not affect chemical destruction (Fernandes et al. 2015; Moreira et al. 2015; Mohajeri et al. 2010a; Hermosilla et al. 2009; Deng and Englehardt 2007; Shao et al. 2006).

Hydroxyl radical (OH[.]) was known as a very influential, unselective and oxidizer agent due to its quick reaction with organic compounds through hydroxylation or dehydrogenation process of nonsaturated bond, succeeding a radical substitution till their comprehensive mineralization, i.e., the modification of inceptive pollutants into carbon dioxide, water and inorganic ions (Boye et al. 2003).

In addition to OH, active chlorine, formed by the oxidation of the existent or added chloride in the wastewater, also oxidizes organic compounds. These compounds are oxidised by assistance of present Cl[-], which is converted to Cl_2, hypochlorite, chlorate, and perchlorate during oxidation process because of electrolysis, and also another oxidizers e.g. O_3, H_2O_2 and peroxides which might be exist because of the electrolyte support oxidation (Chen 2004; Arevalo and Calmano 2007). Existence of chloride ions leads the following chemical reactions to occur (Eqs. 1-3):

$$2Cl^- \rightarrow Cl_2 + 2e^- \tag{1}$$

$$Cl_2 + H_2O \rightarrow HOCl + H^+ + Cl^- \tag{2}$$

$$HOCl \rightarrow H+ + OCl^- \tag{3}$$

These anodic reactions take place concurrently with the following primary cathodic reactions (Eqs. 4-6):

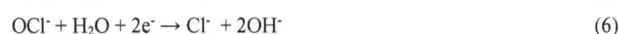

$$2H_2O + 2e^- \rightarrow 2OH^- + H_2 \tag{4}$$

$$2H_2O + O_2 + 4e^- \rightarrow 4OH \tag{5}$$

$$OCl^- + H_2O + 2e^- \rightarrow Cl^- + 2OH^- \tag{6}$$

Hypochlorite (OCl[-]) produced in solution is a powerful oxidizing agent that is able to oxidize aqueous organic compounds (Eq. 7):

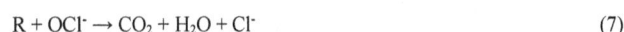

$$R + OCl^- \rightarrow CO_2 + H_2O + Cl^- \tag{7}$$

---

[*]Corresponding author E-mail: zahed51@yahoo.com

Owning to the inexpensive chloride, its nearly high solubility and strong oxidizing attributes of the produced active chlorine, indirect electrochemical oxidation of pollutants in wastewater by electrochemically generated chlorine was utilized. Moreover, it is stated that among chloride, nitrate and sulfate, chloride was the best sustenance electrolyte to oxidize electrochemical of refractory organic pollutants (Chen 2004).

Once the current density and concentration of chloride is augmented; the COD deduction efficiency is enhanced (Szpyrkowicz et al. 2005; Chiang and Chang 2001; Szpyrkowicz et al. 2001). The applied current density and the chloride concentration was significant factors that influence the COD deletion from landfill leachates.

Electrochemical waste destruction reveals numerous profits in terms of cost and safety, operational ease and equipment simplicity. The process generally executes at high electrochemical efficiency and operates necessarily under the same conditions for several types of organic wastes (Oturan et al. 2015). The chief drawback of Electrochemical (EC) systems are due to their high electrical energy consumption and sludge production that depend on the applied operational conditions and electrode material (Bayramoglu et al. 2004). Iron (Fe) electrodes, for example, are particularly notorious for sludge generation. Aluminum electrodes are relatively cheap and are not associated with major sludge generation and disposal problems as with iron electrodes. The aim of this study is to investigate the electrochemical oxidation of landfill leachate using aluminum electrodes. The effect of different operating parameters comprising pH, current density, contact time, electrolyte (NaCl) concentration and agitation rate on the deletion of COD and color was determined.

2. Materials and methods
2.1. Sampling

Sanitary landfill leachate used for the experiments was sampled from Pulau Burung Landfill Site (PBLS), Penang, Malaysia. The site is a semi-aerobic municipal landfill and has been developed as semi-aerobic sanitary landfill Level II by establishing a controlled tipping technique in 1991. Later in 2001, it was upgraded to a sanitary landfill Level III using controlled tipping with leachate (Mohajeri et al. 2010b). The collected landfill leachate was stored at 4°C to keep the wastewater characteristics unaffected before laboratory analysis and electrochemical treatment. Table 1 presents the characteristics of leachate used in the study.

Table 1. Leachate characteristics sampled from PBLS, Malaysia.

Parameters	Range	Average± Standard deviation
pH	8.3-8.8	8.5±0.19
COD (mg/L)	2380-2480	2450±40.62
Color (Pt.Co.)	3020-3150	3100±48.48
Turbidity (FAU)	220-255	240±13.69
TSS (mg/L)	120-135	130±6.12
Conductivity (µs/cm)	22400-25100	23900±994
Chloride (mg/L)	990-1410	1240±161
Sulfate (mg/L)	140-240	184±38
Nitrate (mg/L)	20-95	63±30
Temperature (°C)	27-29	28±0.84

2.2. Electrochemical oxidation

For every run, 500 mL of the leachate was put in the EC reactor. EC oxidation was conducted using prearranged current densities. The electric power needed for the electrolysis test was generated by a laboratory DC power supply 3 A-30 V (DAZHENG, PS-305D, China). Vertically positioned electrodes with a surface area of 15 cm² were placed side by side to each other in the electrolytic cell. They were separated by a distance of 3 cm. A magnetic stirrer maintained at the

desired value was used to mix the electrolytic cell in the solution well. Once the electrode material was disappeared > 10 %, the electrodes were switched. Experiments were carried out at room temperature and atmospheric pressure. The solution was then settled for half an hour and the supernatant was extracted once the run was terminated (Mohajeri et al. 2010a).

2.3. Analytical procedure

Analytical grade sodium chloride (Merck, Darmstadt, Germany) was utilized for electrolyte and source of chloride reactant. Various quantities of NaCl (500–3000 mg/L Cl⁻) were added to the leachate to test the influence of chloride concentration on the efficiency of COD and color removals (Mohajeri et al. 2010a).

The concentration of COD was quantified colorimetrically through a HACH DR/2010 (HACH Co., USA). The same spectrophotometer at 465 nm wavelength was also used to measure the color (APHA, 2017). Eq. (8) was used to discern removal percentage:

$$R = \frac{(C_0 - C_1)100}{C_0} \tag{8}$$

where C_0 presents the initial COD or color and C_1 indicates the residual COD or color after treatment. Experiments and analytical measurements were made in triplicate.

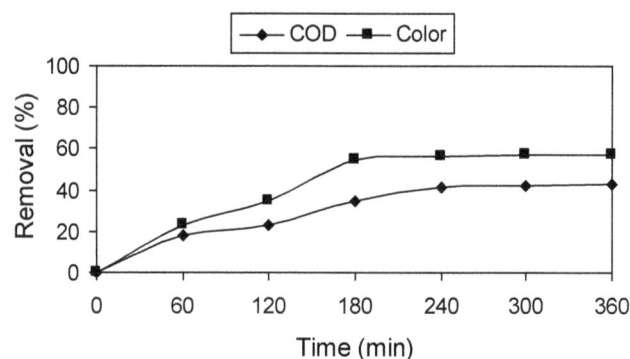

Fig. 1. Effect of reaction time on (a) the COD removal and (b) the color removal, (2000 mg/L NaCl, current density, 60 mA/cm², pH 8).

3. Results and discussion
3.1. Effect of reaction time on removal efficiencies

The initial pH, NaCl concentration and current density were kept constant at 8, 2000 mg/L, 60 mA/cm² respectively. Whereas the reaction time was differed between 30 and 360 min. Fig. 1. shows that color removal was more rapid than COD removal. The ideal operating time was ascertain to be 240 min; at which COD and color removal efficiencies were 42 % and 55 % respectively. Increases in removal efficiencies beyond 240 min were negligible. The choice of COD as the attribute to be degraded indicates that extended processing time will be needed.

3.2. Effect of pH on removal efficiencies

The initial pH of the solution was varied to test the effect of pH on the electrooxidation process. Fig. 2. shows the COD and color removals for samples containing 2000 mg/L NaCl, with pH values (adjusted with sulfuric acid or sodium hydroxide) of 4, 6, 8, 10 and 12, respectively. A current density of 60mA/cm2was employed. The results show that the organic compounds are oxidized preferably in the neutral and alkaline medium. The COD and color removal efficiencies did not change significantly at pH above 8. Alkaline conditions have been suggested to help enhance indirect oxidation (Moreira et al. 2016; Yi et al. 2008).

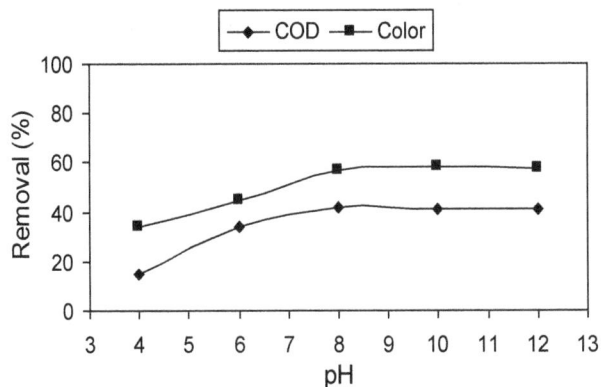

Fig. 2. Effect of initial pH on (a) the COD removal and (b) the color removal. (2000 mg/L NaCl, 240 min, current density, 60 mA/cm^2).

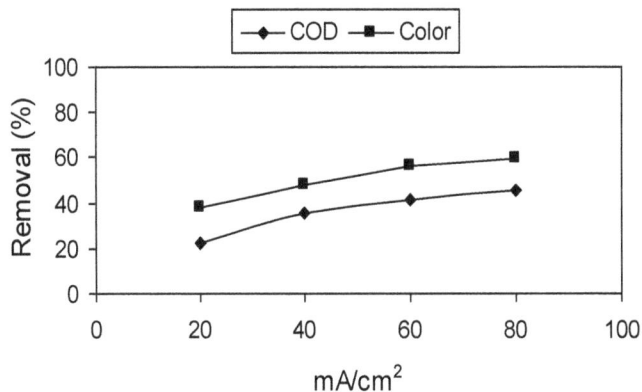

Fig. 3. Effect of current density on (a) the COD removal and (b) the color removal (2000 mg/L NaCl, 240 min)

With the initial pH increasing from 4 to 8, the COD and color removal efficiency also increased after 240 min of electrolysis. Constant COD and color removal efficiencies of about 42% and 55% respectively were attained when initial pH was about 8.0; which indicate that leachate could be effectively degraded at its normal pH. Hence, the pH of leachate was not adjusted in subsequent experiments.

Some researchers reported that the pH effect on the electrooxidation process is not considerable. They emphasized that the initial pH does not affect significantly the degradation of organic pollutants by indirect electrochemical oxidation in the range of 3.0–10.0 (Gotsi et al. 2005). Conversely, other researchers discovered that COD reduced significantly by pH. Some of them proposed acidic and others recommended alkaline condition to increase indirect oxidation (Moreira et al. 2016; Vlyssides et al. 2003). These discrepancies were probably due to the complex compositions of leachates and type of waste.

3.3. Effect e of current density on removal efficiencies

Current density (the electric current per unit surface area of the electrode) plays an important role in the progress of electrochemical elimination of organics owing to its ability to control the reaction rate. Researchers have applied current densities from 5 to 540 mA/cm^2 for electrochemical oxidation of landfill leachate (Deng and Englehardt, 2007). As well as increasing the treatment (electricity) cost, operation at high current densities may cause darkening of leachate and formation of brown precipitates at the anode surface under weak oxidative conditions (Li et al. 2001).

The influence of applied current upon the degradative behavior of landfill leachate was examined by electrolyzing the solution at 20, 40, 60 and 80 mA/cm2. As shown in Fig. 3, COD and color removals of samples both correlated positively with the applied current density. With an increase in the current density from 20 mA/cm2 to 80 mA/cm2, the COD and color removal efficiencies raised from 22% to 45% and 38% to 60%, respectively.

High current density of the anodic oxidation of water accelerates the generation of hydroxyl radicals which could ease the indirect electrochemical oxidation of organic compounds (Fernandes et al. 2015; Un et al. 2008; Yi et al. 2008). It is also possible that an augmented current density throughout electrochemical oxidation increases active chlorine generation, which is liable for the deletion of the pollutants (Krishna et al. 2010). Additionally, some researchers have also reported color removal from leachate fully dependent on current density (Mohajeri et al. 2010c; Gotsi et al. 2005; Moraes and Bertazzoli 2005). Even though a high current density is more advantageous to organic compounds degradation, a current density of 60 mA/cm2 was adopted for the subsequent tests, because of lower energy consumption (increase in current density from 60 mA/cm2 to 80 mA/cm2 resulted in less than 5% increase in COD and color removal efficiencies).

Current efficiency has been utilized to show a total efficacy because of the activity of direct oxidation using hydroxyl radicals and indirect oxidation through electro-generated active chlorine. The COD values calculated using Eq. 9 was utilized to determine the average current efficiency (ACE) percentage of oxidation (Montanaroa and Petrucci, 2009):

$$ACE = \frac{(COD_0 - COD_t)FV}{8It} \times 100 \qquad (9)$$

where COD$_0$ (g O$_2$ L^{-1}) indicates the initial chemical oxygen demand, CODt (g O$_2$ L^{-1}) stands for the chemical oxygen demands at a given time t (s), F for the Faraday constant (96,487 C mol^{-1}), V for the volume of the treated solution (L), I for the current applied (A) and 8 denotes the oxygen equivalent mass (g eq^{-1}). Generally, current density increases might cause a decline in the current efficiency (Mohajeri et al. 2010a). The maximal ACE of 34.7 % was detected for the current density of 20 mA/cm^2 in 240 min experimentation. The phenomenon may occur possibly owning to an enhance in the unwanted reaction of oxygen evolution (Hmani et al. 2009; Panizza and Cerisola, 2009) (Eq. 10):

$$2H_2O \rightarrow O_2 + 4H^+ + 4e^- \qquad (10)$$

These results approve active chlorine role in COD removal.

3.4. Effect of electrolyte concentration on removal efficiencies

A key matter in the application of electrochemical technique is supporting electrolyte usage, which can influence the conductivity of raw leachate because it simplifies the passage of current. Consequently, different concentrations of NaCl (electrolyte) was added to the system and variation in present COD and color removals was recorded.

To find the oxidation effect of active chlorine in electrochemical oxidation of leachate, chloride was added as the supporting electrolyte. Fig. 4. illustrates the electrolysis results regarding the COD and color removals at chloride ion concentrations ranging from 500 to 3000 mg/L. An increase of the Cl$^-$ concentration from 500 to 3000 mg/L raised the COD and color removal efficiencies from 25 to 45 % and 49 to 60% respectively.

The trend might be ascribed to the indirect electro-oxidation of organics by active chlorine (chlorine/hypochlorite), produced from Cl$^-$ at the anode (Un et al. 2008; Körbahti et al. 2007; Iniesta et al. 2001). Thus, the increased electrolyte concentration would improve removal efficiency. The additional NaCl will also decrease the power consumption owing to an elevation in conductivity as described (Mohajeri et al. 2010a). Albeit, addition of further Cl$^-$ generally improves oxidation efficiency, toxic chlorinated organic generated over treatment may impede extensive use of electrochemical oxidation process (Fernandes et al. 2016; Moreira et al. 2015; Mohajeri et al. 2010a).

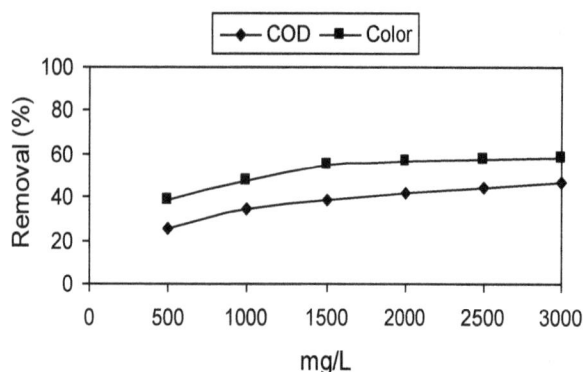

Fig. 4. Effect of electrolyte concentration on (a) the COD removal and (b) the color removal (240 min, current density, 60 mA/cm²).

3.5. Effect of the agitation rate on removal efficiencies

To investigate the effect of mass transfer on the electrochemical treatment of leachate, the influence of agitation rate during electro-oxidation of leachate was studied by changing the magnetic stirrer rate. The rate of agitation was ranged between 0–600 rpm, and the results are shown in Fig. 5. The rest of the process parameters and media conditions remain unaltered.

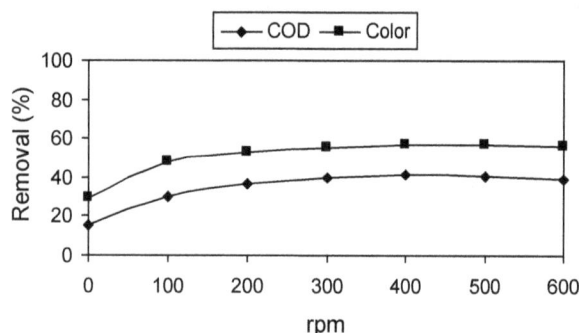

Fig. 5. Effect of agitation rate on (a) the COD removal and (b) the color removal (2000 mg/L NaCl, 240 min, current density, 60 mA/cm²).

Varying the agitation speed affected markedly the COD and color removal efficiencies. High agitation rates led to a more rapid and high efficient electrochemical process. As shown in Fig. 5. via elevation of rotation speed, COD and color removals are increased up to 400 rpm beyond that, there is no substantial increase in COD and color removals by increasing rpm. A similar trend was also observed by Bensalah and co-authors (2009) during their study on the electrochemical treatment of synthetic wastewaters on BDD anodes. These results revealed that stirring the mixture mechanically necessitates to achieve the maximum effect of electro-oxidation. The 400 rpm was selected for all experiments because the set-up may be disrupted in higher speeds.

3.6. Effect of the dilution on removal efficiencies

The effectiveness of dilution on leachate mineralization and decolorization were examined by a series of experiments. When the effect of initial COD concentration was investigated, experimental conditions were as follows: a current density of 60 mA/cm2, electrolyte concentration of 2000 mg/L and the original pH of the leachate solution. The leachate samples were diluted 50, 100, 150, and 200% with distilled water. Fig. 6, shows the influence of dilution during electrolyzes of leachate. It was found that dilution affects significantly the removal

efficiencies in the electrochemical oxidation process. At first, the degradation efficiency increased with decreasing the initial COD of leachate and then reached a maximum value of 57.1%, at a dilution of 50%. Further decreases of the initial COD lead to a sharp decrease of the removal efficiency from 57.1 % down to 36.6 % at dilution percentage of 200 %. It can be observed that high dilution provided low COD and color reduction. Oxidation of organic matters in leachate can be affected by dilution since the concentration of organics is a key factor. Like other chemical reactions, oxidation is limited in low concentrations.

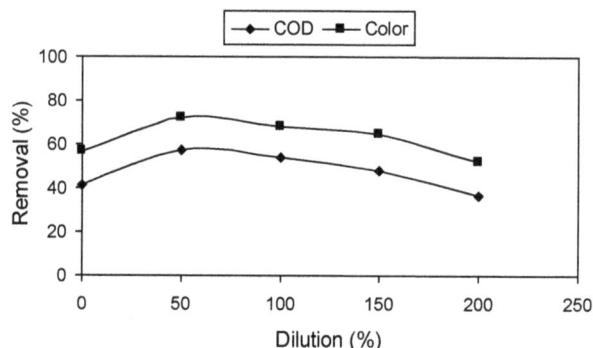

Fig. 6. Effect of dilution on (a) the COD removal and (b) the color removal (2000 mg/L NaCl, 240 min, current density, 60 mA/cm²).

3.7. Economic evaluation

To evaluate the economy of the electrochemical process, the required electrical energy consumption (EEc) per volume of leachate is calculated as kWh/m3 according to (El-Ashtoukhya et al. 2009) using Eq. (11):

$$EE_c = \frac{VCt}{S_v} \tag{11}$$

where V is voltage (volt), C is current (Ampere), t is treatment time (h) and Sv is sample volume (L). In this study the operational parameters of the experiment including electrolysis current, cell voltage, and the reaction time were determined to be 2 A, 8V and 240 min, respectively. The volume of the leachate treated was 0.5 L, therefore energy consumption was determined as 128 kWh/m³ for Pulau Burung semi-aerobic landfill leachate.

3.8. Comparison with other techniques

Other researchers (e.g. Feki et al. 2009; Ilhan et al. 2008; Moraes and Bertazzoli 2005) also utilized electrochemical oxidation technique for the treatment of landfill leachate. Moraes and Bertazzoli (2005) applied electrochemical treatment for landfill leachate in a pilot scale flow reactor, using oxide-coated titanium anode. The maximum COD removal of 73 % was acquired at a current density of 116.0 mA/cm² and 180 min of processing. In a study performed by Cabeza et al. (2007), complete removal of both COD and ammonium were obtained after 6–8 h using a boron-doped diamond (BDD) anode, to treat biologically and physicochemically treated leachates from a municipal landfill site of Meruelo in Cantabria, Spain. Ilhan and coauthors (2008) reported that more than 59 % of COD and 14 % of ammonia removal were achieved within 30 min reaction time for a current density of 631 mA/cm² using aluminum electrode. Feki et al. (2009) also found that under the optimal operational conditions (t = 1 h, J = 4 A dm⁻², Ti/Pt electrode), COD, color and heavy metal concentrations of the final effluent meet the Tunisian discharge standards in the sewer. In addition, Bashir et al. (2009) used graphite carbon electrodes and sodium sulfate as the electrolyte for electrochemical treatment of landfill. The maximum COD removal of 68 % was obtained when the reaction time was 4 h and the current density was 79.9 mA/cm^2 whilst the original COD was 1414 mg/L. In comparison to the previous literatures, we obtained specific conditions for electrochemical treatment of highly polluted leachate, indicating that electrochemical degradation could be used as a step of joint treatment. Our results here call for the more treatment of leachate to reach higher COD and color removals compliance with the environmental regulations.

4. Conclusions

Electrochemical oxidation process, using aluminum electrodes, was applied on mineralization and decolorization of landfill leachate.

The effects of operating parameters i.e., pH, reaction time, current density, electrolyte concentrations, agitation rate and dilution were tested. The results demonstrated that electrochemical oxidation could effectively eliminate COD and color from municipal landfill leachate under proper conditions. The highest deletion of COD and color were obtained as 41.6 % and 56.4 % respectively. By dilution of 50 % within this condition, COD and color removal raised to 57.1% and 72.0 % respectively. Electrochemical treatment may be a viable option for the degradation and remediation of landfill leachate. We therefore require additional treatments of leachate for increase of COD and color removals to comply with the environmental obligations.

References

APHA Standard Methods for the Examination of water and Wastewater. 23rd ed. American Public Health Association, Washington DC, pp. 1-733 (2017).

Arevalo E., Calmano W., Studies on the electrochemical treatment of wastewater contaminated with organotin compounds, Journal of Hazardous Materials 146 (2007) 540-545.

Bashir M.J.K., Isa M.H., Kutty S.R.M., Awang Z.B., Aziz H.A., Mohajeri S. and Farooqi I.H., Landfill leachate treatment by electrochemical oxidation, Waste Management 29 (2009) 2534-2541.

Bayramoglu M., Kobya M., Can O.T., Sozbir M., Operating cost analysis of electrocoagulation of textile dye wastewater, Separation and Purification Technology 37 (2004)117-125.

Bensalah N., Quiroz Alfarob M.A., Martínez-Huitle C.A., Electrochemical treatment of synthetic wastewaters containing Alphazurine A dye, Chemical Engineering Journal 149 (2009) 348-352.

Boye B., Dieng M.M., Brillas E., Anodic oxidation, electro-Fenton and photoelectro-Fenton treatments of 2, 4, 5-trichlorophenoxyacetic acid, Journal of Electroanalytical Chemistry 557 (2003) 135-146.

Chen G., Electrochemical technologies in wastewater treatment, Separation and Purification Technology 38 (2004) 11-41.

Chiang L.C., Chang J.E., Electrochemical oxidation combined with physical-chemical pretreatment processes for the treatment of refractory landfill leachate, Environmental Engineering Science 18 (2001) 369-379.

Deng Y., Englehardt J.D., Electrochemical oxidation for landfill leachate treatment, Waste Management 27 (2007) 380-388.

Deng Y., Englehardt J.D., Electrochemical oxidation for landfill leachate treatment, Waste Management 27 (2007) 380-388.

El-Ashtoukhya E.S.Z., Amina N.K., Abdelwahab O., Treatment of paper mill effluents in a batch-stirred electrochemical tank reactor, Chemical Engineering Journal 146 (2009) 205-210.

Fan Y., Ai Z., Zhang L., Design of an electro-Fenton system with a novel sandwich film cathode for wastewater treatment, Journal of Hazardous Materials 176 (2010) 678-684.

Feki F., Aloui F., Feki M., Sayadi S., Electrochemical oxidation post-treatment of landfill leachates treated with membrane bioreactor, Chemosphere 75 (2009) 256-260.

Fernandes, A., Pacheco, M. J., Ciríaco, L., & Lopes, A., Review on the electrochemical processes for the treatment of sanitary landfill leachates: Present and future, Applied Catalysis B: Environmental 176 (2015) 183-200.

Fernandes A., Santos D., Pacheco M.J., Ciríaco L., Lopes A., Electrochemical oxidation of humic acid and sanitary landfill leachate: Influence of anode material, chloride concentration and current density. Science of The Total Environment 541(2016) 282-291.

Gotsi M., Kalogerakis N., Psillakis E., Samaras P., Mantzavinos D., Electrochemical oxidation of olive oil mill wastewaters, Water Research 39 (2005) 4177-4187.

Hermosilla D., Cortijo M., Huang C.P., Optimizing the treatment of landfill leachate by conventional Fenton and photo-Fenton processes, Science of the Total Environment 407 (2009) 3473-3481.

Hmani E., Chaabane Elaoud S., Samet Y., Abdelhédi R., Electrochemical degradation of waters containing O-Toluidine on PbO$_2$ and BDD anodes, Journal of Hazardous Materials 170 (2009) 928-933.

Ilhan F., Kurt U., Apaydin O., Gonullu M.T., Treatment of leachate by electrocoagulation using aluminum and iron electrodes, Journal of Hazardous Materials 154 (2008) 381-389.

Iniesta J., Gonzalez-Garcia J., Exposito E., Montiel V., Aldaz A., Influence of chloride ion on electrochemical degradation of phenol in alkaline medium using bismuth doped and pure PbO2 anodes, Water Research 35 (2001) 3291-3300.

Körbahti B.K., Aktaş N., Tanyolaç A., Optimization of electrochemical treatment of industrial paint wastewater with response surface methodology, Journal of Hazardous Materials 148 (2007) 83-90.

Krishna B.M., Murthy U.N., Manoj Kumar B., Lokesh K.S., Electrochemical pretreatment of distillery wastewater using aluminum electrode, Journal of Applied Electrochemistry 40 (2010) 663-673.

Li M., Wang M., Jiao Z.K., Chen Z.Y., Study on electrolytic oxidation for landfill leachate treatment, China Water and Wastewater 17 (2001) 14-17.

Mohajeri S., Hamidi A.A., Isa M.H., Zahed M.A., Landfill Leachate Treatment through electro-Fenton oxidation, Pollution 5 (2019) 199-209.

Mohajeri S., Aziz H.A., Isa M.H., Bashir M.J.K., Mohajeri L., Adlan M.N., Influence of Fenton reagent oxidation on mineralization and decolorization of municipal landfill leachate, Journal of Environmental Science and Health - Part A Toxic/Hazardous Substances and Environmental Engineering 45 (2010b) 692-698.

Mohajeri S., Aziz H.A., Isa M.H., Zahed M.A., Adlan M.N., Statistical optimization of process parameters for landfill leachate treatment using electro-Fenton technique, Journal of Hazardous Materials 176 (2010c) 749-758.

Mohajeri S., Aziz H.A., Isa M.H., Zahed M.A., Bashir M.J.K., Adlan M.N., Application of the central composite design for condition optimization for semi aerobic landfill leachate treatment using electrochemical oxidation, Water Science and Technology 61 (2010a) 1257-1266.

Montanaroa D., Petrucci E., Electrochemical treatment of Remazol Brilliant Blue on a boron-doped diamond electrode, Chemical Engineering Journal 153 (2009) 138-144.

Moraes P.B., Bertazzoli R., Electrodegradation of landfill leachate in a flow electrochemical reactor, Chemosphere 58 (2005) 41-46.

Moreira F.C., Soler J., Fonseca A., Saraiva I., Boaventura R.A., Brillas E., Vilar V.J., Incorporation of electrochemical advanced oxidation processes in a multistage treatment system for sanitary landfill leachate, Water Research 81 (2015) 375-87.

Moreira F.C., Soler J., Fonseca, A., Saraiva, I., Boaventura R. A., Brillas E., Vilar V.J., Electrochemical advanced oxidation processes for sanitary landfill leachate remediation: Evaluation of operational variables, Applied Catalysis B: Environmental 182 (2016) 161-171.

Oturan N., Van Hullebusch E.D., Zhang H., Mazeas L., Budzinski H., Le Menach K., Oturan M.A., Occurrence and removal of organic micropollutants in landfill leachates treated by electrochemical advanced oxidation processes, Environmental Science & Technology 49 (2015) 12187-12196.

Panizza M., Cerisola G., Electrochemical degradation of gallic acid on a BDD anode, Chemosphere 77 (2009) 1060-1064.

Rivas F.J., Beltrán F., Carvalho F., Gimeno O., Frades J., Study of Different Integrated Physical–Chemical+Adsorption Processes for Landfill Leachate Remediation, Industrial & Engineering Chemistry Research 44 (2005) 2871-2878.

Shao L., He P., Xue J., Li G., Electrolytic degradation of biorefractory organics and ammonia in leachate from bioreactor landfill, Water Science and Technology 53 (2006) 143-150.

Szpyrkowicz L., Kaul S.N., Neti R.N., Satyanarayan S., Influence on the anode material on electrochemical oxidation for the treatment of tannery wastewater, Water Research 39 (2005) 1601-1613.

Szpyrkowicz L., Kelsall G., Kaul S.N., De Faveri M., Satyanarayan S., Performance of electrochemical reactor for treatment of tannery wastewaters, Chemical Engineering Science 56 (2001) 1579-1586.

Un U.T., Altay U., Koparal A.S., Ogutveren U.B., Complete treatment of olive mill wastewaters by electrooxidation, Chemical Engineering Journal 139 (2008) 445-452.

Vlyssides A.G., Karlis P.K., Mahnken G., Influence of various parameters on the electrochemical treatment of landfill leachates, Journal of Applied Electrochemistry 33 (2003) 155-159.

Wang Y., Gong B., Lin Z., Wang J., Zhang J., Zhou J., Robustness and microbial consortia succession of simultaneous partial nitrification, ANAMMOX and denitrification (SNAD) process for mature landfill leachate treatment under low temperature, Biochemical Engineering Journal 10 (2018) 1-15.

Yi F., Chen S., Yuan C., Effect of activated carbon fiber anode structure and electrolysis conditions on electrochemical degradation of dye wastewater, Journal of Hazardous Materials 157 (2008) 79-87.

Yusoff M.S., Aziz H.A., Zamri M.F.M.A., Abdullah A.Z., Basri N.E.A., Floc behavior and removal mechanisms of cross-linked Durio zibethinus seed starch as a natural flocculant for landfill leachate coagulation-flocculation treatment, Waste Management 74 (2018) 362-372.

25

A comparative study on reaction kinetic of textile wastewaters degradation by UV/TiO$_2$ and UV/ZnO

Mojtaba Ahmadi*, Pegah Amiri

Chemical Engineering Department, Faculty of Engineering, Razi University, Kermanshah, Iran.

ARTICLE INFO	ABSTRACT
	A comparative kinetic study of photocatalytic degradation of textile wastewater by UV/TiO$_2$ and UV/ ZnO for removal of chemical oxygen demand (COD), and color was carried out. The effects of some parameters such as the initial concentration of catalyst, initial COD concentration and light intensity on the photocatalytic process were also examined. It was demonstrated that the COD removal by TiO$_2$/UV and ZnO/UV was about 49 %, 33.3 %, and color removal was 30 % and 10 %, respectively. The experiment demonstrated that the photo-degradation efficiency of TiO$_2$ was significantly higher than that of ZnO. On the other hand, the kinetic study shows that decomposition of chemical oxygen demand follows a first-order for processes. The rate of degradation is highly dependent on the initial concentration of TiO$_2$, ZnO and light intensity. A comparison between experimental and calculated degradation rate constants shows that TiO$_2$/UV process gives better results than photocatalytic treatment. Maximum degradation rate was achieved for TiO$_2$/UV at optimum concentrations of TiO$_2$.

Keywords:
Kinetics
Photocatalytic degredation
Textile wastewater
Titanium dioxide
Zinc oxide

1. Introduction

Textile industry is one of the largest industries in various parts of the world. Water pollution by the textile mills is the result of wastewater discharge from various stages of production like desizing, scouring, bleaching, mercerizing, dyeing, and printing (Barka et al. 2011). The colored wastewater discharged from a textile industry exhibits low biochemical oxygen demand (BOD), high values of COD, changeable pH, suspended solids, and organic chlorine compounds (Lee and Yoon. 2004). An important characteristic of the textile mills is the use of different types of dye (Aslam et al. 2004). Thus, removal of color is the most difficult constituent of the textile wastewater (Oguz and Keskinler. 2008). The physical, chemical and mostly biological technologies that have been widely used to treat textile effluents are: advanced oxidation methods, physic-chemical methods like adsorption (Belaid et al. 2013; Rangabhashiyam et al. 2013), electro-catalytic treatment (Ibrahim et al. 2013), and biological sludge methods (Wu et al. 2008).

Advanced oxidation processes (AOP) such as ozonation, UV and ozone/UV combined oxidation, photo catalysis (UV/TiO$_2$), Fenton's reagent, and ultrasonic oxidation are based on the production of hydroxyl radicals as oxidizing agents to mineralize organic chemicals (Blanco et al. 2012). The reaction mechanism of organic compounds with hydroxyl radicals is very complex, so that the mechanism can be briefly described in three stages including: 1. initiation reaction: in this process, free radicals are produced. 2. propagation reaction: the radicals generated in the previous step are converted to other free radicals, and 3. termination Reaction: in this process, the free radicals produce a stable compound (Lin et al. 1998). Among the different existing methods, heterogeneous photocatalysis has been shown to be successful and beneficial for degradation of wastewater organic pollutants (Yahiat et al. 2011). In photocatalysis systems, combination of semiconductors (e.g. TiO$_2$, ZnO, ZnS, WO$_3$, CdS and SrTiO$_3$) and UV or visible lights can be used. Semiconductors are characterized by two separate energy bands: a low-energy valence band and a high-energy conduction band. Photon energy, hv, with a wavelength of 387.5 nm can be used to excite an electron from the valence band into the conduction band (Al-Momani et al. 2002; Ren et al. 2010). The hole generated can degrade organic pollutants in wastewater to CO$_2$ and H$_2$O$_2$ (Aguedach et al. 2005). TiO$_2$ and ZnO among the semiconductors that have been most studied, present good features of stability, non-toxicity and insolubility (Yeber et al. 2000). In addition, since ozone is a

powerful oxidant in textile wastewater treatment, it generally used to remove stable organic compounds and dyes from industrial effluents (Langlais et al. 1991). With dissolving ozone in water, it oxidizes the organic ~compounds in two different ways: one way is by a direct reaction of molecular ozone, and the other way is through free radicals as OH˙ radicals produced by ozone decomposition in water (Perkowski et al. 1996; Baig and Liechti. 2001; Soares et al. 2006; Somensi et al. 2010). Advanced oxidation processes have a significant impact on the ability to reduce COD and treatment of non-biodegradable pollutants. As a result of better access to these methods, the processes can be combined with biological treatment (Tabrizi and Mehrvar. 2004; Wang et al. 2009).

Among the advanced oxidation processes the homogeneous AOPs employing Titanium dioxide photocatalysis process (Ghezzar et al. 2009; Foo and Hameed. 2010), UV/TiO$_2$ (Yu et al. 2010; Botía et al. 2012; Gupta et al. 2012), ZnO (Chakrabarti and Dutta. 2004; Ahsan Habib et al., 2012) and ozone (Soares et al. 2006; Oguz and Keskinler. 2008; Wu et al. 2008; Somensi et al. 2010) have been found to be very effective to degrade dye and pollutants. Ghezzar et al. (2009) studied the bleaching and textile effluent degradation with the plasma-catalytic process. The best conditions of experimental parameters for the removal of textile wastewaters have been determined and the wastewater was completely decolorized after only 30 min (Ghezzar et al. 2009). The research of Akyol et al. (2004) was conducted using ZnO for degradation of Remazol Red RR dye (Akyol et al. 2004). Peralta-Zamora et al. reported the treatment of effluent from the cellulose and textile industries by applying a heterogeneous photocatalytic procedure using ZnO, TiO$_2$ by supported ZnO on the photoassisted (Peralta-Zamora et al. 1998). The results showed that the treatment of textile effluents by UV-irradiation in the presence of free TiO$_2$ and ZnO or silica gel supported ZnO, the color fades progressively to reach substantial decolorization ratios. Using of ozone as pretreatment of textile wastewater in a pilot-scale plant was investigated by Somensi et al. (2010) ozonation was enhanced the biodegradability of textile wastewater (BOD$_5$/COD ratios) by a factor of up to 6.8-fold (Somensi et al. 2010). Kinetic study of decolorization C.I. Basic Blue 3 with immobilized TiO$_2$ nanoparticles photocatalytic was studied by Khataee and Mirzajani et al. (2010). The kinetic characteristics of the photocatalytic degradation of BB3 by using immobilized titanium dioxide nanoparticles were experimentally investigated (Khataee and Mirzajani. 2010). Gupta et al. (2012) investigated mineralization of toxic

*Corresponding author Email: m_ahmadi@razi.ac.ir

dye amaranth on TiO_2/UV. The decolorization and degradation kinetics were investigated followed a pseudo first order kinetics with regards to the substrate concentration under the experimental conditions (Gupta et al. 2012). The research of Bensaadi et al. (2014) was conducted by using photodegradation with TiO_2 for mineralization of acebutolol. The results showed that the photodegradation follows a pseudo-first-order kinetic.

The aim of this study is to compare photocatalysts techniques by TiO_2 and ZnO for reduction of color and COD textile wastewater. The primary focus is to study the effect of TiO_2 and ZnO concentration, light intensity, initial concentration of COD in on the photocatalytic system. In the literature, only a limited number of studies have so far been focused on the kinetic analysis of degradation textile wastewater so the experimental data were also analyzed by using the first-order kinetic model.

2. Materials and methods
2.1. Material

The used wastewater for the laboratory-scale was obtained from the textile plant. The characteristics of the untreated wastewater are given in Table 1. The performance of two catalysts was compared in the photocatalytic Treatment. TiO_2 (P-25) photocatalyst was purchased from Degussa Co. Ltd., Germany (anatase 75 %, rutile 25 %, BET specific surface area 48 m^2/g, and mean particle size 25 nm) and ZnO purchased from Merk Co.Ltd., Germany. The absorbance at specific wavelengths of the supernatants was concluded by using ultra violet visible spectrophotometer, (model UV-2100 UV/Vis Spectrophotometer). Decolorization efficiency (DE) was estimated from a mathematical equation based on calculation of decolorization used before:

$$DE = \frac{(absorbance)_0 - (absorbance)_t}{(Absorbance)_0} \times 100$$

where $(absorbance)_0$ and $(absorbance)_t$ are the absorbance before irradiation and the absorbance at time t, respectively.

2.2. Experimental set-up
2.2.1. Photoreactor

The photocatalytic experiments used to photodegradation of textile wastewater were performed in a Pyrex glass vessel photoreactor equipped with an aeration system, as shown in Fig. 1. The photoreactor was illuminated by a 125 W mercury lamp with a peak light intensity at 254 nm located in the distance of 12 cm from center of suspension surface. The temperature of the suspension in the photoreactor was kept constant at 30 0C by circulating water in bathroom water, and the irradiation time was 2 h. The photocatalytic degradation researches were done by loading 500ml of the wastewater solutions in the photocatalytic reactor. The effects of initial concentration of textile wastewater and photocatalyst concentrations were examined by using different initial COD concentrations (500–1650 mg/l) and varying

amount of the photocatalyst from 0.125 to 1 mg/l with an initial COD concentration of 1650 mg/l, respectively. Tests under two 125 W and 250 W light sources were separately done and also sampled once every 30 minutes. The samples were taken from the reactor at scheduled times, and COD was measured.

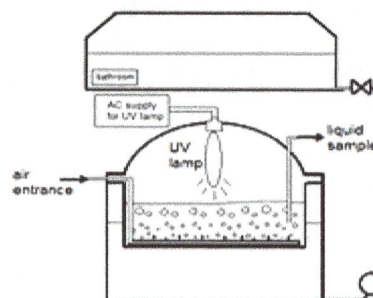

Fig. 1. Schematic of photocatalytic reactor.

3. Results and discussion
3.1. UV/ TiO_2 photocatalytic process
3.1.1. Effect of initial textile wastewater concentration

The effect of initial concentration of the textile wastewater on the COD removal is shown in Fig. 2. The figure indicates that the initial concentration of the textile wastewater is an effective parameter on removal efficiency. The decrease in photooxidative degradation with increasing initial textile wastewater concentration has been observed. As increasing of the concentration from 500 to 1650 mg/l, the decolorization efficiency decreased from 80 % to 49 %. The slope of the plots increases with decreasing initial concentration, i.e., when initial concentration increased, the catalyst efficiency decreased. It was noticed that the maximum percentage of removal occurred when initial concentration was 500 mg/L. The higher textile wastewater concentration, and the more absorption of UV light are the presumed reasons. Photoreactor solution is more resistant to UV radiation (Sleiman et al. 2007; Barka et al. 2010).

Table 1. Characteristics of textile wastewater.

Parameters	Value
COD (mg O_2 L^{-1})	1650
BOD (mg O_2 L^{-1})	390
BOD/COD	0.23
pH	11
Color (A_{335})	0.81

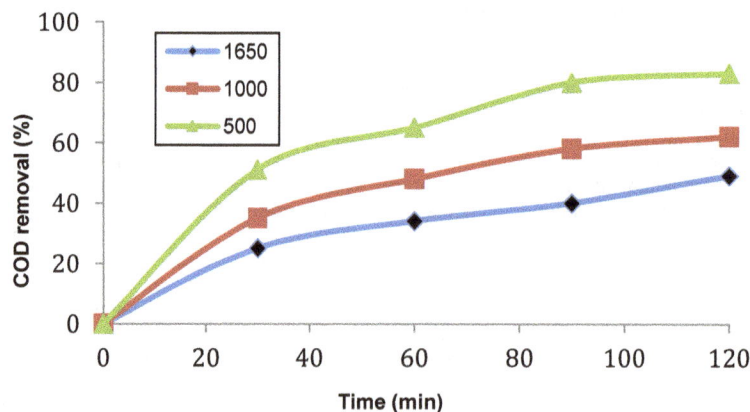

Fig. 2. Effect of initial COD concentration on TiO_2/UV treatment (TiO_2 concentration: 0.25 g/l, Light source: 250 W).

3.1.2. Effect of initial concentration of TiO$_2$

The initial concentration of TiO$_2$ was found to be an important parameter for the photooxidative photocatalysis process of textile wastewater in the UV/TiO$_2$ process. The effect of initial TiO$_2$ concentration on photodegradation efficiency has been expressed in Fig. 3. As the concentration of TiO$_2$ increased, the photocatalytic efficiency increased proportionally. Also, at higher concentrations, this beneficial effect tends to level off and then decreases because of the screening effect of excess particles which prevent of the photosensitive surface active sites and impediment the UV irradiation(Sleiman et al. 2007). On the other hand, the photocatalytic efficiency on color removal was observed 27 and 30 %   for light sources 125 and 250 W respectively. As seen in Fig. 3, the COD removal increases with increasing light power. It is hoped that kinetic modeling and studies will explain the effect of the light intensity on the photocatalytic efficiency.

3.1.3. Kinetics analyses

The kinetics of the photocatalytic degradation rate of most organic com pounds is described by pseudo-first order kinetics:

$$-\frac{dC(t)}{dt} = K_{app}C(t) \tag{1}$$

where t is time, C is the COD concentration, and K$_{app}$ is the apparent rate constant. By integrating the equations (1) with the initial condition C(0)=C0 , the following equations could be obtained:

$$\ln\left(\frac{C}{C_0}\right) = -K_{app}t \tag{2}$$

Regression analysis based on the first-order reaction kinetics for the photocatalytic degradation of textile wastewater in the UV/TiO$_2$ process was conducted, and the results were shown in Fig. 4. The results indicated that the photodegradation kinetics of textile wastewater followed the first-order kinetics well. The rate constants at different initial TiO$_2$ concentration were obtained and the results were shown in Table 2. Increasing TiO$_2$ particles prevents UV light from the light source, so COD removal was reduced. The results indicate that 0.25 g/l of TiO$_2$ is sufficient for the maximum rate of decolorization and COD removal. The proposed kinetic model is also in good agreement with our experimental data. Increasing the light intensity from 125 to 250 W could increase the kinetic rate constants of UV/ TiO$_2$ photocatalytic process. This observation agrees to the results of other studied (Khataee et al. 2011). It is reasonable, since the light intensity increased, more hydroxyl radicals are available to attack the dye pollutants and the photocatalytic reaction rate constant increases.

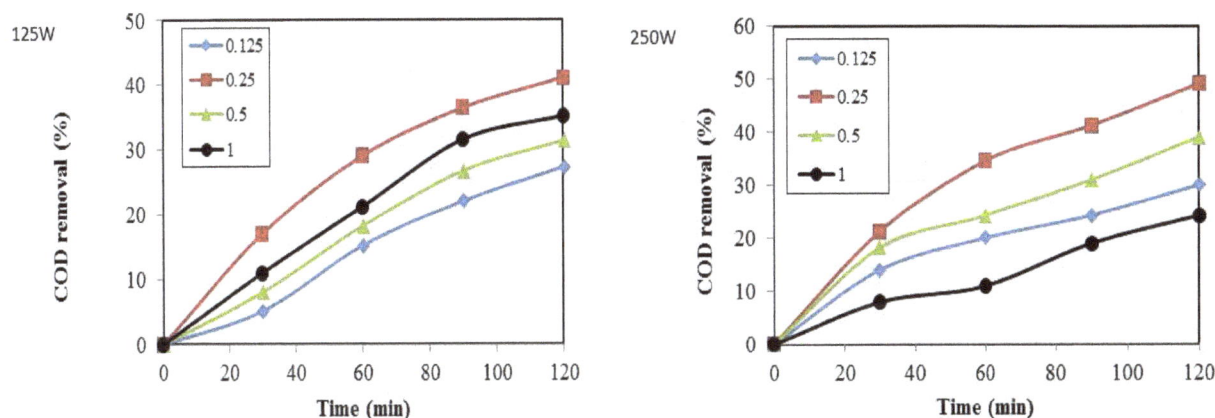

Fig. 3. Effect of initial concentration of TiO$_2$ for different light intensity (initial COD: 1650mg/l, temperature: 30 °C).

3.2. UV/ ZnO photocatalytic process
3. 2.1. Effect of initial concentration of ZnO

Effect of catalyst mass on COD removal efficiency of real textile industrial wastewater has also been investigated by employing different masses of ZnO varying from 0.125 to 1 g/l under UV irradiation, and the presented results in Table 3 are plotted in Figs. 5 and 6 shows results curves of rate constant. K and R^2 calculated values are given in Table 4. As can be seen, the removal rate increased with increasing of ZnO particles. The results indicate that 1g/l of ZnO is sufficient for the maximum rate of removal COD increasing of initial concentration of ZnO increases the catalyst surface and light absorption by the catalyst surface. Akyol et al. (2004) studied the effect of the catalyst loading in photodegradation of Remazol Red RR in aqueous ZnO suspensions. They commented that the decolorization efficiency increases by increasing loading of ZnO.

The results of the effect of titanium dioxide (anatase and rutile) and zinc oxide masses on the photocatalytic decolorization of real and simulated textile dyeing wastewater shows that the decolorization efficiency is increased with increasing in the mass of catalysts, and it becomes constant at a certain mass, and then starts to decrease with increasing in the mass of catalyst further. The increasing of catalytic activity with increasing in the mass of catalyst was explained due to increasing availability of photocatalysts sites, and the decreasing of catalytic activity after the plateau region is related to increasing of light scattering due to the excess of catalyst.

3.2.2. Effect of light intensity

The rate of photocatalysis activation, and electron–hole formation in photochemical reaction is strongly dependent on light intensity (Cassano and Alfano. 2000). For studying the influence of UV intensity on photocatalytic efficiency, different UV source lamps with powers of 125 and 250 W were used. All experiments were created in same conditions, so the obtained results are shown in Figs. 3 and 5 obviously, with increasing the UV light intensity from 125 to 250 W the removal COD efficiency was increased. Generally, the light intensity of the solution had a significant effect on kinetics of the reactions. It was observed that by an increase in light intensity, an increase in rate constant occurred. This can be due to leading to more catalyst activation and more photolcatalytic degradation of textile wastewater.

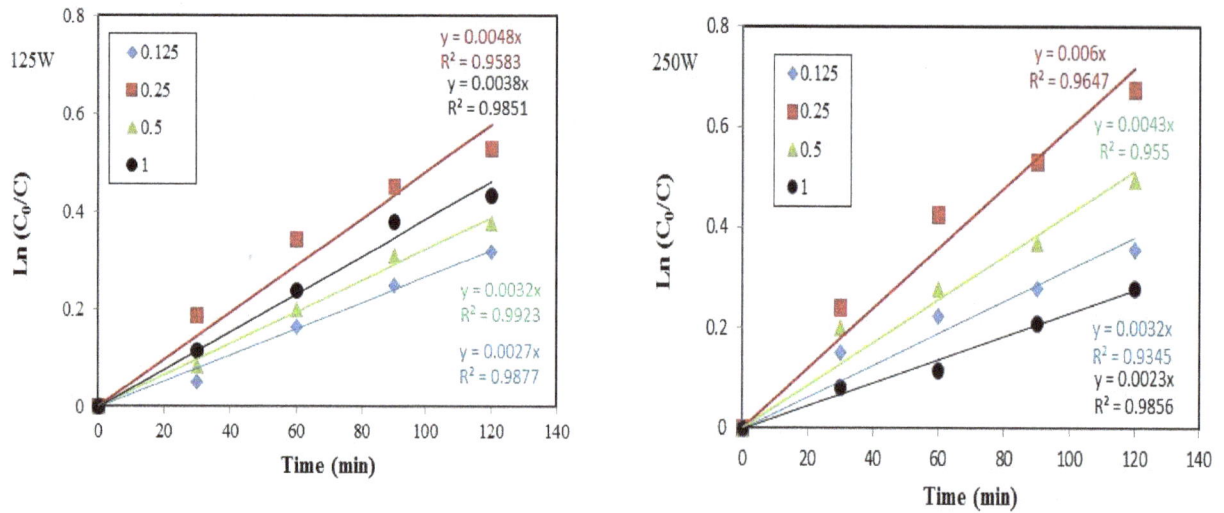

Fig. 4. Effect of initial TiO_2 concentration on decomposition rate constant (initial COD: 1650 mg/l, temperature: 30 °C).

Table 2. The first-order kinetic rate constants of UV/ TiO_2 photocatalytic process at 30 °C and under different initial TiO_2 concentration.

TiO_2 (g/l)	UV source lamp (W)			
	125		250	
	K (1/min)	R^2	K (1/min)	R^2
0.125	0.0027	0.987	0.0032	0.934
0.25	0.0048	0.958	0.0060	0.965
0.5	0.0032	0.992	0.0043	0.955
1	0.0038	0.985	0.0023	0.985

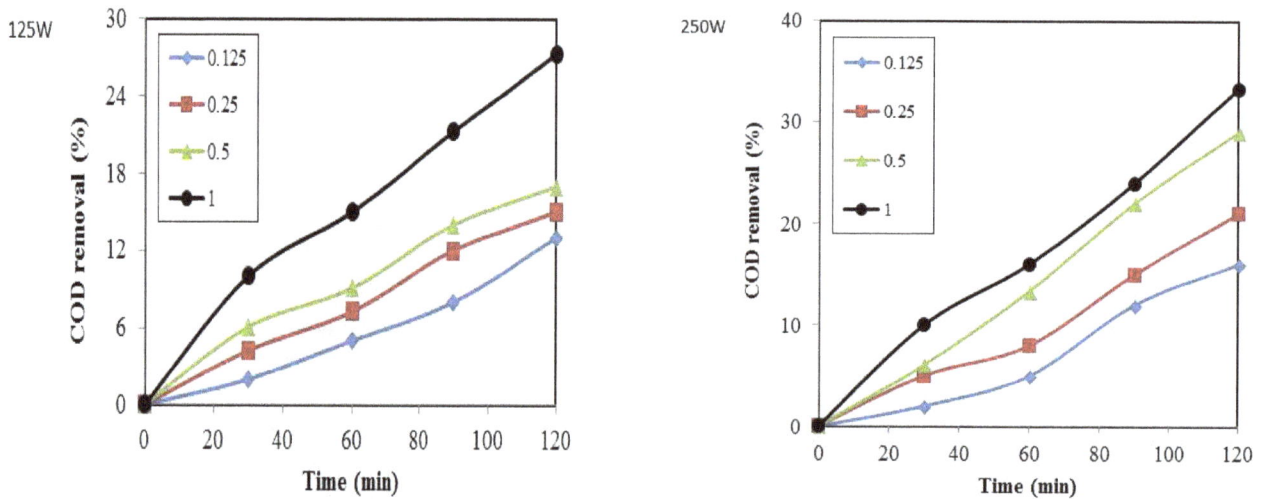

Fig. 5. Effect of initial concentration of ZnO for different light intensity.

Table. 3. Color removal percent after ZnO/UV process.

	Color (A_{335})	Color removal (%)
Raw wastewater	0.81	
After (ZnO/UV: 250 watt)	0.72	10
After (ZnO/UV:125 watt)	0.74	8

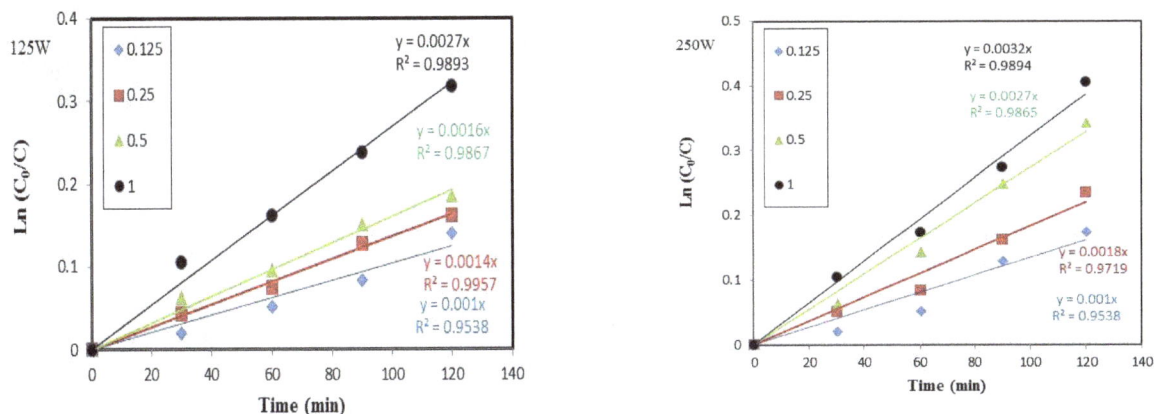

Fig. 6. Effect of initial ZnO concentration on decomposition rate constant.

Table 4. Rate constants of UV/ ZnO photocatalytic process at 30 °C and under different initial ZnO.

TiO_2 (g/l)	UV source lamp (W)			
	125		250	
	K (1/min)	R^2	K (1/min)	R^2
0.125	0.001	0.953	0.0010	0.954
0.25	0.0014	0.995	0.0018	0.971
0.5	0.0016	0.987	0.0027	0.986
1	0.0027	0.989	0.0032	0.989

4. Conclusions

In this study, photocatalytic processes for the removal of dyes and organic load of textile effluent were examined. The results show that the degree of photodegradation of textile industrial wastewater are obviously influenced by several parameters. Under optimal conditions, the extent of decolorization and COD removal were achieved about 80 % by using TiO_2, and 37 % by using ZnO at 303 K, respectively. The kinetic study shows that COD removal follows first-order models

adequate to describe the kinetic of biodegradation process. It is clear that mineralization of the total organic content of the textile wastewater is not economical during photocatalytic in a pilot-scale, but partial oxidation altering the original product can have beneficial effects by reducing toxicity effects of wastewater, and enhancing the biodegradability of organic wastewater contents in a sequential microbiological treatment of textile wastewaters.

References

Aguedach A., Brosillon S., Morvan J., Lhadi E.K., Photocatalytic degradation of azo-dyes reactive black 5 and reactive yellow 145 in water over a newly deposited titanium dioxide, Applied Catalysis B: Environmental 57 (2005) 55-62.

Ahsan Habib M., Ismail I.M.I., Mahmood A.J., Rafique Ullah M., Photocatalytic decolorization of Brilliant Golden Yellow in TiO_2 and ZnO suspensions, Journal of Saudi Chemical Society 16 (2012) 423-429.

Akyol A., Yatmaz H.C., Bayramoglu M., Photocatalytic decolorization of Remazol Red RR in aqueous ZnO suspensions, Applied Catalysis B: Environmental 54 (2004) 19-24.

Aslam M.M., Baig M., Hassan I., Qazi I.A., Malik M., Saeed H., Textile wastewater characterization and reduction of its COD and BOD by oxidation, EJEAF CHE 3 (2004) 804-811.

Baig S., Liechti P., Ozone treatment for biorefractory COD removal, Water Science & Technology 43 (2001) 197-204.

Barka N., Abdennouri M., Makhfouk M.E.L., Removal of Methylene Blue and Eriochrome Black T from aqueous solutions by biosorption on Scolymus hispanicus L.: Kinetics, equilibrium and thermodynamics,

Journal of the Taiwan Institute of Chemical Engineers 42 (2011) 320-326.

Barka N., Qourzal S., Assabbane A., Nounah A., and Ait-Ichou Y., Photocatalytic degradation of an azo reactive dye, Reactive Yellow 84, in water using an industrial titanium dioxide coated media, Arabian Journal of Chemistry 3 (2010) 279-283.

Belaid K.D., Kacha S., Kameche M., Derriche Z., Adsorption kinetics of some textile dyes onto granular activated carbon, Journal of Environmental Chemical Engineering 1 (2013) 496-503.

Bensaadi Z., Yeddou-Mezenner N., Trari M., and Medjene F., Kinetic studies of β blocker photodegradation on TiO_2, Journal of Environmental Chemical Engineering 2 (2014) 1371-1377.

Blanco J., Torrades F., De la Varga M., García-Montaño J., Fenton and biological-Fenton coupled processes for textile wastewater treatment and reuse, Desalination 286 (2012) 394-399.

Botía D.C., Rodríguez M.S., Sarria V.M., Evaluation of UV/TiO_2 and UV/ZnO photocatalytic systems coupled to a biological process for the treatment of bleaching pulp mill effluent, Chemosphere 89 (2012) 732-736.

Cassano A.E., Alfano O.M., Reaction engineering of suspended solid heterogeneous photocatalytic reactors, Catalysis Today 58 (2000) 167-197.

Chakrabarti S., Dutta B.K., Photocatalytic degradation of model textile dyes in wastewater using ZnO as semiconductor catalyst, Journal of Hazardous Materials 112 (2004) 269-278.

Cristina Yeber M., Rodríguez J., Freer J., Durán N., Mansilla H.D., Photocatalytic degradation of cellulose bleaching effluent by supported TiO_2 and ZnO, Chemosphere 41 (2000) 1193-1197.

Foo K.Y., Hameed B.H., Decontamination of textile wastewater via TiO_2/activated carbon composite materials, Advances in Colloid and Interface Science 159 (2010) 130-143.

Ghezzar M.R., Abdelmalek F., Belhadj M., Benderdouche N., Addou A., Enhancement of the bleaching and degradation of textile wastewaters by Gliding arc discharge plasma in the presence of TiO_2 catalyst, Journal of Hazardous Materials 164 (2009) 1266-1274.

Gupta V.K., Jain R., Mittal A., Saleh T.A., Nayak A., Agarwal S., Sikarwar S., Photo-catalytic degradation of toxic dye amaranth on TiO_2/UV in aqueous suspensions, Materials Science and Engineering: C 32 (2012) 12-17.

Ibrahim D.S., Praveen Anand A., Muthukrishnaraj A., Thilakavathi R., and Balasubramanian N., In situ electro-catalytic treatment of a Reactive Golden Yellow HER synthetic dye effluent, Journal of Environmental Chemical Engineering 1 (2013) 2-8.

Khataee A.R., Mirzajani O., UV/peroxydisulfate oxidation of C. I. Basic Blue 3: Modeling of key factors by artificial neural network, Desalination 251 (2010) 64-69.

Khataee A.R., Fathinia M., Aber S., Kinetic study of photocatalytic decolorization of C.I. Basic Blue 3 solution on immobilized titanium dioxide nanoparticles, Chemical Engineering Research and Design 89 (2011) 2110-2116.

Langlais B., Reckhow D.A., Brink D.R., Ozone in water treatment: Application and engineering: Cooperative research report, CRC Press (1991).

Lee C., Yoon J., Application of photoactivated periodate to the decolorization of reactive dye: reaction parameters and mechanism, Journal of Photochemistry and Photobiology A: Chemistry 165 (2004) 35-41.

Lin B., Yamaguchi R., Hosomi M., Murakami A., A new treatment process for photo-processing waste using a sulfur-oxidizing bacteria/granular activated carbon system followed by fenton oxidation, Water Science and Technology 38 (1998) 163-170.

Neamtu M., Siminiceanu I., Yediler A., Kettrup A., Kinetics of decolorization and mineralization of reactive azo dyes in aqueous

solution by the UV/H_2O_2 oxidation, Dyes and Pigments 53 (2002) 93-99.

Oguz E., Keskinler B., Removal of colour and COD from synthetic textile wastewaters using O_3, PAC, H2O2 and HCO3⁻, Journal of Hazardous Materials 151 (2008) 753-760.

Peralta-Zamora P., Moraes S.G., Pelegrini R., Freire Jr M., Reyes J., Mansilla H., and Durán N., Evaluation of ZnO, TiO_2 and supported ZnO on the photoassisted remediation of black liquor, cellulose and textile mill effluents, Chemosphere 36 (1998) 2119-2133.

Perkowski J., Kos L., Ledakowicz S., Application of Ozone in Textile Wastewater Treatment, Ozone: Science & Engineering 18 (1996) 73-85.

Rangabhashiyam S., Anu N., Selvaraju N., Sequestration of dye from textile industry wastewater using agricultural waste products as adsorbents, Journal of Environmental Chemical Engineering 1 (2013) 629-641.

Ren C., Yang B., Wu M., Xu J., Fu Z., lv Y., Guo T., Zhao Y., Zhu C., Synthesis of Ag/ZnO nanorods array with enhanced photocatalytic performance, Journal of Hazardous Materials 182 (2010) 123-129.

Sleiman M., Vildozo D., Ferronato C., Chovelon J., Photocatalytic degradation of azo dye Metanil Yellow: Optimization and kinetic modeling using a chemometric approach, Applied Catalysis B: Environmental 77 (2007) 1-11.

Soares O.S.G.P., Órfão J.J.M., Portela D., Vieira A., Pereira M.F.R., Ozonation of textile effluents and dye solutions under continuous operation: Influence of operating parameters, Journal of Hazardous Materials 137 (2006) 1664-1673.

Somensi C.A., Simionatto E.L., Bertoli S.L., Wisniewski Jr A., Radetski C.M., Use of ozone in a pilot-scale plant for textile wastewater pre-treatment: Physico-chemical efficiency, degradation by-products identification and environmental toxicity of treated wastewater, Journal of Hazardous Materials 175 (2010) 235-240.

Tabrizi G.B., Mehrvar M., Integration of Advanced Oxidation Technologies and Biological Processes: Recent Developments, Trends, and Advances, Journal of Environmental Science and Health, Part A, 39 (2004) 3029-3081.

Wang X., Chen S., Gu X., Wang K., Pilot study on the advanced treatment of landfill leachate using a combined coagulation, fenton oxidation and biological aerated filter process, Waste Management 29 (2009) 1354-1358.

Wu J., Doan H., Upreti S., Decolorization of aqueous textile reactive dye by ozone, Chemical Engineering Journal 142 (2008) 156-160.

Yahiat S., Fourcade F., Brosillon S., Amrane A., Photocatalysis as a pre-treatment prior to a biological degradation of cyproconazole, Desalination 281 (2011) 61-67.

26

Rapid removal of mercury ion (II) from aqueous solution by chemically activated eggplant hull adsorbent

Ali Ahmadpour[1,*], Mohammad Zabihi[2], Tahereh Rohani Bastami[3], Masoomeh Tahmasbi[4], Ali Ayati[3]

[1]Department of Chemical Engineering, Faculty of Engineering, Ferdowsi University of Mashhad, Mashhad, Iran.
[2]Chemical Engineering Faculty, Sahand University of Technology, Sahand New Town, Tabriz, Iran.
[3]Department of Chemical Engineering, Quchan University of Advanced Technology, Quchan, Iran.
[4]Department of Chemistry, Ferdowsi University of Mashhad, Mashhad, Iran.

ARTICLE INFO

Keywords:
Eggplant hull
Food waste
Chemical Treatment
Adsorption
Mercury

ABSTRACT

Eggplant hull (EH) as a waste material was introduced as a new adsorbent through activation process followed by a chemical treatment method using H_2O_2 and NH_3. For the first time, the EH adsorption was used for the adsorptive removal of mercury ion (Hg^{++}) from aqueous solutions and the effects of different parameters, including pH of solution, contact time, temperature, and initial concentration, on the adsorption efficiency were studied. The results revealed that the EH, as an effective adsorbent, has high adsorption activity in higher pH values. The temperature study indicated the endothermic nature of the adsorption processes. Also, the Hg (II) adsorption by EH follows the pseudo first order kinetic model and both Langmuir and Freundlich isotherm models. However, Freundlich adsorption model fitted the experimental data better than Langmuir model. The maximum adsorption capacity of EH was obtained 147.06 mg/g which was higher than most conventional adsorbents. The surface activity of EH was estimated using iodine method as 300 mg/g.

Nomenclature

q	Amount adsorbed
C_0	Initial metal ion concentration
C_e	Equilibrium concentration
V	Volume of the solution
q_e	Equilibrium amount adsorbed
K_f	Freundlich constants related to adsorption capacity
n	Freundlich constants related to adsorption intensity
q_m	Langmuir constant related to the maximum adsorption capacity
b	Langmuir constant related to the energy or net enthalpy of adsorption
q_t	Amount adsorbed at time t

1. Introduction

Water pollution caused by metal ions is considered one of the major environmental problems in many countries which can threaten human health and other living organisms in higher contents than admissible sanitary standards (Ayati et al. 2016). Mercury, as one of the top ten toxic metal ion, is mainly present in the wastewater effluents of various process industries such as paper, pulp and paint, oil refining, rubber processing, and fertilizer plants (Johari et al. 2016). The converting of mercury to methyl mercury, as its organic form, and its bioaccumulation in food chain makes it high toxicity which can constitute a threat to the aquatic life and living creatures (Yu et al. 2014). The exposure of human cells to methyl mercury can lead to neurological effect, mood swings, memory loss, damage of gastrointestinal tract, and the kidneys (Adams et al. 2010). Accordingly, its MCLG (Maximum Contaminant Level Goals) was specified at low level of 0.2 ppb by Environmental Protection Agency (EPA). Thus, many researchers focused on the mercury removal by different methods (Fu and Wang 2011) such as membrane technology (Fard and Mehrnia 2016), chemical precipitation (Nguyen et al. 2016), ion exchange (Oehmen et al. 2014), coagulation (Samrani et al. 2008), and adsorption (Inbaraj and Sulochana 2006; Zabihi et al. 2010). However, some of these methods such as reverse osmosis and ion exchange are excellent techniques, but adsorption was found a superior to other methods, due to its low cost, simplicity and flexibility, ease of operation (Rafatullah et al. 2010) and removal efficiency. A wide range of adsorbents and biosorbents have been used for mercury ion removal from aqueous solutions, such as activated carbon, as commonly used sorbent (Bhatnagar et al. 2013; De et al. 2013), waste rubber (Gupta et al. 2012), carbon fibres (Nabais et al. 2006), chitosan (Reddy and Lee 2013), polymer composites (Say et al. 2008) and etc. In last decades, many researches have been done to find eco-friendly, low cost, highly active and locally available adsorbents (Rangabhashiyam et al. 2013). In this regard, agricultural wastes, including peanut hull (Liao et al. 2011), sago waste (Kadivelu et al. 2004), Walnut shell (Zabihi et al. 2009), have attracted significant interest as emerging low cost adsorbents of mercury.

In the present work, eggplant hull (EH), as one of the food wastes, was activated by a chemical treatment method using H_2O_2 and NH_3 and introduced as a novel low cost biosorbent for the ion removal from aqueous solution under varied experimental condition. Moreover, the adsorption kinetics and isotherms were studied by the well-known models.

2. Materials and methods

Mercury chloride was obtained from Merck in analytical grade. The sorbent was the eggplant hull as a waste material. It was washed with distilled water to remove impurities and dried at 180 °C for 20 h. The sample was ground and chemical treated by the solution of hydrogen peroxide (2 % vol) and ammonia (2 % vol) for 30 min, in order to extract the soluble organic compounds of hulls and enhance chelating efficiency.

2.1. Batch adsorption

For the batch adsorption studies, 0.1 g of EH adsorbent (fraction with 0.2 mm particles in size) was added to 50 mL of Hg(II) solution with the desired concentration (in the range of 11.2-105.6 mg/L) at pH of 5 in several conical flasks. The pH was adjusted using HCl and NaOH solutions. The suspensions were agitated on mechanical stirrer (720 rpm) for specific time intervals at room temperature. Finally, the adsorbents were separated through microporous filter paper and the solutions were analyzed to obtain Hg (II) ion concentrations by Varin atomic absorption spectrophotometer (spectra-110-220/880, Australia Pty. Ltd.) equipped with a Zeeman atomizer.

2.2. Characterization of adsorbent (eggplant hull)

The surface activity of EH toward iodine was determined by using the DIN 53582 standard method. The iodine No. of the sample was obtained as 300 mg/g. The apparent density of this sorbent was measured as 25 g/cm^3 using ASTM (D 2845-89).

3. Results and discussion
3.1. Influence of pH

The effect of the solution pH on mercury ion adsorption removal efficiency is illustrated in Fig.1, while the initial concentrations of Hg(II) ions was fixed at 52.6 mg/L. The results show that hydrogen ion concentration is an important parameter in the Hg(II) ions adsorption by EH. So that, the Hg(II) adsorptive removal percentage increases in the range of 2 to 5 and followed a plateau trend for higher pH values. The surface charge is a key factor for the metal ions adsorption (Rio and Delebarre 2003; Budinov et al. 2008). As in our previous study, the surface of food and agricultural wastes has oxygen –functional groups due to presence of lactonic, carboxylic, and phenolic groups on their surface (Zabihi et al. 2010). Furthermore, the surface of EH was treated by the solution of hydrogen peroxide (2 % vol) and ammonia (2 % vol). So it is assumed that, the oxygen-containing functional groups on the EH surface is responsible for mercury adsorption. On the other hand, the Hg(II) ions adsorption may also include surface complexes formation with the functional groups precipitation on the surface of adsorbent. In the presence of Hg (II) in solution, the following surface complexes may be formed (Boehm et al. 1966; Boehm 1994):

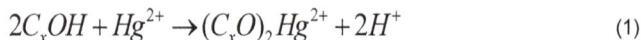

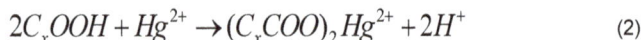

$$2C_xOH + Hg^{2+} \rightarrow (C_xO)_2 Hg^{2+} + 2H^+ \tag{1}$$

$$2C_xOOH + Hg^{2+} \rightarrow (C_xCOO)_2 Hg^{2+} + 2H^+ \tag{2}$$

At pH<4 (acidic medium), the reverse reactions may occurred and mercury ion dissolved in the solution.

3.2. Effect of temperature and contact time

Fig. 2 shows the effect of contact time on the Hg(II) ions adsorption at different temperature. As can be seen, the process is very fast and the ion concentration sharply decreased to attain equilibrium just within 2 to 7 min. This rapid uptake of ions by EH clarifies that it can be considered to apply in economical wastewater treatment. Also, the Hg(II) ions concentration decreased with the increase in temperature indicating an endothermic nature of the adsorption processes. The adsorption capacity enhanced by increasing the temperature suggesting that more surface active sites are available for adsorption at higher temperature which might be attributed to the pore size change as well as diffusion rate enhancement in an endothermic process.

3.3. Effect of initial Hg(II) ion concentration

In this investigation, the effect of initial Hg (II) concentration on the EH adsorption performance was studied using solutions with various concentrations of 11.2, 24.6, 52.6 and 105.6 mg/l at 25 °C (Fig.3). It is clear that the Hg (II) ions adsorption amount by EH adsorbent increases with increasing the initial ion concentration. It was very similar to the obtained results of many other Hg(II) ion adsorptive removal studies (Carrott et al. 1998; Namasivayam and Kadirvelu. 1999; Zhang et al. 2004). It might be due the higher probability of collisions between ions and EH particles by increasing the ions concentration.

As can be seen, the ions solutions were equilibrated at first 5 min. Therefore, it was obtained as equilibrium time in other batch experiments. The amount of Hg(II) ions adsorptive uptake (q, mg/g) was calculated by:

$$q = \frac{(C_o - C_e)V}{m} \tag{3}$$

where C_0 and C_e are the initial and equilibrium mercury ions concentrations (mg/L), respectively, V is the solution volume (L), and m is the mass of adsorbent (g). In the presence of 0.1g adsorbent and pH of 5, the equilibrium adsorbed amounts of Hg (II) at 11.2, 24.6, 52.6 and 105.6 mg/L initial concentrations were measured as: 1.8, 8.3, 22.05 and 48.1 mg/g, respectively.

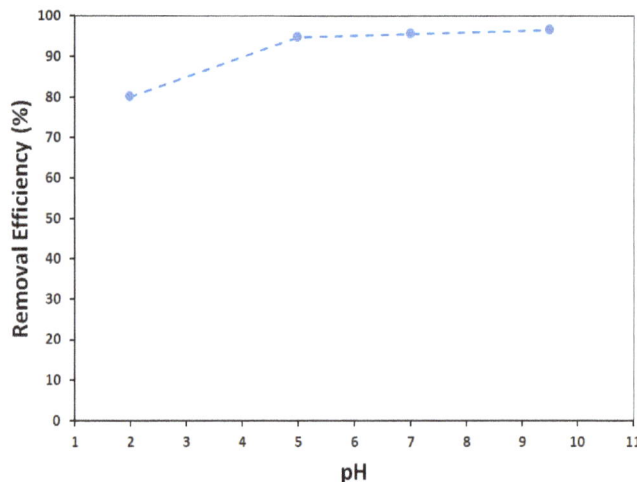

Fig.1. Effect of pH on the removal efficiency of Hg(II) ion (EH dosage=0.1 g, C$_0$=52.6 mg/l, T=25 °C, t=5 min).

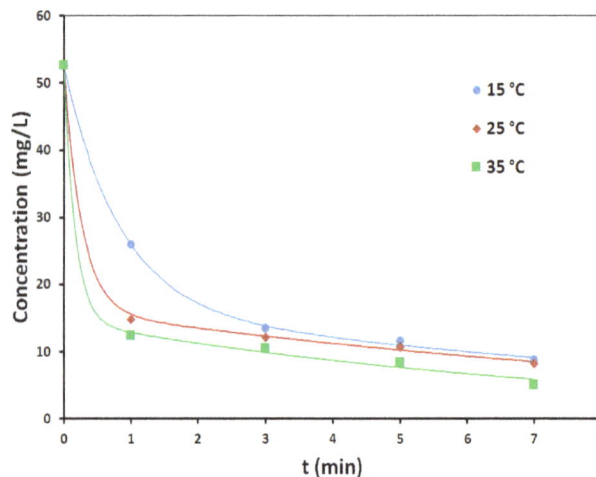

Fig. 2. Effect of contact time on the Hg(II) ion removal at different temperature (C$_o$=52.6 mg/l, sorbent dose= 0.1 g, T=25 °C, pH= 5).

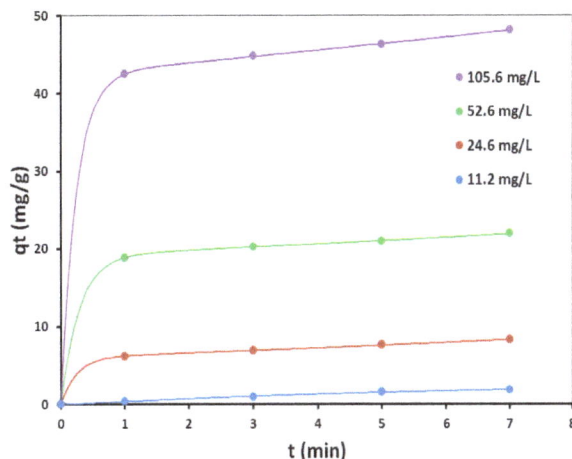

Fig. 3. Effect of initial concentration on the mercury adsorption (EH dose=0.1g, T=25 °C, pH=5).

3.4. Adsorption Isotherm

The adsorption isotherms study of Hg(II) ions removal by EH was carried out by varying the initial Hg(II) ions concentration in the range of 11.2 to 105.6 mg/L at room temperature. The obtained data were correlated with two well-known Freundlich and Langmuir isotherm models for further investigation.

3.4.1. Freundlich model

The Freundlich model, which often gives better fit for adsorption from liquid phases, can be expressed as (Freundlich. 1906):

$$\log q_e = \log K_f + \frac{1}{n}\log c_e \qquad (4)$$

where q_e is the equilibrium capacity (mg/g), C_e is the equilibrium concentration of solution (mg/L) and K_f and n are the Freundlich constants. The Freundlich plot of Hg(II) adsorption on EH is shown in Fig. 4a. The plot gives a good fit to the data with correlation coefficient of $R^2 = 0.9994$. Usually, good adsorbent illustrates 1<n<10. Smaller value of n attributes to better adsorption performance and formation of relatively strong bonds between the adsorbate and adsorbent surface.

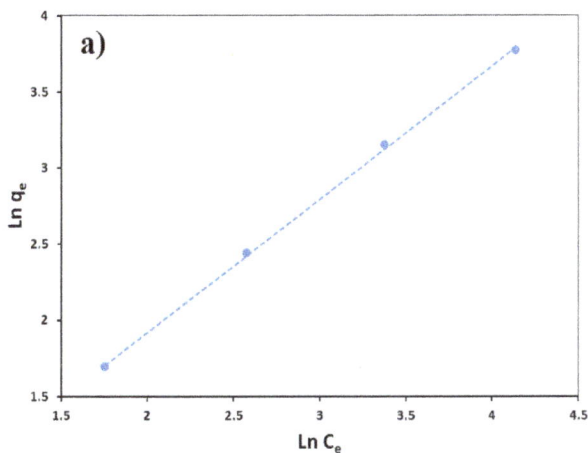

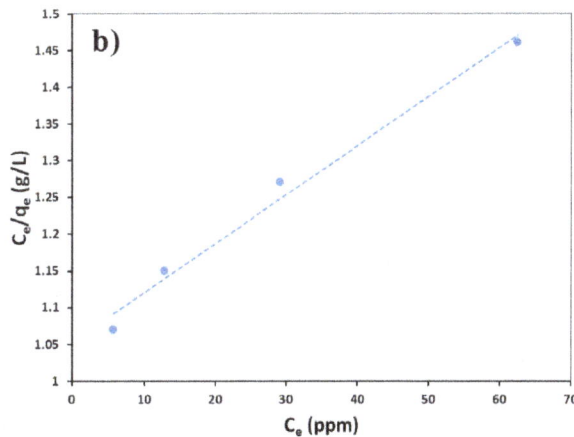

Fig. 4. a) Freundlich isotherm, and b) for adsorption of mercury by EH at 25 °C.

3.4.2. Langmuir model

The Langmuir adsorption isotherm was used as follow (Langmuir 1918):

$$\frac{C_e}{q_e} = \frac{1}{q_m b} + \frac{1}{q_m}C_e \qquad (5)$$

where q_m is the maximum monolayer capacity of adsorbent (mg/g), obtained by complete monolayer coverage) and b is the Langmuir constant (1/mg). In the present work, we found that the Langmuir isotherm model (shown in Fig. 4b) gives a fairly good fit the experimental data of mercury ions adsorption by EH (R^2 =0.9849). The fitted data of these two models are summarized in Table 1.

Table 1. Measured parameters of the isotherm models for adsorption of Hg (II) onto EH.

Langmuir model			Freundlich model		
q_m (mg/g)	b (l/mg)	R^2	K_f (mg/g)	N	R^2
147.06	0.00648	0.98	1.160	1.136	0.99

From the results, the Freundlich isotherm model predicted the experimental data slightly better than the other isotherm model. The monolayer adsorption capacity of Hg (II) ions on EH was calculated 147.06 mg/g.

3.5. Kinetic studies

The kinetic adsorption data was modeled with pseudo- first- order (Eq. 6) (Lagergren 1898) and pseudo- second –order (Eq. 7) (Ho and McKay 2000) equations which are given as:

$$\log(q_e - q_t) = \log q_e - \frac{k_1 t}{2.303} \tag{6}$$

$$\frac{t}{q_t} = \frac{1}{k_2 q_e^2} + \frac{t}{q_e} \tag{7}$$

where q_e and q_t are the amounts of Hg (II) ions uptake by adsorbent (mg/g) at equilibrium and t time (min), respectively, and k_1 is the first-order rate constant (1/min) and k_2 is the second-order rate constant (g/mg min).

The values of different parameters determined from these two kinetic models with their corresponding correlation coefficients are presented in Table 2. The correlation coefficients of the second-order model at all temperatures are more than 0.99, showing that this model can well explain the kinetic adsorption data of mercury on EH.

3.6. Comparison with other adsorbents

The adsorption capacity to EH toward the Hg(II) ion removal was compared with the adsorption capacities of other adsorbents from the literature, as given in Table 3, It shows that the adsorption capacity of eggplant hull for Hg (II) ion is comparable with the many conventional adsorbents.

Table 2. Comparison of the parameters of the (a) pseudo-first-order and (b) pseudo-second-order kinetic models at different temperatures.

(a)

T (°C)	K_1 (1/min)	q_e (mg/g), Exp.	q_e (mg/g), Cal.	%D	R^2
15	0.3388	21.80	24.15	10.77	0.91
25	0.2538	22.05	21.27	3.53	0.99
35	0.2347	23.55	22.27	5.43	0.98

(b)

T (°C)	K_2 (g/mg min)	q_e (mg/g), Exp.	q_e (mg/g), Cal.	% D	R^2
15	0.0508	21.80	24.04	10.27	0.99
25	0.3664	22.05	26.54	20.36	0.99
35	0.3937	23.55	29.06	23.39	0.99

Table 3. Monolayer adsorption capacity of various adsorbents for mercury.

Adsorbent type	q_m (mg/g)	Ref.
Activated carbon (fertilizer waste)	3.62×10^{-3}	(Mohan et al. 2001)
Fuller's earth	1.145	(Oubagaranadin et al. 2007)
Silico-aluminous ashes	3.2	(Rio and Delebarre. 2003)
Sulfo-calcic ashes	4.9	(Rio and Delebarre. 2003)
Carbon aerogel	34.96	(Kadirvelu et al. 2008)
Coal adsorbents (Bolluca)	37	(Ekinci et al. 2002)
Sago waste carbon	55.6	(Kadivelu et al. 2004)
Activated carbon	69.44	(Oubagaranadin et al. 2007)
Coal adsorbent (Mengen)	92	(Ekinci et al. 2002)
Activated carbon (Indian almond)	94.43	(Inbaraj and Sulochana. 2006)
Coal adsorbents	105	(Ekinci et al. 2002)
Activated carbon (antibiotic waste)	129	(Oubagaranadin et al. 2007)
Eggplant hull (EH)	147.06	The present study
Furfural	174	(Mohan et al. 2001)
Carbon fibers (Acrylic, Kynol)	290-710	(Nabais et al. 2006)

4. Conclusions

The foregoing study has revealed the feasibility of using a new sorbent derived from eggplant hull (EH) for the rapid removal of Hg(II) ions from aqueous solutions. The adsorption capacity of the EH was obtained 147.06 mg/g at pH of 5.0 for the adsorbent particles with 0.88 mm diameter. The adsorption kinetics and equilibrium data for this process were well described by pseudo-second-order and Freundlich models respectively. The EH was found a cheap and available adsorbent which gives a rapid sorption of mercury ion.

Acknowledgements

The authors gratefully acknowledge financial support of the project from Khorasan Razavi Regional Water Organization.

References

Adams D.H., Sonne C., Basu N., Dietz R., Nam D.H., Leifsson P.S., Jensen A.L., Mercury contamination in spotted seatrout, Cynoscion nebulosus: An assessment of liver, kidney, blood, and nervous system health, Science of the Total Environment 408 (2010) 5808-5816.

Aguado J., Arsuaga J.M., Arencibia A., Adsorption of aqueous mercury(ii) on propylthiol-functionalized mesoporous silic an obtained by cocondensation, Industrial & Engineering Chemistry Research 44 (2005) 3665-3671.

Ayati A., Tanhaei B., Sillanpää M., Lead(II)-ion removal by ethylenediaminetetraacetic acid ligand functionalized magnetic chitosan-aluminum oxide-iron oxide nanoadsorbents and microadsorbents: Equilibrium, kinetics, and thermodynamics, Journal of Applied Polymer Science 134 (2016) 44360.

Bhatnagar A., Hogland W., Marques M., Sillanpää M., An overview of the modification methods of activated carbon for its water treatment applications, Chemical Engineering Journal 219 (2013) 499-511.

Boehm H.P., Some aspects of the surface chemistry of carbon blacks and other carbons, Carbon 32 (1994) 759-769.

Boehm H.P., Eley D.D., Pines H., Weisz P.B. (1966). Advances in Catalyses and Related Subjects, New York, Academic Press.

Budinov T., Petrov N., Parra J., Baloutzov V., Use of an activated carbon from antibiotic waste for the removal of Hg(II) from aqueous solution, Journal of Environmental Management 88 (2008) 165-172.

Carrott P.J., Carrott M.M.L.R., Nabais J.M.V., Influence of surface ionization on the adsorption of aqueous mercury chlorocomplexes by activated carbons, Carbon 36 (1998) 11-17.

Chiarle S., Ratto M., Rovatti M., Mercury removal from water by ion exchange resins adsorption, Water Research 34 (2000) 2971-2978.

De M., Azargohar R., Dalai A.K., Shewchuk S.R., Mercury removal by bio-char based modified activated carbons, Fuel 103 (2013) 570-578.

Ekinci E., Budinova T., Yardim F., Petrov N., Razvigorova M., Minkova V., Removal of mercury ion from aqueous solution by activated carbons obtained from biomass and coals, Fuel Processing Technology 77 (2002) 437- 443.

Fard G.H., Mehrnia M.R., Investigation of mercury removal by Micro-Algae dynamic membrane bioreactor from simulated dental waste water, Journal of Environmental Chemical Engineering In Press. (2016).

Freundlich H.M.F., Over the adsorption in solution, The Journal of Physical Chemistry 57 (1906) 384-470.

Fu F., Wang Q., Removal of heavy metal ions from wastewaters: A review, Journal of Environmental Management 92 (2011) 407-418.

Gupta V.K., Ganjali M.R., Nayak A., Bhushan B., Agarwal S., Enhanced heavy metals removal and recovery by mesoporous adsorbent prepared from waste rubber tire, Chemical Engineering Journal 197 (2012) 330-342.

Herrero R., Lodeiro P., Rey-Castro C., Vilarino T., Vicente M.E.S.d., Removal of inorganic mercury from aqueous solutions by biomass of the marine macroalga Cystoseira baccata, Water Research 39 (2005) 3199-3210.

Ho Y.S., McKay G., The kinetics of sorption of divalent metal ions onto sphagnum moss peat, Water Research 34 (2000) 735-742.

Inbaraj B.S., Sulochana N., Mercury adsorption on a carbon sorbent derived from fruit shell of Terminalia catappa, Journal of Hazardous Materials 133 (2006) 283-290.

Johari K., Saman N., Tien S.S., Chin C.S., Kong H., Mat H., Removal of elemental mercury by coconut pith char adsorbents, Procedia 148 (2016) 1357 - 1362.

Kadirvelu K., Goel J., Rajagopal C., Sorption of lead, mercury and cadmium ions in multi-component system using carbon aerogel as adsorbent, Journal of Hazardous Materials 153 (2008) 502-507.

Kadivelu K., Kavipriya M., Karthika C., Vennilamani N., Pattabhi S., Mercury (II) adsorption by activated carbon made from sago waste, Carbon 42 (2004) 745-752.

Lagergren S., Zur theorie der sogenannten adsorption gelöster stoffe, Kungliga Svenska Vetenskapsakademiens, Handlingar. 24 (1898) 1-39.

Langmuir I., The Adsorption of Gases on Plane Surfaces of Glass, Mica and Platinum, Journal of the American Chemical Society 40 (1918) 1361-1368.

Liao S.-W., Lin C.-I., Wang L.-H., Kinetic study on lead (II) ion removal by adsorption onto peanut hull ash, Journal of the Taiwan Institute of Chemical Engineers 42 (2011) 166-172.

Mohan D., Gupta V.K., Srivastava S.K., Chander S., Kinetics of mercury adsorption from wastewater using activated carbon derived from fertilizer waste, Colloids and Surfaces A 177 (2001) 169-181.

Nabais J.V., Carrott P.J. M., Carrott M.M.L.R., Belchior M., Boavida D., Diall T., Gulyurtlu I., Mercury removal from aqueous solution and flue gas by adsorption on activated carbon fibres, Applied Surface Science 252 (2006) 6046-6052.

Namasivayam C., Kadirvelu K., Uptake of mercury (II) from wastewater by activated carbon from an unwanted agricultural solid by-product: coirpith, Carbon 37 (1999) 79-84.

Nguyen D.L., Kim J.Y., Shim S.-G., Ghim Y.S., Zhang X.-S., Shipboard and ground measurements of atmospheric particulate mercury and total mercury in precipitation over the Yellow Sea region, Environmental Pollution 129 (2016) 262-274.

Oehmen A., Vergel D., Fradinho J., Reis M.A.M., Crespo J.G., Velizarov S., Mercury removal from water streams through the ion exchange membrane bioreactor concept, Journal of Hazardous Materials 264 (2014) 65-70.

Oubagaranadin J.U.K., Sathyamurthy N., Murthy Z.V.P., Evaluation of Fuller's earth for the adsorption of mercury from aqueous solutions: A comparative study with activated carbon, Journal of Hazardous Materials 142 (2007) 165-174.

Rafatullah M., Sulaiman O., Hashim R., Ahmad A., Adsorption of methylene blue on low-cost sorbents: a review, Journal of Hazardous Materials 177 (2010) 70-80.

Rangabhashiyam S., Anu N., Selvaraju N., Sequestration of dye from textile industry wastewater using agricultural waste products as adsorbents, Journal of Environmental Chemical Engineering1 (2013) 629-641.

Reddy D.H.K., Lee S.M., Application of magnetic chitosan composites for the removal of toxic metal and dyes from aqueous solutions, Advances in Colloid and Interface Science 201-202 (2013) 68-93.

Rio S., Delebarre A., Removal of mercury in aqueous solution by fluidized bed plant fly ash, Fuel 82 (2003) 153-159.

Samrani A.G.E., Lartiges B.S., Villiéras F., Chemical coagulation of combined sewer overflow: Heavy metal removal and treatment optimization, Water research 42 (2008) 951-960.

Say R., Birlik E., Erdemgil Z., Denizli A., Ersöz A., Removal of mercury species with dithiocarbamate-anchored polymer/organosmectite composites, Journal of Hazardous Materials 150 (2008) 560-564.

Yu C.-T., Chen Y.-L., Cheng H.-W., Development of an Innovative Layered Carbonates Material for Mercury Removal Sorbents, Energy Procedia 61 (2014) 1270-1274.

Zabihi M., Ahmadpour A., Asl A.H., Removal of mercury from water by carbonaceous sorbents derived from walnut shell, Journal of Hazardous Materials 167 (2009) 230-236.

Zabihi M., Asl A.H., Ahmadpour A., Studies on adsorption of mercury from aqueous solution on activated carbons prepared from walnut shell, Journal of Hazardous Materials 174 (2010) 251-256.

Zhang F.S., Nriagu J.O., Itoh H., Photocatalytic removal and recovery of mercury from water using TiO_2-modified sewage sludge carbon, Journal of Photochemistry and Photobiology A: Chemistry167 (2004) 223-228.

27

Application of artificial neural networks for the prediction of Gaza wastewater treatment plant performance-Gaza strip

Mazen Hamada[1], Hossam Adel Zaqoot[2,*], Ahmed Abu Jreiban[3]

[1]Department of Chemistry, Faculty of Science, Al Azhar University, Gaza Strip, Palestine.
[2]Environment Quality Authority, Gaza Strip, Palestine.
[3]Institute of Water and Environment, Al Azhar University, Gaza Strip, Palestine.

ARTICLE INFO	ABSTRACT
Keywords: Artificial neural network BOD COD Gaza wastewater treatment plant Prediction TSS	This paper is concerned with the use of artificial neural network and multiple linear regression (MLR) models for the prediction of three major water quality parameters in the Gaza wastewater treatment plant. The data sets used in this study consist of nine years and collected from Gaza wastewater treatment plant during monthly records. Treatment efficiency of the plant was determined by taking into account of influent input values of pH, temperature (T), biological oxygen demand (BOD), chemical oxygen demand (COD) and total dissolved solids (TSS) with effluent output values of BOD, COD and TSS. Performance of the model was compared via the parameters of root mean squared error (RMSE), mean absolute percentage error (MAPE) and correlation coefficient (r). The suitable architecture of the neural network model is determined after several trial and error steps. Results showed that the artificial neural network (ANN) performance model was better than the MLR model. It was found that the ANN model could be employed successfully in estimating the BOD, COD and TSS in the outlet of Gaza wastewater treatment plant. Moreover, sensitive examination results showed that influent TSS and T parameters have more effect on BOD, COD and TSS predicting to other parameters.

1. Introduction

The wastewater usually is exposed to many changed processes that can remove maximum of the pollutants such as organic substances, ammonium, phosphorus, nitrogen and other residuals from industrial surroundings and urban or rural areas community. Wastewater treatment processes are very complex, intensely nonlinear and considered by uncertainties concerning to its parameters (Henze et al. 1996). Boogaard and Eslamian. (2015); Hamdy and Eslamian. (2015) were focused on wastewater monitoring, wastewater treatment, sustainable reuse and recycling for treated urban wastewater to conserve the water resources for management purpose. Sewage management–including the collection, treatment and disposal of sewage has been a major environmental challenge in the Gaza Strip for several decades. Recent reports showed that about 60% of the population lives in areas with sewage networks, while the rest uses septic tanks and cesspits (Ashour et al. 2009). Due to low per capita water consumption, the sewage in the Gaza Strip is highly concentrated, with characteristic influent levels of biological oxygen demand (BOD) of up to 600 mg/l as compared to 250 mg/l, which is the standard for urban sewage. Given that the Gaza wastewater treatment plant function only irregularly, little sewage is treated and most returned to lagoons, wadis and the sea. Sewage systems were effected in several ways during the aggressions. First, as the electricity supply collapsed, transfer pumps ceased to function, resulting in sewage being diverted to the nearest available lagoons, including infiltration lagoons. Second, the limited treatment that had been taking place in sewage treatment plants also ceased due to electricity shortages. The effluent leaving sewage treatment plants to be disposed of in the sea or by infiltration in the groundwater was therefore entirely untreated. Recent

data (CMWU. 20093) on Gaza wastewater treatment plant shows an inflow BOD of 415 mg/l and effluent BOD of 172 mg/l, with 58% efficiency. Evaluation of water quality parameters is necessary to enhance the performance of an assessment operation and develop better management and planning for water resources (Abyaneh 2014). The traditional modelling techniques may possibly present relatively good predictions for water quality variables; yet such models need large data and group of input data sets that are often unknown. The wastewater treatment method is quite complex. However, the advances in intelligent methods make them conceivable to use in complex systems modeling (Hanbay 2008). Artificial neural network (ANN) can be used for better prediction of the process performance owing to their high accuracy, adequacy and quite promising applications in engineering, water sciences and environmental fields (Govindaraju 2002; Maier & Dandy. 2000; Maier and Dandy. 2000; Neelakantan et al. 2001). There are definite key descriptions of parameters, which can be used to evaluate the wastewater treatment plant performance. These parameters include chemical oxygen demand (COD), BOD and total suspended solids (TSS). Until now, most of the current studies for modeling wastewater treatment plants (WWTPs) used these parameters. The ANN established models find out acceptable results. ANN model was developed for BOD removal process in horizontal subsurface flow constructed wetlands by Akratos et al. (2008). Mjalli et al. (2007) used neural network with single input and multi-input layers and gave comparable predictions of the plant performance criteria. Prediction of BOD and suspended solid (SS) concentrations based on ANN were presented by Hamed et al. (2004). TSS is an indication of plant performance. A simple prediction models based on neural network for TSS was demonstrated by (Belanche 2000). Many other ANN models for wastewater treatment performance prediction have been proposed either in the past (Zhua et al. 1998; Choi and Park 2000; Choi and Park 2000; Oliveira-Esquerre et al. 2001; El-Din and Smith 2002; Geissler et al. 2005; El-Din et al. 2004; Pai et al. 2012) or more recently (Zhang and Hu. 2012; Vyas et al. 2011; Nasr et al. 2012; Djeddou 2014; Yordanova e t al. 2014; Guo et al. 2014; Bagheri et al. 2015; Kundu et al. 2014; Gholikandi et al. 2014; Pakrou et al. 2015; Guo et al. 2015; Djeddou and Achour 2015; Djeddou and Achour 2015). Due to numerous problems in the recording and measurement of wastewater quality such as BOD and COD, the main objective of the present paper is to find the optimized topology of the ANN and compare the obtained prediction results with multiple linear regression (MLR) model for prediction of complex wastewater quality data; to select the best method in prediction of the wastewater quality data, and to evaluates the results of the multilayer perceptron and radial basis function type of ANN in prediction of BOD, COD and TSS and selecting the optimized topology. As a first case study for predicting the performance of Gaza wastewater treatment plant it is hoped that, the results of this study would contribute in assisting the local authorities in developing plans and policies to reduce the pollution generated from the wastewater treatment plant and improve its performance.

2. Materials and methods
2.1. Study area

The study area is the Gaza wastewater treatment plant, which lies to the southwest of Gaza city. The exact location inside the plant is the drying lagoons, which are being used as filtration basins. Treated wastewater and produced sludge are disposed to open areas a few meters beside the plant itself. The plant is close to less urbanized and agricultural areas. Fig. 1 shows the location of the Gaza wastewater treatment plant. The area has a long history of exposure to wastewater and sludge. Large areas have been used for the disposal of raw sewage effluents and untreated sludge from 1977 up to date. The plant was designed to treat about 42,000 m^3 per day but now facing a daily inflow of more than 90,000 m^3. This has overcome the biological stage of the treatment process. As an emergency measure to stop sewage from overflowing, scarcely treated wastewater is currently piped to the coast, where the dark grey liquid can be seen, and smelled, flowing along the beach of Gaza (Shomar 2011). The GWWTP comprises two sections including: operation and laboratory. The operation section is responsible for the daily operations of the plant and to monitor the performance of the different mechanical facilities in the plant and to record the daily activities while the laboratory is responsible for monitoring the quality of influent and effluent coming to the plant or discharging to the sea or infiltration ponds.

2.2. Data collection

Historical monthly database describing the operation of the wastewater treatment plant in Gaza city for a period of approximately 9 years (2007-2015) with a total of 108 data vectors were obtained from (Gaza wastewater treatment plant operators). These variables include influent and effluent temperature (T), pH, BOD, COD and TSS variables. ANN input and output variables of GWWTP has to be chosen based on engineering judgment on which input and output may have a significant effect in predicting effluent BOD, COD and TSS. Using MATLAB software proper training validating and testing is done and branded constructive algorithm is applied to the network.

2.3. Artificial Neural Networks (ANNs)

ANNs are sensitive to the composition of the training data set and to the initial network parameters (Talib et al. 2009), it comprised of three independent layers, the input layers, where the data introduce to the ANN, the hidden layers, where data are processed that can be either multiple layers or a single layer, and output layers, where the result of ANN are produced (Diamantopoulou et al. 2005). Each layer consists of several processing neurons. Each neuron in a layer operates in logical similarity. Information is transmitted from one layer to others in serial operations. The most widely used training algorithm for neural networks is the back propagation algorithm (Civelekoglu et al. 2007). The multilayer perceptron (MLP) is an example of an artificial neural network that is used extensively to solve a number of different problems, including pattern recognition and interpolation (Haykin 2005), (Musavi and Golabi. 2008) that feed the input data to the neural layer to produce desire output (Talib and Amat. 2012). Each layer is composed of neurons. In each neuron, a specific mathematical function

called the activation function accepts input from previous layers and generates output for the next layer. Each layer is interconnected with each other by weights. In the experiment, the activation function used is the hyperbolic tangent sigmoid transfer function (Fausett 1994). This paper demonstrating the application of artificial neural networks to predict the performance of the Gaza wastewater treatment plant through predicting the major indicators of the wastewater quality. Two types of feedforward networks are used to construct the ANN predictive model. They are MLP and RBF neural networks; both are trained on the collected data for building predictive models for the wastewater quality parameters prediction. The chosen MLP one hidden layer was trained using the backpropagation incorporated with LM algorithm. The RBF network was trained using Orthogonal Least Squares (OLS) algorithm. Before running all models data sets were normalized to be included within the interval {0, 1} (Saen 2009). The methodology used to train, validate and test ANNs models is described below. Five models are developed (MLP networks) to choose the best model for predicting wastewater quality parameters including: BOD, COD and TSS and then to compare results with RBF and MLR statistical model.

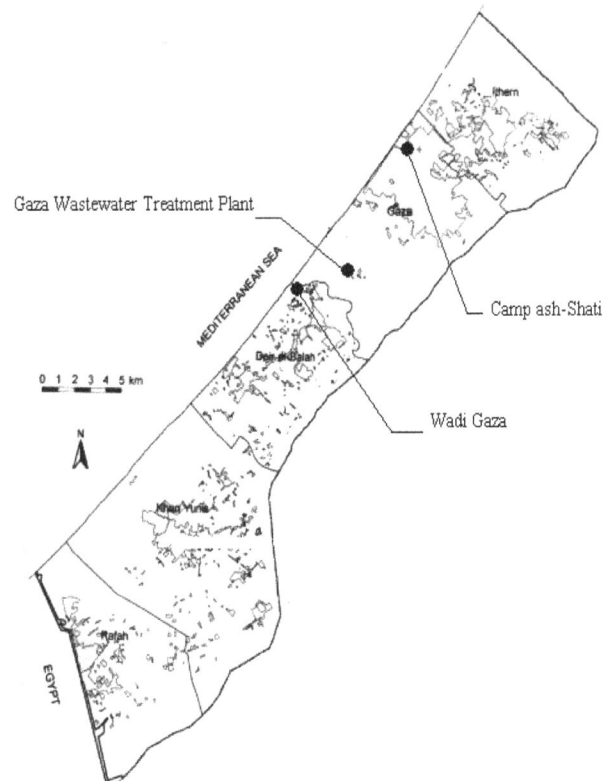

Fig. 1. Map shows the location of Gaza wastewater treatment plant

2.4. Data statistical and multiple linear regression

The collected data were entered as Microsoft Excel sheets, uploaded to Statistical Package for Social Sciences (SPSS) and analyzed using min, max, average, standard deviation tools. In addition, the Pearson correlation coefficient (a measure of linear association) is used to measure the linear association among the selected parameters. The training, validation and testing of the developed ANN models were carried out using neural network toolbox in the MATLAB. Two types of feedforward networks are used. They are multilayer perceptron and radial basis function neural networks. Root mean squared error, mean absolute percentage error and correlation statistics were calculated using MATLAB software. Statistical methods, such as MLR models, are good tools used to investigate any relationship between dependent and independent parameters of small sample numbers (Razi and Athappilly 2005). In this study the MLR is a method used to model the linear relationship between a dependent parameter and one or more independent parameters. MLR is based on least squares. In the best model, sum of square error between observed and predicted parameters should be minimum value. However, in this paper MLR statistical model is used for comparison purpose with the predictions of ANN developed model.

2.5. Data processing and training

Computationally competent deterministic approach, first-order gradient method (back propagation) because of its increased ability to find global optima in the error surface, was used to conduct the ANN training. The aim of model (ANN training) is to find a set of model parameters that enables a model with a given functional form to best represent the desired input/output relationship (Glorot and Bengio 2010). At early stage of the Gaza wastewater treatment plant performance prediction, inlet and outlet wastewater quality data, over a period of nine years beginning from 2007 to 2015 were collected. The all collected data (108 readings) in this study are combined in one set to examine the possibility for developing a neural network model for predicting the effluent wastewater quality parameters including: BOD, COD and TSS. The main obtainable selected influent wastewater quality parameters including: T, pH, BOD, COD and TSS. The MLP network had attained good results when trained using the backpropagation incorporated with Levenberg Marquardt algorithm. The tangent hyperbolic function is used as the activation function in the hidden layer neurons. The linear activation function is used in the output layer neurons (Haykin 2005). The RBF network is trained using the backpropagation incorporated with the orthogonal least squares algorithm and the Gaussian radial basis function is used as the activation function in the hidden layer. The linear activation function is used in the output layer (Chen et al. 1991). Before running ANN networks, the data set was divided into three data sets 60% of the data used for training purpose, 20% used for validation and 20% used for testing the networks performance and then data was normalized to be included within the interval {0,1} (Saen 2009). The MLP network training procedure started with utilizing 10 neurons in the hidden layer, then gradually the number of neurons increased till 18 neurons and at 14 neurons the performance of the developed network was good. The architecture of the developed MLP neural network for predicting BOD, COD and TSS contains three layers, 5 neurons in the input, 14 neurons in the hidden layer and 3 neurons in the output layer. The input neurons made from five influent important parameters including: T, pH, BOD, COD and TSS. For training RBF neural network same input neurons was used as utilized for MLP network. During training process, the RBF neural network performed good at 17 neurons in the hidden layer. Referring to Rounds (2002) study linear regression can be considered as a different instance of ANN model, which uses linear transfer functions and certainly not hidden layers. If the linear model attains as well as other complex ANN, then using the nonlinear neural networks may not be realistic and so the linear models are appropriate as a basis for comparison. However, in this study the linear regression is used to compare the prediction results attained from both developed MLP and RBF neural networks.

2.6. Evaluation criteria for ANN and MLR prediction

In order to determine which network structure is optimal, the performance of a calibrated model was evaluated against one or more criteria. In this paper, the ANN model performance was assessed using a quantitative error metric. The employed metrics belonging to this category include mean absolute percentage error, root mean square

error and R correlation. Once a model structure has been chosen and the network trained, the selected model needs to be evaluated. In practice, the accuracy of a model is determined by the 'goodness of fit' between outputs of the model and the system given the same input. Hence, some validation tests need to be considered. Generally, the accuracy of a model must be evaluated for three sets of data samples. These data sets are: training data that express the effectiveness of learning, validation data set that used to save the model from overfitting problem, and the testing data set that measure the generalization capability of the network. There is a need to point out that the testing data set should ideally not have previously been presented to the network and it must represent the entire operation range (Cawley and Talbot 2010). In this study, the mean absolute percentage error (MAPE), root mean square error (RMSE) and correlation coefficient (r) have been considered as evaluation criteria.

$$MAPE = \frac{100}{N} \times \sum_{k=1}^{N} \left| \frac{X_k - Y_k}{X_k} \right| \qquad (1)$$

$$RMSE = \frac{1}{N} \sum_{k=1}^{N} (X_k - Y_k)^2 \qquad (2)$$

$$r = \frac{(X_k - \bar{X})(Y_k - \bar{Y})}{\sqrt{\sum_{k=1}^{N}(X_k - \bar{X})^2 \sum_{k=1}^{N}(Y_k - \bar{Y})^2}} \qquad (3)$$

where X_k is the is the actual observation time series values; Y_k is the predicted time series values, N is the number of values in the data set; $\bar{X}$ is the mean of observed values and $\bar{Y}$ is the mean of predicted values.

3. Results and discussion
3.1. Data set statistical analysis

The data of wastewater quality were made and used for training of artificial neural networks for predicting the BOD, COD and TSS of wastewater effluents in Gaza wastewater treatment plant. All collected data were entered as Microsoft Excel sheets, uploaded to SPSS software, and analyzed using min, max, average, standard deviation statistical and coefficient of variance (CV) tools. Additionally, the correlation coefficient (a measure of linear association) were used to measure the linear association among the wastewater quality parameters. Table.1 shows the statistical analysis summary of wastewater quality parameters in the Gaza wastewater treatment plant. The treatment process in Gaza wastewater treatment plant showed good performance when it is compared with the wastewater treatment plant in Yazd-Iran. The measured average of BOD, COD and TSS concentrations of raw wastewater in Yazd treatment plant, were around 272.08, 577.13, 258.66 mg/L, where the concentrations of treated wastewater were around 135.18, 307, and 139.75 respectively (Farzadkia et al. 2014).

Table 1. Statistical analysis summary of GWWTP wastewater quality parameters.

Parameters	Ranges	Data Statistics Average	S.D	CV
Input Layer				
T (ºC)	13.8-32	22.71	4.30	0.188
pH	6.75-8.56	7.81	0.24	0.031
BOD (mg/l)	380-840	495	79.46	0.160
COD (mg/l)	720-1520	991	163.42	0.164
TSS (mg/l)	363-80	501	88.40	0.176
Output layer				
BOD (mg/l)	40-230	103	34.75	0.339
COD (mg/l)	53-412	230	79.24	0.344
TSS (mg/l)	42-300	113	41.38	0.367

The study of correlation coefficient is mostly measures the association between two or more functionally independent variables. The values of correlation coefficient during this study are calculated using SPSS software. Pearson's correlation was used to detect linear associations between various variables. Influent BOD is inversely correlated with effluents of Temp, BOD, COD, TSS, T influent and positively with influent of COD, TSS and pH. Influent COD is inversely correlated with effluent of T, BOD, COD, TSS, T influent and positively

with influent of pH, BOD, TSS and effluent of pH. Influent TSS is positively correlated with influent of pH, BOD, COD, pH effluent and inversely is correlated with effluent of T, BOD, COD, TSS and T influent. Effluent BOD is inversely correlated with influent T, pH, BOD, COD, TSS, effluent T, pH and positively correlated with effluent COD and TSS. Effluent COD is inversely correlated with influent T, pH, BOD, COD, TSS and effluent Temp and pH and positively correlated with effluent BOD and TSS. Effluent TSS is positively correlated with effluent BOD, COD and pH and inversely correlated with influent T, pH, BOD, COD, TSS and effluent T. Effluent BOD is found to be strongly correlated with effluent COD and TSS (r=0.89 and 0.77) and moderately to weakly correlated with influent T, pH, BOD, COD, TSS and effluent T and weakly with effluent pH. Effluent COD is correlated moderately to weakly with influent T, BOD, COD, TSS and effluent

Temp and poorly with influent pH and effluent pH and correlated strongly with effluent BOD and TSS (r=0.89 and 0.72). Effluent TSS is found to be strongly correlated with effluent BOD and COD (r=0.77 and 0.72) is correlated moderately to weakly with influent Temp and effluent T and poorly with influent pH, BOD, COD, TSS and effluent pH.

Table 2. Variation in R value with change in number of neurons and hidden layers.

Models	Architecture	Train R	Valid R	Test R	All R	Remarks
M1-MLP	5-10-3	0.8858	0.6360	0.7837	0.8018	Architectures with 14 neurons in
M2-MLP	5-12-3	0.8713	0.7859	0.6831	0.8032	the hidden layer is selected the
M3-MLP	5-14-3	0.8220	0.7855	0.8181	0.8134	optimum model for prediction of
M4-MLP	5-16-3	0.8207	0.7183	0.7318	0.7725	BOD, COD and TSS. M3 is the
M5-MLP	5-18-3	0.8574	0.7107	0.6540	0.7811	best model.

3.2. Performance of the ANN

Because of the lack of theoretical foundations, training a neural network requires a long trial and error process, experimenting different combinations of learning rates, momentum terms, transfer functions, and network architectures. The determination of the learning rates and other network parameters is fundamental to train the network successfully. So as to overcome this difficulty, LM algorithm is used in this work to reduce the random nature of the determination of the training parameters, improving the training process and, therefore, the forecasting performance of the network. During training, the weights of the neural network was adjusted in order to minimize the error between the network output and the target value for all of the records in the training set. To ensure that the network does not over-fit the training data (by learning patterns specific only to the training set), the performance of the network on the validation set was periodically evaluated. When performance on the validation set begins to degrade, training was stopped. To predict the future concentrations effluent of BOD, COD, and TSS (one month a head) in the wastewater quality of GWWTP, feedforward MLP and RBF neural networks are employed. For this purpose, several algorithms are used during training process including: Backpropagation (BP), Levenberg Marquardt (LM), Conjugate gradients (CG), Resilient backpropagation (rprop) and Gradient descent (GD). During the MLP network training process five models were developed to choose the optimal model for predicting the GWWTP performance. ANN model architecture refers to the layout of neurons and the number of hidden layers. The feed forward backpropagation training algorithm is a supervised training mechanism and is normally adopted in most of science and the engineering applications. The primary goal of training is to minimize the error at the output layer by searching for a set of connection strengths that cause the ANNs to produce outputs that are equal to or closer to the targets (Tarke et al. 2016). Atypical ANN model with a backpropagation incorporated with LM algorithm is constructed to predict BOD, COD and TSS concentrations. From the candidate models described above; because of the reason that the R values are closer to each other for the train, validation and test sets, is minimal in comparison, the MLP one hidden layer based BOD, COD and TSS prediction model with 14 neurons in the hidden layer is selected as optimum model for predicting the performance of GWWTP. The analysis of the performance statistics is supported by plot of the measured values of BOD, COD and TSS against the predicted values for the five setting of MLP network are presented in Table 2. Figs. 2 and 3 show the performance of MLP and RBF neural networks.

Table 2 shows the variation of R value with varying hidden layer neurons number. Hyperbolic tangent transfer function with 14 numbers of neurons in one hidden layer showed the best model performance. In this study the best network training results was achieved at 9 epochs using LM learning algorithm. Hence, the suitable optimum architecture for prediction was determined to consist of an input layer with five neurons, a hidden layer with 14 neurons and an output layer with three neurons.

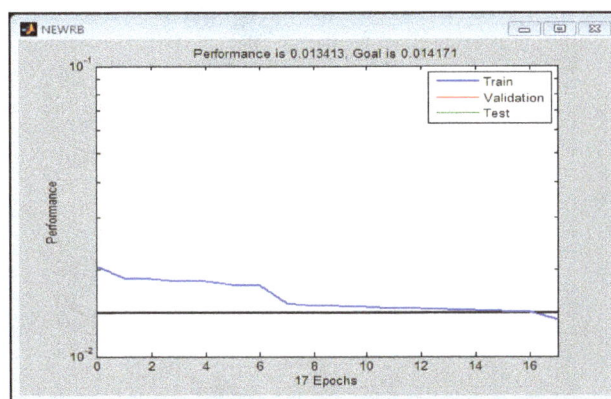

Fig. 3. RBF network training performance.

3.3. Prediction of ANN

In a WWTP, there are certain key descriptive variables which can be used to assess the plant performance. These variables include biological oxygen demand, chemical oxygen demand, and total suspended solids. Most of the available literature on the application of ANNs for modelling WWTPs utilized these variables and found that the ANN-based models provide an efficient and robust tool in predicting WWTP performance. For modelling WWTPs using ANN, Hamoda et al. (1999) found a correlation index of 0.74 for BOD prediction; Belanche et al. (1999) found 0.504 for COD prediction; Häck and Köhne (1996) found 0.92 and 0.82 for COD and nitrate prediction, Abyaneh (2014) found RMSE = 25.1 mg/L, r = 0.83 for the prediction of BOD and for the prediction of COD found RMSE = 49.4 mg/L, r = 0.81, and Nasr et al. (2012) found that the ANN can predict the plant performance with a correlation coefficient between the observed and predicted output variables reached up to 0.90, respectively. This paper addresses the problem of how to capture the complex relationships that exist between process variables and to diagnose the dynamic behavior of Gaza WWTP by applying an ANN model. Nonthreatening operation and control of the plant can be achieved by developing an ANN model for predicting the plant performance based on past observations of certain key product quality parameters. The regression button in the training window of network in MATLAB performs a linear regression between the network outputs and the corresponding targets. Fig.4 shows the best model regression results. It is observed that the output tracks the targets very well using MLP for training (R-value= 0.82202), validation (R-value= 0.78551) and testing (R-value= 0.8181). These values can be equivalent to a total response of R-value= 0.81343. It is observed that the output tracks the targets very well using RBF for training (R-value= 0.81559), validation (R-value= 0.76837) and testing (R-value= 0.70076). These values can be equivalent to a total response of R-value= 0.79024. The prediction results of MLP model found to be slightly better than RBF in training, validation and testing data set.

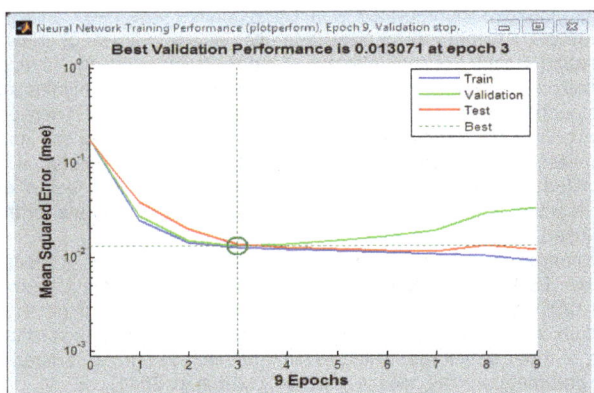

Fig. 2. MLP network training performance.

There are many statistical tools for model validation, but the primary tools for most process modeling applications include correlation coefficient (r), root mean squared error (RMSE) and mean absolute percentage error (MAPE). Summary of these statistical tools used in the evaluation of developed models prediction results as well as multiple regression model are given in Table 3. It can be understood from the results presented in Table 3 which shows the coefficient correlations between the observed and predicted values of BOD, COD and TSS using MLP, RBF and MLR for training, validating and testing the developed models. The correlations between the predicted and actual values of BOD, COD and TSS for MLP, RBF model are found to be near strong and better than MLR model whereas coefficient correlation values are [0.7846, 0.7594, 0.7655], [0.7162, 0.7062, 0.7183] and [0.5425, 0.5263, 0.5372] respectively. It also understood that in all developed models predictions of the BOD with ANN and MLR models were found to be better than TSS and COD. In this case the achieved results reveal that the developed MLP and RBF (neural network models) have satisfactory competence and accuracy in predicting BOD, COD and TSS concentrations in the water quality of Gaza wastewater treatment plant as shown in Table 3.

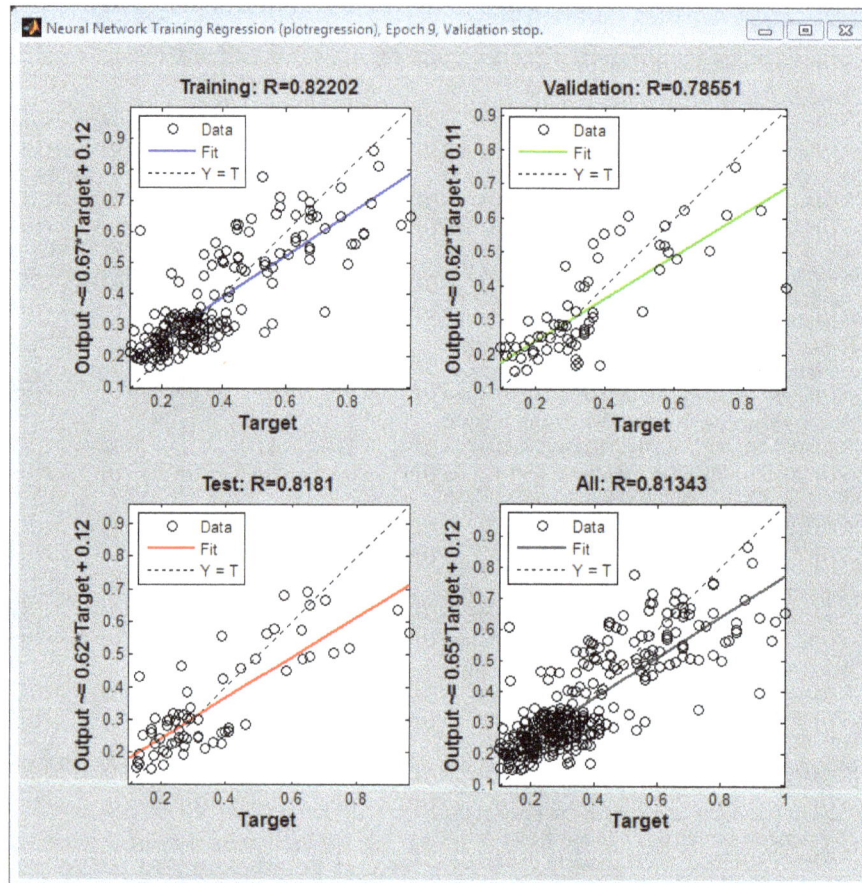

Fig. 4. The best ANN network regression.

Table 3. Analytical comparison between MLP, RBF and MLR prediction results.

Parameters	Models	Performance		
BOD		**R**	**RMSE**	**MAPE (%)**
	MLP	0.7846	29.69	25.45
	RBF	0.7162	31.46	27.85
	MLR	0.5425	32.61	30.53
COD				
	MLP	0.7594	59.48	26.29
	RBF	0.7062	63.39	28.77
	MLR	0.5263	68.92	32.66
TSS				
	MLP	0.7655	37.38	26.33
	RBF	0.7183	38.13	28.15
	MLR	0.5372	38.28	29.48

3.4. Comparison between ANNs and MLR predictions

Figs. 5, 6 and 7 show comparisons of MLP and RBF neural networks prediction results of effluent BOD, COD and TSS with the conventional method (MLR model) predictions. From the figures it can be seen that the performance of MLP is slightly better than RBF and both are better than MLR.

This good predictions result obtained from the ANN models is due to the good correlation between the selected input and output data. From the above shown figures it can be understood that ANN predictions are better than conventional methods. The good prediction results prove that the chosen approach is adept and appropriate for modeling the performance of Gaza wastewater treatment plant.

3.5. Sensitivity of input parameters

Additional analysis about the sensitivity of the prediction results against the input factors are done on the results of ANN approximation for the optimal network. Thus, to determine the sensitivity and impact of different input factors, MLP network with 1 hidden layer and trained with backpropagation incorporated with LM algorithm was used. First, 5 parameters of wastewater quality were used as a primary input of ANN developed model. For the selection of the most important ANNs input parameters the periodic remove method was used. Then, by eliminating any input parameter, the structure of enhanced artificial neural network was run. With comparing neural network output by eliminating any input parameter, the network sensitivity to any input parameter was calculated. Sensitivity grade or impact of each of the input factors on the prediction results of the training, validation and testing data for the developed model results are presented in Fig. 8

It can be seen that TSS > Temp > BOD5 > COD > pH. It means that the TSS concentration is the most influential factor and Temp, BOD, COD, and pH are in the next levels. It can be concluded that ANN structure with 5 parameters in some places had a greater error than those of other structures, which means that increasing the number of input parameters is not always effective (Zare et al. 2011).

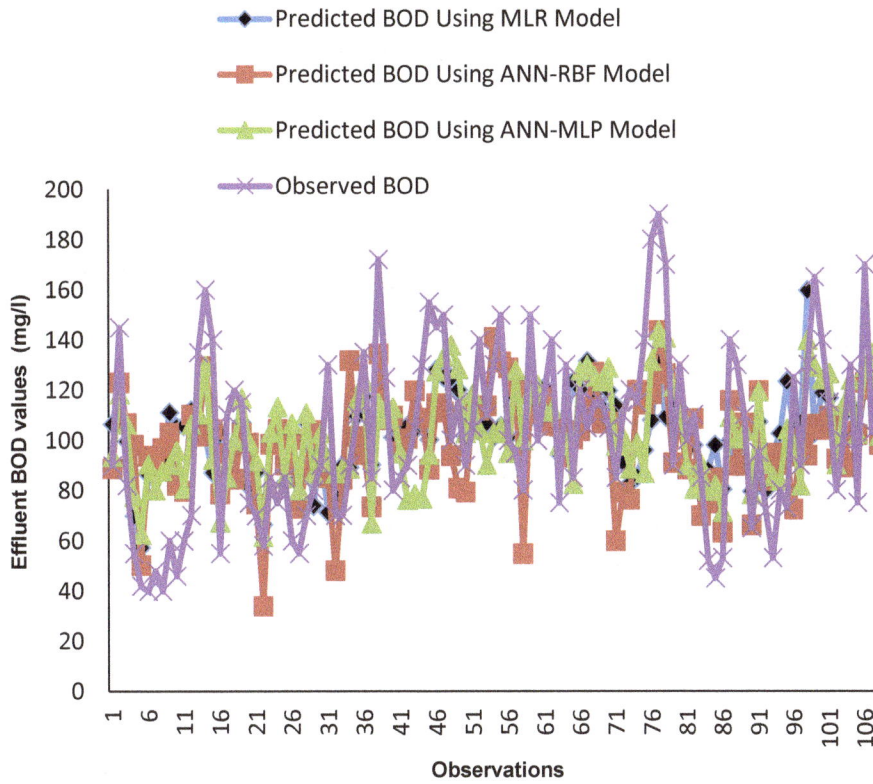

Fig. 5. Comparisons between MLP, RBF and MLR for predicting BOD.

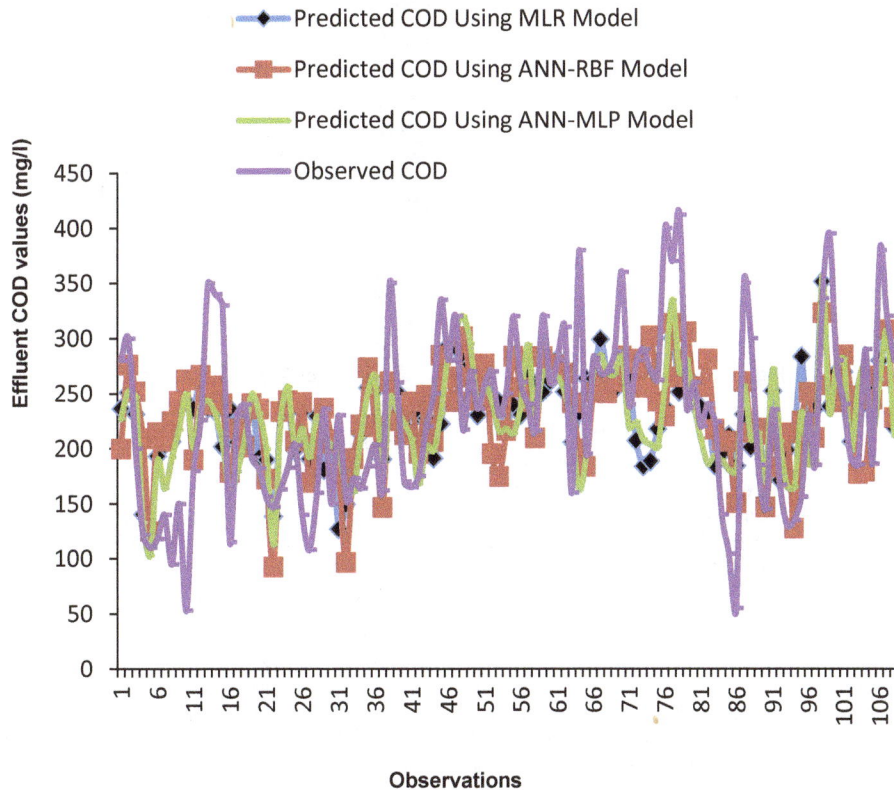

Fig. 6. Comparisons between MLP, RBF and MLR for predicting COD.

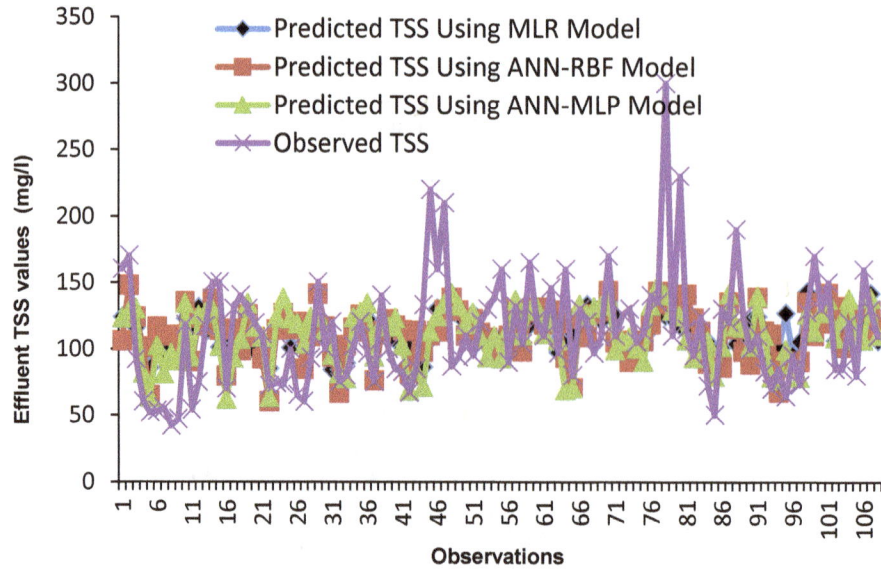

Fig. 7. Comparisons between MLP, RBF and MLR for predicting TSS.

Artificial neural network prediction sensitivity

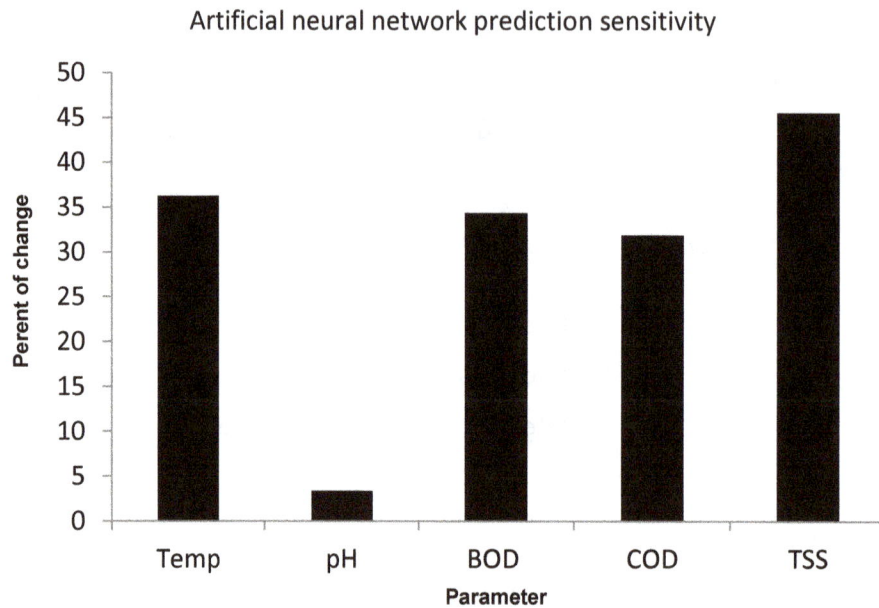

Fig. 8. Sensitivity of each input index for prediction of GWWTP water quality.

4. Conclusions

In the present study MLP and RBF neural networks were successfully developed to predict one month a head values of BOD, COD and TSS for modelling the Gaza wastewater treatment plant performance. Several scenarios were used to train MLP and RBF networks for choosing the best model for predicting the water quality of Gaza wastewater treatment plan. Performance of the models was evaluated using coefficient of correlation (r), RMSE, and MAPE. The results indicated that the ANN model with minimum input parameters, temperature (T), pH, BOD5, COD and TSS could be successfully used for predicting BOD5, COD and TSS effluent concentrations. It was found in the present study that ANN model trained with LM algorithm is an effective adsorbent for the prediction of BOD, COD and TSS concentrations. The choice structure had the highest correlation value (r = 0.81) and the least error (RMSE = 0.1374 mg/L for normal data). Comparison of the ANN and MLR models showed that the ANN model performed much better than the MLR. The results provided sufficient assessment of each model performance (for BOD predictions MLP model r= 0.78 and RMSE = 29.69 mg/L, RBF model r=0.71 and RMSE=31.46 and MLR model in contrast r= 0.54 and RMSE= 32.61

mg/L, for COD predictions MLP model r= 0.75 and RMSE= 59.48 mg/L, RBF model r=0.71 and RMSE=63.39 and MLR model in contrast r=0.52 and RMSE = 68.92 mg/L and TSS predictions MLP model r=0.76 and RMSE=37.38, RBF model r=0.718 and RMSE=38.13 and MLR in contrast r=0.537 and RMSE 38.28). In the three developed models, predictions of the BOD, COD and TSS concentrations with MLP found to be better than RBF and MLR models. In all developed models predictions of the BOD with ANN and MLR models were found to be better than TSS and COD. Further sensitivity analysis of the input factors effects on the developed ANN models were made for the best network. Sensitivity degree or impact of each of the input factors on the outcomes of the training, validation and testing data for predicting BOD, COD and TSS model result presented in the following order: TSS > Temp > BOD > COD > pH. It can be concluded that the concentration of TSS has the highest influence on the developed ANN model.

5. Future perspective

Further research efforts are to be suggested and directed towards an improved understanding of ANN performance in predicting other water quality parameters such as nitrate and heavy metals. It is also suggested to investigate the possibility of using support vector machine (SVM) for the prediction of Gaza wastewater treatment plant performance and then to compare the obtained results with the ANN technology.

Acknowledgement

The authors would like to thank the Middle East Desalination Research Centre (MEDRC) and Palestinian Water Authority (PWA) Project No: 15-DC-014 for their support and funding of the study.

References

Abyaneh H.Z., Evaluation of multivariate linear regression and artificial neural networks in prediction of water quality parameters, Journal of Environmental Health Science & Engineering 1 (2014) 12:40.

Akratos C.S., Papaspyros, J.N.E., Tsihrintzis V.A., An artificial neural network model and design equations for BOD and COD removal prediction in horizontal subsurface flow constructed wetlands, Chemical Engineering Journal 143 (2008) 96–10.

Ashour F., Ashour B., Zynski K.M., Nassar Y., Kudla M., Shawa N., Henderson G., A brief outline of sewage infrastructure and public health risks in the Gaza Strip, Technical Report Submitted to the World Health Organization, (2009).

Bagheri M., Mirbagheri S.A., Ehteshami M., Bagheri Z., Modeling of a sequencing batch reactor treating municipal wastewater using multi-layer perceptron and radial basis function artificial neural networks, Process Safety and Environmental Protection 93 (2015) 111-123.

Belanche L., Valde´s J.J, Comas J., Roda I.R. Poch M., Prediction of the bulking phenomenon in wastewater treatment plants, Artificial Intelligence in Engineering 14 (2000) 307–317.

Blanche L.B., Valdes J.J., Comas J., Towards a model of input–output behaviour of wastewater treatment plants using soft computing techniques, Environmental Modelling & Software 14 (1999) 409-419.

Cawley G.C., Talbot N.L.C., On over-fitting in model selection and subsequent selection bias in performance evaluation, Journal of Machine Learning Research 11 (2010) 2079–2107.

Chen S, Cowan C.F.N., Grant, P.M., Orthogonal least squares learning algorithm for radial basis function networks, IEEE Transactions on neural networks 2 (2001) 302–309.

Choi D.J., Park H., A hybrid artificial neural network as a software sensor for optimal control of a wastewater treatment process, Water research 35 (2001) 3959-3967.

Civelekoglu G., Yigit N.O., Diamadopoulos E., Kitis M., Prediction of bromate formation using multi-linear regression and artificial neural networks, Journal of Ozone Science and Engineer 5 (2007) 353-362.

Coastal Municipalities Water Utility (CMWU)., Gaza Emergency Water Project II. Second Quarter Technical Report in Water Quality in Gaza Strip, Palestinian Authority, (2009) 1-19.

Diamantopoulou M.J., Antonopoulos V.Z., Papamichail, D.M., The use of a neural network technique for the prediction of water quality parameters of Axios River in Northern Greece, European Water, 11 (2005) 55-62.

Djeddou M., Achour B., Wastewater treatment plant reliability prediction using artificial neural networks, 12th IWA Specialised Conference on Design, Operation and Economics of Large Wastewater Treatment Plants, Prague, Czech Republic, (2015) 242-245.

Djeddou M., Rate failure prediction in wastewater treatment plant using artificial neural networks, PhD thesis, Civil Engineering and Hydraulic Department, University Mohamed Khider of Biskra, Algeria, (2014) 195 pages.

El-Din A.G., Smith D.W., A neural network model to predict the wastewater inflow incorporating rainfall events, Water Research 36 (2002) 1115–1126.

El-Din A.G., Smith D.W. El-Din M.G., Application of artificial neural networks in wastewater treatment, Journal of Environmental Engineering and Science 3 (2004) S81-S95.

Fausett L., Fundamentals of neural networks architecture. Algorithms and applications, Pearson Prentice Hall, USA (1994).

Geissler S., Wintgens T., Melin T., Vossenkaul K., Kullmann C., Modelling approaches for filtration processes with novel submerged capillary modules in membrane bioreactors for wastewater treatment, Desalination 178 (2005) 125-134.

Gholikandi G.B., Jamshidi S., Hazrati H., Optimization of anaerobic baffled reactor (ABR) using artificial neural network in municipal wastewater treatment, Environmental Engineering and Management Journal 13 (2014) 95-104.

Glorot, X., Bengio Y., Understanding the difficulty of training deep feedforward neural networks. Appearing in Proceedings of the 13th International Conference on Artificial Intelligence and Statistics (AISTATS) 2010, Chia Laguna Resort, Sardinia, Italy. Volume 9 of JMLR: W&CP 9.

Govindaraju R.S., Artificial neural network in hydrology. II: hydrologic application, ASCE task committee application of artificial neural networks in hydrology, Journal of Hydrologic Engineering 5 (2000) 124–137.

Guo Y.M., Liu Y.G., Zeng G.M., Hu X.J., Xu W.H., Liu Y.Q., Huang, H.J., An integrated treatment of domestic wastewater using sequencing batch biofilm reactor combined with vertical flow constructed wetland and its artificial neural network simulation study, Ecological Engineering 64 (2014) 18-26.

Guo H., Jeong K., Lim J., Jo J., Kim Y.M., Park J., Kim J.H., Cho K.H., Prediction of effluent concentration in a wastewater treatment plant using machine learning models, Journal of Environmental Sciences 32 (2015) 90-101.

Hack M., Köhne M., Estimation of wastewater process parameters using neural networks, Water Science and Technology 33 (1996) 101–115.

Hamed M.M., Khalafallah M.G., Hassanien E.A., Prediction of wastewater treatment plant performance using artificial neural networks. Environmental Modelling & Software 19 (2004) 919–928.

Hamoda M.F., Al-Gusain I.A., Hassan A.H., Integrated wastewater treatment plant performance evaluation using artificial neural network, Water Science and Technology, 40 (1999) 55-69.

Hanbay D.I., Turkoglu I., Demir Y., Prediction of wastewater treatment plant performance based on wavelet packet decomposition and neural networks, Expert Systems with Applications 34 (2008) 1038-1043.

Haykin S., Neural networks: A comprehensive foundation second edition, Pearson Prentice Hall, (2005) Delhi India.

Henze M., Harremoës P., Jansen J., Arvin E., Wastewater Treatment Biological and Chemical Processes, 2nd ed., (1996) Berlin: Springer Verlag.

Kundu P., Debsarkar A., Mukherjee S., Kumar S., Artificial neural network modelling in biological removal of organic carbon and nitrogen for the treatment of slaughterhouse wastewater in a batch reactor, Environmental Technology 35 (2014) 1296-1306.

Maier H.R., Dandy G.C., Neural Networks for the prediction and forecasting of water resources variables: a review of modeling issues and applications, Environmental Modeling and software 15 (2000) 101-124.

Mjalli F.S., Al-Asheh S., Alfadala H.E., Use of artificial neural network black-box modeling for the prediction of wastewater treatment plants performance, Journal of Environmental Management 83 (2007) 329-338.

Musavi S.H., Golabi M., Application of artificial neural networks in the river water quality modeling: Karoon River, Iran, Journal of Applied Sciences, Asian Network for Scientific Information (2008) 2324-2328.

Nasr M.S., Moustafa M.A.E., Seif H.A.E., El Kobrosy J., Application of artificial neural network (ANN) for the prediction of EL-AGAMY wastewater treatment plant performance-EGYPT, Alexandria Engineering Journal 51 (2012) 37–43.

Neelakantan T.R. Brion T.R., Lingireddy S., Neural network modeling of cryptosporidium and giardia concentrations in Delware River, USA, Water Science and Technology 43 (2001) 125–132.

Oliveira-Esquerre K.M., Seborg D.E., Mori M., Bruns R.E., Application of steady state and dynamic modeling for the prediction of the BOD of an aerated lagoon at a pulp and paper mill Part II. Nonlinear approaches, Chemical Engineering Journal 105 (2004) 61–69.

Pai T.Y., Chuang S.H., Ho H.H., Predicting performance of grey and neural network in industrial effluent using online monitoring parameters, Process Biochemistry 43 (2008) 199-205.

Pakrou S., Mehrdadi N., Baghvand A., ANN modeling to predict the COD and efficiency of waste pollutant removal from municipal wastewater treatment plants. Current World Environment. 10 (2015) 873-881.

Razi M.A., Athappilly, K., A comparative predictive analysis of neural networks (NNs), nonlinear regression and classification and regression tree (CART) models, Expert Systems with Applications 29 (2005) 65–74.

Rounds S.A., Development of a neural network model for dissolved oxygen in the Tualatin River, Oregon, Presented at the Proceedings of the Second Federal Interagency Hydrologic Modeling Conference (Sub-Committee on Hydrology of the Interagency Advisory Committee on Water Information), Las Vegas, Nevada, July 29 to August 1, 2002.

Saen F.R., The use of artificial neural networks for technology selection in the presence of both continuous and categorical data. World Applied Sciences Journal 6 (2009) 1177-1189.

Shomar B., The Gaza Strip: Politics and environment, Water Policy 13 (2011) 28-37.

Talib A., Amat M.I., Prediction of Chemical Oxygen Demand in Dondang River Using Artificial Neural Network, International Journal of Information and Education Technology 2 (2012) 259-261.

Talib A., Abu Hasan Y., Abdul Rahman N.N., Predicting Biochemical Oxygen Demand as Indicator Of River Pollution Using Artificial Neural Networks, present at the World IMACS / MODSIM Congress, Cairns, Australia, July 13-17, 2009.

Tarke P.D., Sarda P.R., Sadgir P.A., Performance of ANNs for Prediction of TDS of Godavari River, India, International Journal of Engineering Research, 5 (2016) 115-118.

Vyas M., Modhera B., Vyas V., Performance forecasting of common effluent treatment plant parameters by artificial neural network, ARPN, Journal of Engineering and Applied Sciences 6 (2011) 38-42.

Yordanova S., Petrova R., Noykova N., Tzvetkov P., Neuro-fuzzy modeling in anaerobic wastewater treatment for prediction and control, International Journal of Computing 5 (2014) 51-56.

Zare A.H., Bayat V.M., Daneshkare A.P., Forecasting nitrate concentration in groundwater using artificial neural network and linear regression models, International Agrophysics 25 (2011) 187-192.

Zhang R., Hu, X., Effluent quality prediction of wastewater treatment system based on small-world ANN, Journal of Computers 7 (2012) 2136-2143.

Zhua J., Zurcher J., Raoc M., An on-line wastewater quality predication system based on a time-delay neural network, Engineering Applications of Artificial Intelligence 11 (1998) 747-758.

28

Biosorption of crystal violet from aqueous solution by pearl millet powder: Isotherm modelling and kinetic studies

Palanikumar Selvapandian[1], Kanakkan Ananthakumar[2,*], Arulsamy Cyril[1]

[1]Department of Chemistry, Raja Doraisingam Govt. Arts College, Sivagangai – 630 561. Tamil Nadu, India.
[2]Department of Chemistry, Kamarajar Govt. Arts College, Surandai – 627 859. Tamil Nadu, India.

ARTICLE INFO

Keywords:
Adsorption isotherm
PMP
Endothermic
Pseudo second order kinetics
Thermodynamic parameters

ABSTRACT

Environmental pollution caused by industrial effluents is an important issue. Biosorption of crystal violet (CV) from aqueous solution using pearl millet powder (PMP) was investigated in a batch mode. The use of pearl millet powder was as an additional substitution of activated carbon for the adsorption of CV from its aqueous solution. The adsorbent pearl millet powder was productively used for the biosorption of dye from its aqueous solution. The effect of initial concentration for CV, sorption time, dose of adsorbent, pH and temperature on dye removal were studied. The equilibrium sorption isotherms have been analyzed by the Freundlich, Langmuir and Temkin models. Sorption kinetic is quick and the data agree well with pseudo-first order kinetic model. But the kinetics studies were provided with Pseudo second order. The adsorption capacity (Q_o) of PMP was found to be 48.535 mg/g. Thermodynamic parameters such as the free energy change ($\Delta G°$), enthalpy change ($\Delta H°$) and entropy change ($\Delta S°$) were determined from Van't Hoff plot. The data reported shows the adsorption process is endothermic in nature.

1. Introduction

Recent years, Environmental pollution caused by industrial effluents is an important issue. Different types of dyes are common pollutants usually found with textile effluents in many countries (Carliell et al. 1995). Environmental pollution could be termed as ecological crisis which has posed threat on basic amenities such as air, water, and soil. Among all, the water pollution (Sharma et al. 1994) is one of the major serious problems, which is highly significant. Most of the dyes are toxic and carcinogenic compounds which cause serious problems to human and animal which is not only limited to themselves but may be passed onto further generations (Crini. 2008) Large amounts of dyes are annually produced and consumed by textile, cosmetics, paper, lather, food and other industries (Hocking. 2005). Dyes from the wastewater of those factories must be removed before discharging into water. Even a small amount of dye when discharged into water can affect aquatic life, food webs and also on humans due to the carcinogenic and mutagenic effects of synthetic dyes (Allen et al. 2004). Many treatment processes have been applied for the removal of dyes from wastewater such as photocatalytic degradation, Sonochemical degradation, miceller enhanced ultrafiltration, cation exchange membranes, electrochemical degradation, adsorption/precipitation processes, integrated chemical-biological degradation, and adsorption on activated carbon (Bockris. 1997).

As synthetic dyes in wastewater cannot be efficiently decolorized by traditional methods, the adsorption of synthetic dyes on inexpensive and efficient solid supports was considered as a simple and economical method for their removal water and wastewater (Rafatullah et al. 2010). Adsorption processes (Forgacs. 2004) provide a feasible treatment especially if the adsorbent is inexpensive and readily available. Earlier works showed that the activated carbon prepared from rice husk (Nawar et al. 1989) and wood of Ailanthus Altissima (Bangash et al. 2009) were effectively used as adsorbent for the removal of dyes. The carbons are prepared from locally available cheap materials like, palm nutshells, cashew nut shells and broom sticks (Rajavel et al. 2003) have

been used for the removal of dark green PLS dye from textile industrial waste. The adsorptions of the dye on the carbons were in between 92 and 95 %. The removal of textile dyes from textile dye effluent using TBAB based aqueous biphasic systems has been studied (Meghna et al. 2005). Adsorption of acid dyes from aqueous solution onto the surface activated kammoni leaf powder has been studied (Ubale et al. 2010). Dyeing is the finishing and the important process in the textile, paper and leather industries. Color is a visible pollutant and color pollution caused by the discharge of effluents from pulp, paper, textile and leather industries containing heavy metals, organic and other micro-toxic pollutants may affect and alter the aquatic eco systems, creating un-aesthetic conditions also. Crystal violet base (methylrosanilide, tris [4-(dimethylamino) phenyl] methanol, $C_{25}H_{31}N_3O$) is taken as dye for the present study. However, in large quantities, gentian violet may lead to ulceration of a baby's mouth and throat and is linked with mouth cancer (14). Gentian violet has also been linked to cancer in the digestive tract of other animals. Fig. 1 displays chemical structure of Crystal Violet (CV).

A careful literature survey has revealed that no studies have been made on the removal of CV from aqueous solution by *Pearl Millet* powder as an adsorbent. This has prompted us to take this investigation.

2. Materials and methods

The stock solution (500 ppm) was diluted to the required initial concentration of dye with double distilled water. 50 ml of the dye solution was taken in each 250 ml leak proof corning reagent bottles. Required amount of adsorbent of fixed particle size was exactly weighed and then transferred into each one of these bottles. The bottles were placed in a mechanical shaker and shaken vigorously for a required period of time (15 minutes). After shaking for a certain period, the flask was set aside for the adsorbent to settle out and the sample taken from the flask was centrifuged. The filtrate equilibrium concentration can be obtained from the standard curve as usual.

Corresponding author Email: rakeshraahul@gmail.com

The optical density of each solution was measured by using UV-Visible spectrophotometer (Systronic Spectrophotometer, model no: 104) at 582 nm (λ_{max}). A plot of optical density versus concentration results a straight line for this particular dye. The optical density for dye solution before and after adsorption is obtained using the spectrophotometer. The corresponding relative concentration can be obtained from the standard curve.

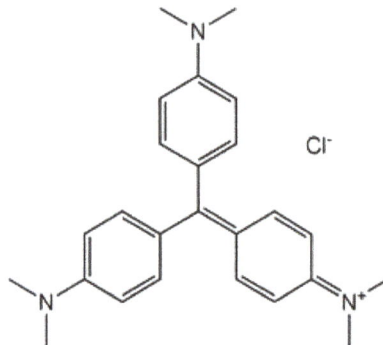

Fig. 1. Chemical structure of CV dye.

3. Results and Discussion
3.1. Effect of Initial dye concentration

For PMP of 2 g/L, the percentage removal of dye decreases from 80 to 71 as the dye concentration increases from 15 to 50 mg/L. (Fig. 2). But, the amount of CV adsorbed per unit mass of adsorbent increased with increasing in CV dye concentration. This may be due to all CV there in solution could intermingle with the binding sites at lower concentration and thus the percentage adsorption was higher than those at higher initial dye concentration. At higher concentrations lower adsorption yield is due to the saturation of adsorption sites. At low concentrations, sorption sites took up the available CV molecule more rapidly while at higher concentrations the rate of diffusion became slow (Zafar et al. 2006).

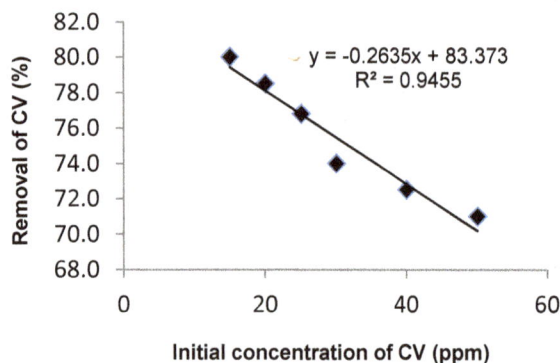

Fig. 2. Effect of Initial CV concentration.

3.2. Effect of adsorbent dosage

The effect of changing adsorbent dosage on the removal of CV was studied by varying the adsorbent dosage from 0.5 to 4 g/L while keeping the other experimental conditions as constant (Fig. 3). An increase in percentage color removal of CV with increasing adsorbent dosage was observed whereas the adsorption capacity for the adsorbents decreased (Garg et al. 2003). This is due to the larger surface area and availability of more surface functional groups at higher concentration of adsorbent. The decrease in adsorption capacity can be explained with the reduction in effective surface area of the adsorbent (Ozer and Dursun. 2007).

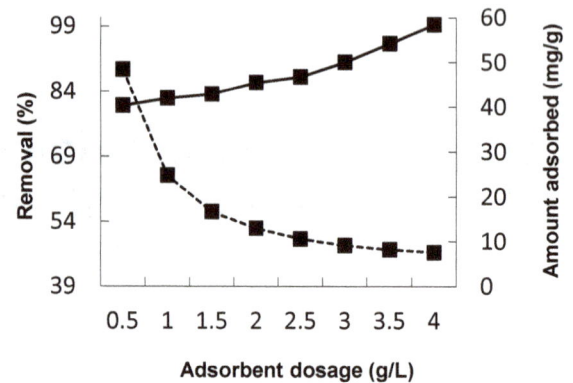

Fig. 3. Effect of adsorbent dosage for the adsorption of CV onto PMP.

3.3. Effect of pH

Fig. 4 shows the effect of pH on the percentage removal of CV by varying the pH of the dye solution from 3.6 to 11. The maximum percentage of dye removal was observed at pH 11. The result indicated that at higher pH the removal of dye was maximum and at low pH the removal of dye was minimum. The hydrogen ion concentration (pH) primarily affects the degree of ionization of the dyes and the surface properties of the adsorbents. The percentage of dye adsorbed increased from 70 to 98 % with increase in pH from 3.6 to 11 at 30 °C. At higher pH, the surface of PMP becomes negative which attract the dye molecules and hence the dye adsorption increases. The adsorbents can also interact with dye molecules via hydrogen bonding and hydrophobic-hydrophilic mechanisms (Al-Degs et al. 2008).

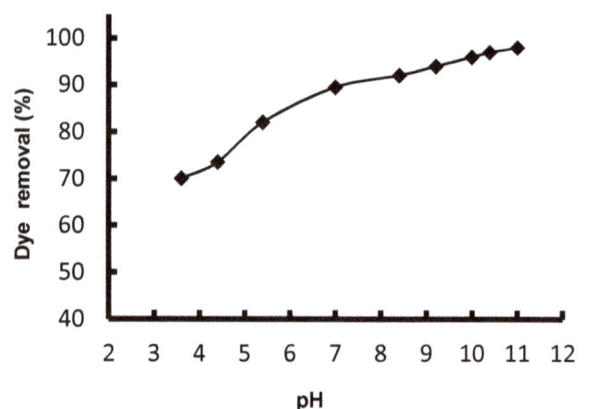

Fig. 4. Effect of pH for the adsorption of CV onto PMP.

3.4. Effect of contact time

The effect of contact time on the removal of CV onto PMP for different concentration of dye is shown in Fig. 5. The results revealed that the rate of CV removal is higher at the beginning by the adsorbent, which was due to the large available surface area. As the surface adsorption sites become exhausted, the uptake rate is controlled by the rate at which the adsorbate is transported from exterior to interior sites of the adsorbent. Almost 70 % of the total dye molecule was adsorbed in the initial 10 minutes period and remained almost unchanged after 1 hr, indicating that the adsorption process has reached equilibrium. **3.5.**

Adsorption Isotherms

Three Langmuir, Freundlich and Temkin isotherm were tested to describe the experimental data for the adsorption of CV onto PMP. Figs. 6a-c show Freundlich, Langmuir and Temkin isotherms respectively. The calculated isotherm constants and their correlation coefficients are given in Table 1.

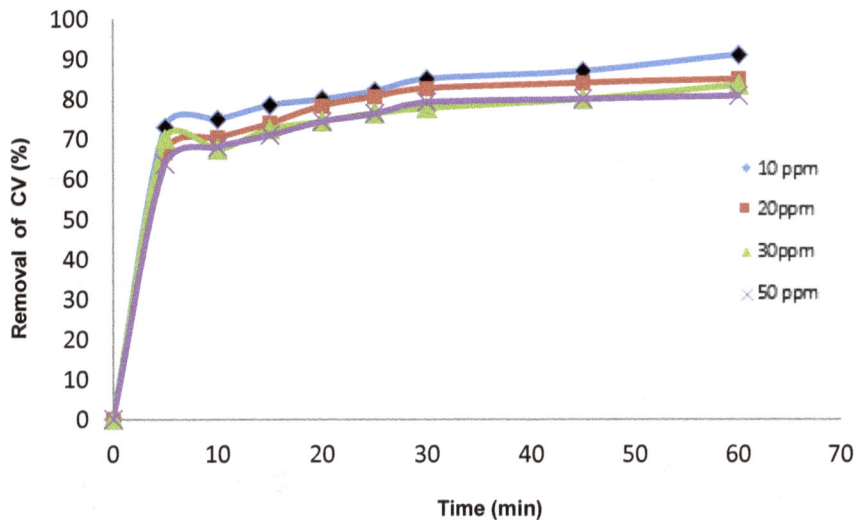

Fig. 5. Effect of contact time for the adsorption of CV onto PMP.

From the experimental results, it can be seen that the Langmuir and Freundlich models fitted better than the Temkin model. The constants for all the isotherms studied and their correlation coefficients are given in Table 1. PMP showed the highest adsorption capacity (48.535 mg/g).

The high adsorption capacity of PMP is due to the abundant formation of oxygen-containing functional groups and adequate pore size distribution.

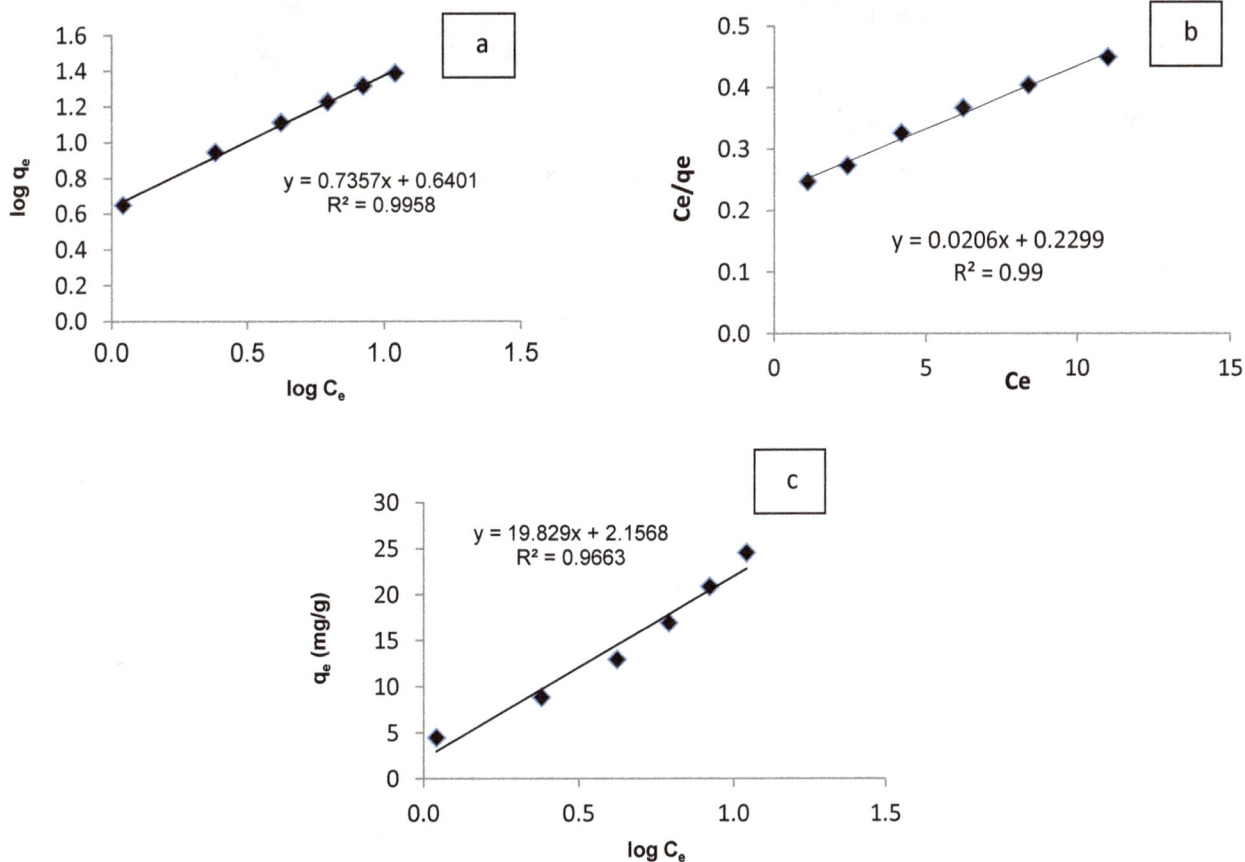

Fig. 6. Isotherm, (a) Freundlich, (b) Langmuir, (c) Temkin.

Table 1. Isotherm constants for the adsorption of CV onto PMP.

Adsorbent	Freundlich isotherm			Langmuir isotherm				Temkin isotherm		
	K_F (mg/g)	n	R^2_F	Q_m (mg/g)	K_L (L/mg)	R_L	R^2_L	K_T (L/mg)	B_1	R^2_E
PMP	4.365	0.735	0.995	48.535	0.0857	0.286	0.9914	0.5173	19.82	0.966

3.6. Characterization of PMP

The SEM image of PMP (Fig. 7a) also confirmed the presence of heterogeneous pores on the surface. From the SEM and FT-IR studies (Figs. 7a and b, 8a and b), it is clear that PMP has a considerable number of pores and surface functional groups and therefore there is a good possibility for dye molecule to be adsorbed and trapped into these pores.

The FT-IR spectra of PMP before and after biosorption were recorded in the range of 400-4000 cm^{-1}. Figs. 7a-b show the FTIR spectra of the PMP before the adsorption and after loaded with the CV dye. The intense absorption bands at 3423 cm^{-1} is assigned to O–H bond stretching. The two CH_2 stretching bands at 2924 and 2856 cm^{-1} are assigned to asymmetric and symmetric stretching of CH_2 groups which present the same wave numbers before and after the adsorption,

indicating that these groups did not participate in the adsorption process. Sharp intense peaks observed at 1645 cm^{-1}, before and after absorption are assigned to the aromatic C—C ring stretching. The wave numbers of these bands were not different before and after the adsorption of CV. Bands ranging from 1116 to 1022 and 1122 to 1024 cm^{-1} before and after adsorption, respectively, are assigned to C–O stretching vibrations. Comparing Figs. 8 and 9, we can conclude that some of these peaks are shifted or disappeared and new peaks are also detected. These changes observed in the spectrum, indicated the possible involvement of those functional groups on the surface of the PMP in adsorption process. Surface morphology of CV loaded adsorbent (Fig. 7b) shows that the surface of PMP is covered with dye molecules.

Fig. 7. (a) SEM Image of PMP before adsorption of CV, (b) After adsorption of CV.

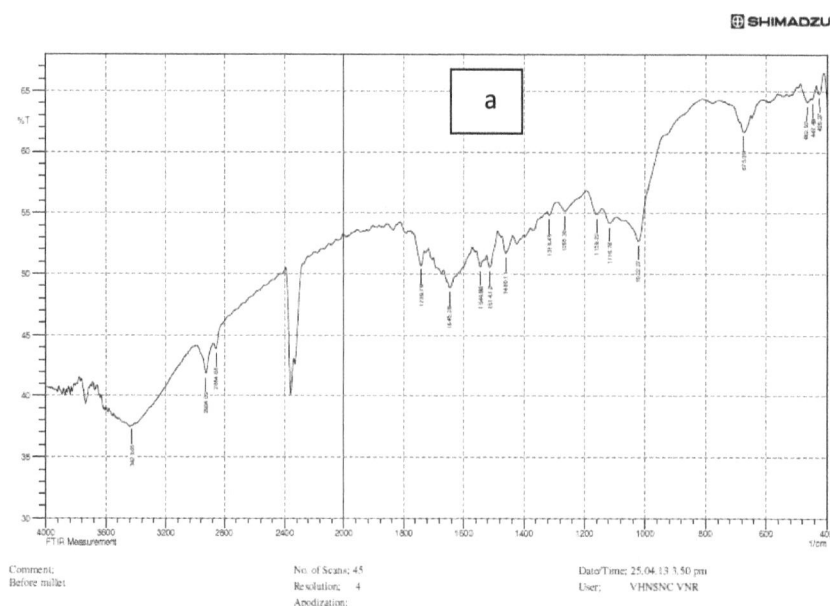

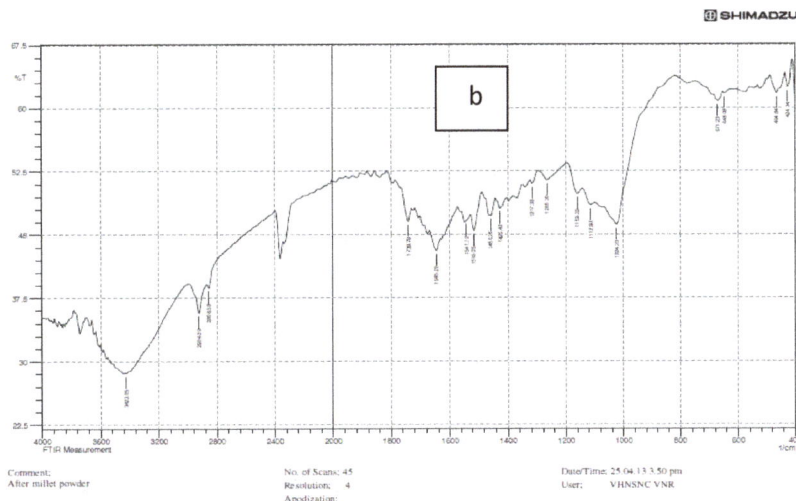

Fig. 8. (a) FTIR spectra of the PMP before the adsorption of CV, (b) FTIR spectra of the PMP after the adsorption of CV.

3.7. Adsorption Kinetics

The experimental kinetic data were analyzed to four kinetic models like pseudo-first order, pseudo-second order, and intra-particle diffusion to evaluate the adsorption mechanism. Figs. 9a-b and 10 represent the pseudo-first order, pseudo-second order and intra-particle diffusion model respectively, for the adsorption of CV onto PMP. The constants associated with these kinetic models are given in Table 2. The correlation coefficients values are greater than 0.99 in the linear plot of pseudo–second order model. The value of equilibrium capacity is found to be 25.64 mg/g which is close to the experimental value of 24.85 mg/g for the initial dye concentration of 50ppm, indicating pseudo–second order nature.

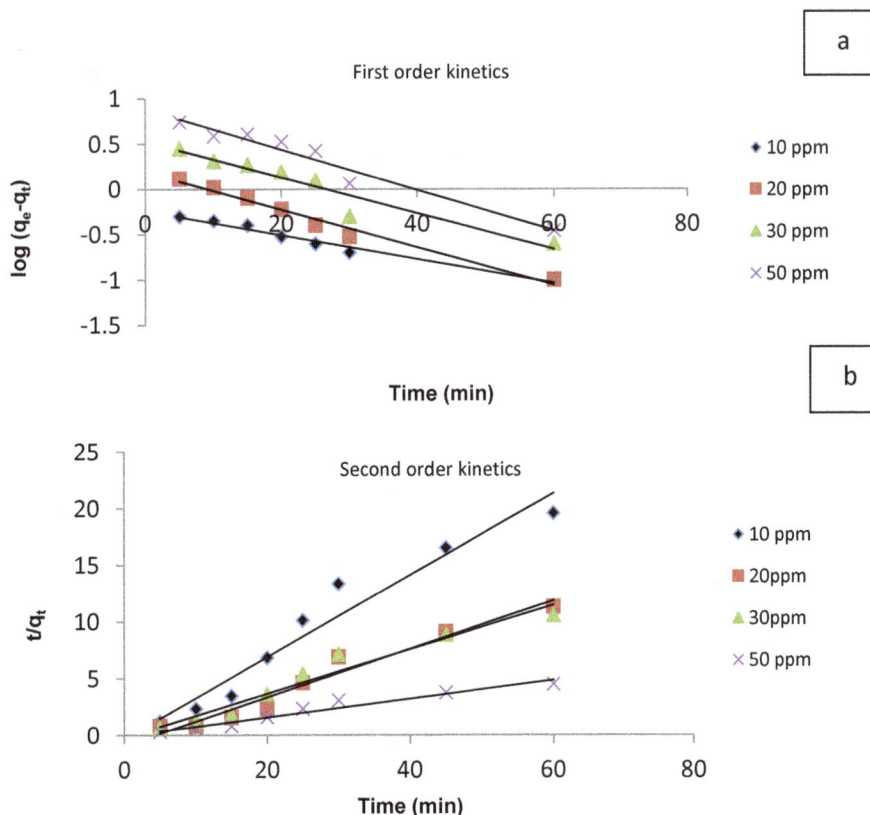

Fig. 9. (a) The pseudo-first order kinetic model for the adsorption of CV onto PMP, (b) The pseudo-second order kinetic model for the adsorption of CV onto PMP.

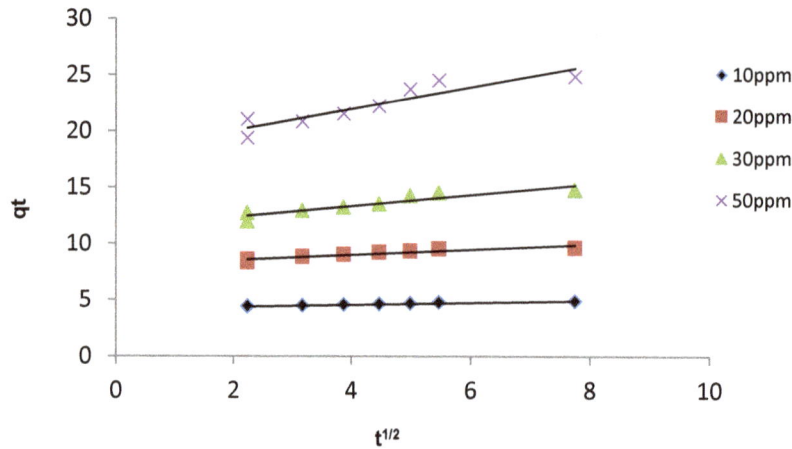

Fig. 10. Intra-particle diffusion model for the adsorption of CV onto PMP.

Table 2. Kinetic parameters for the adsorption of CV onto PMP.

C_0 (mg/L)	Pseudo-first order				Pseudo-second order			Intra-particle diffusion		
	q_e	q_f (mg/g)	k_f (min^{-1})	R^2_f	q_s (mg/g)	k_s x 10^{-2} (g/mg/ min)	R^2_s	k_{id} (mg/g/ min$^{1/2}$)	I	R^2_{id}
10	4.85	1.74	0.0299	0.988	4.901	1.12	0.999	0.071	4.208	0.980
20	9.6	1.54	0.0483	0.989	16.12	11.42	0.941	0.232	8.041	0.933
30	14.75	3.341	0.046	0.959	15.15	623.03	0.999	0.382	11.491	0.941
50	24.85	7.62	0.0506	0.975	25.64	165.80	0.999	0.769	18.755	0.945

3.8. Thermodynamic Studies

Fig. 11 shows the relationship between K_c and T for the adsorption of CV onto PMP. The calculated thermodynamic constants are given in Table 3. The negative value of $\Delta G°$ indicates the spontaneous nature of CV adsorption onto PMP However, the value of $\Delta G°$ decreased with an increase in temperature, indicating that the spontaneous nature of adsorption is inversely proportional to the temperature (Bulut and Tez. 2007). The positive values of $\Delta H°$ show that the adsorption is endothermic nature. The adsorption process in the solid-liquid system is a combination of two processes: i) desorption of the molecules of solvent (water) previously adsorbed, and ii) the adsorption of adsorbate species. The CV ions have to displace more than one water molecule for their adsorption and thus results in the endothermicity of the adsorption process.

The positive value of $\Delta S°$ suggests increased randomness at the solid /solution interface during the adsorption of CV onto PMP. The enhancement of adsorption at higher temperatures may be attributed to the enlargement of pore size and/or activation of the adsorbent surface (Gaballah and Kilbertus. 1998).

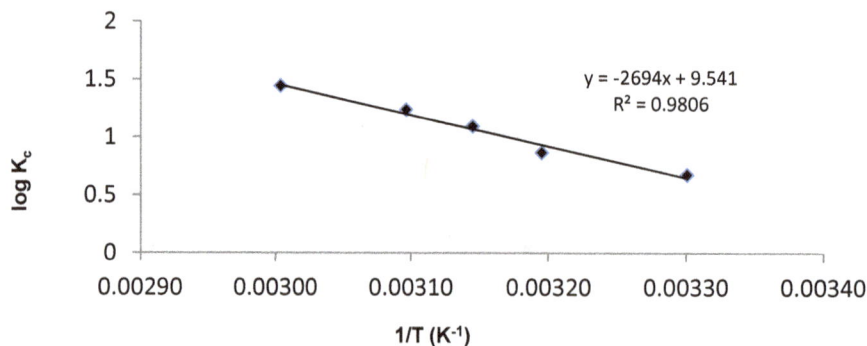

Fig. 11. Plot of log K_c versus 1/T for the adsorption of CV onto PMP.

Table 3. Thermodynamic parameters for the adsorption of CV onto PMP.

Adsorbents	$\Delta G°$(kJ/mol) at 313 K	$\Delta H°$ (kJ/mol)	$\Delta S°$ (kJ/mol)
PMP	-5.6031	51.582	0.1827

4. Conclusion

We have introduced an appropriate new adsorbent in the place of activated carbon. The current investigation shows that PMP is an effective adsorbent for the removal of CV from aqueous solution. From the studies it is observed that:

- ➢ The adsorption capacity (Q_o) of PMP was found to be 48.535mg/g
- ➢ Langmuir and Freundlich models fitted better than the Temkin model.

- ➢ Sorption kinetic is quick and the data agree well with pseudo-second order kinetic model.
- ➢ The data reported shows the adsorption process is endothermic in nature.

References

Al-Degs Y.S., El-Barghouthi M.I., El-Sheik A.H., Walker G.M., Effect of solution pH, ionic strength, and temperature on adsorption behavior of reactive dyes on activated carbon, Dyes Pigments 77 (2008) 16-23.

Allen S.J., Mckay G., Porter J.F., Optimisation of adsorption isotherm models for the prediction of basic dye adsorption by peat in single and binary component systems, Journal of Colloid and Interface Science 280 (2004) 322-333.

Bangash F. K., Alam S., Adsorption of acid blue 1on activated carbon produced from the wood of Ailanthus altissima, Brazilian Journal of Chemical Engineering 26 (2009) 275-285.

Bockris J.O. M., Environmental chemistry, plenum press, New York, (1997).

Bulut Y. and Tez Z., Adsorption studies on ground shells of hazelnut and almond, Journal of Hazardous materials 149 (2007) 35-41.

Carliell C.M., Barclay S.J., Naidoo N., Buckly C.A., 'Microbial decolourisation of a reactive azo dye under anaerobic condition' Water SA 21 (1995) 61-69.

Crini G., Non-conventional low-cost adsorbents for dye removal: a review, Dyes pigments 77 (2008) 415-426.

Forgacs E., Serhati T.C., Oros G., Removal of synthetic dyes from wastewaters: a review, Environment International 30 (2004) 953-971.

Gaballah I., Kilbertus G., Recovery of heavy metal ions through decontamination of synthetic solutions and industrial effluents using modified barks, Journal of Geochemistry Explorer 62 (1998) 241-286.

Garg V.K., Gupta R., Bala Yadav A. Kumar R., Dye removal from aqueous solution by adsorption on treated sawdust, Bioresource Technology 89 (2003) 121-124.

Hocking M.B, Hand Book of chemistry technology and pollution control, Acadamic press, Ed. 3, (2005).

Dilip M., Venkateswaran P., Palanivelu K., Removal of textile dyes from textile dye effluent using TBAB based aqueous biphasic systems, Journal of Environmental Science and Engineering 47 (2005) 176 – 181.

Nawar S.S., Doma H.S., Removal of dyes from effluents using low-cost agricultural by-products, Science of the Total Environment 79 (1989) 271-279.

Ozer A., Dursun G., Removal of methylene blue from aqueous solution by dehydrated wheat bran carbon, Journal of Hazardous materials 146 (2007) 262-269.

Paul Drinkwater, "Gentian violet – is it safe" The Australian and New Zealand, Journal of Obstetrics and Gynaecology, 30 (1990) 65-66,

Rafatullah M., Sulaiman O., Adsortion of methylene blue on low-cost adsorbents-A review, Journal of Hazardous materials 177 (2010) 70-80.

Rajavel G., Anathanarayanan C., Prabhakar L.D., Palanivel C., Removal of dark green PLS dye from textile industrial waste through low cost carbons, Indian Journal of Environmental Health 45 (2003) 195-202.

Sharma B K., Kaur H., Air pollution Goel publishing house, Meerut, (1994).

Ubale M., Shelke R., Bharad J., Madje B., Adsorption of acid dyes from aqueous solution onto the surface of acid activated kammoni leaf powder: A case study, Journal of Chemical Pharmlogical Research 2 (2010) 747-753.

Zafar M.N., Nadeem R., Hanif M.A., Biosorption of Nickel from Protonated Rice Bran, Journal of Hazardous materials 143 (2006) 478-485.

Permissions

All chapters in this book were first published in JARWW, by Razi University; hereby published with permission under the Creative Commons Attribution License or equivalent. Every chapter published in this book has been scrutinized by our experts. Their significance has been extensively debated. The topics covered herein carry significant findings which will fuel the growth of the discipline. They may even be implemented as practical applications or may be referred to as a beginning point for another development.

The contributors of this book come from diverse backgrounds, making this book a truly international effort. This book will bring forth new frontiers with its revolutionizing research information and detailed analysis of the nascent developments around the world.

We would like to thank all the contributing authors for lending their expertise to make the book truly unique. They have played a crucial role in the development of this book. Without their invaluable contributions this book wouldn't have been possible. They have made vital efforts to compile up to date information on the varied aspects of this subject to make this book a valuable addition to the collection of many professionals and students.

This book was conceptualized with the vision of imparting up-to-date information and advanced data in this field. To ensure the same, a matchless editorial board was set up. Every individual on the board went through rigorous rounds of assessment to prove their worth. After which they invested a large part of their time researching and compiling the most relevant data for our readers.

The editorial board has been involved in producing this book since its inception. They have spent rigorous hours researching and exploring the diverse topics which have resulted in the successful publishing of this book. They have passed on their knowledge of decades through this book. To expedite this challenging task, the publisher supported the team at every step. A small team of assistant editors was also appointed to further simplify the editing procedure and attain best results for the readers.

Apart from the editorial board, the designing team has also invested a significant amount of their time in understanding the subject and creating the most relevant covers. They scrutinized every image to scout for the most suitable representation of the subject and create an appropriate cover for the book.

The publishing team has been an ardent support to the editorial, designing and production team. Their endless efforts to recruit the best for this project, has resulted in the accomplishment of this book. They are a veteran in the field of academics and their pool of knowledge is as vast as their experience in printing. Their expertise and guidance has proved useful at every step. Their uncompromising quality standards have made this book an exceptional effort. Their encouragement from time to time has been an inspiration for everyone.

The publisher and the editorial board hope that this book will prove to be a valuable piece of knowledge for researchers, students, practitioners and scholars across the globe.

List of Contributors

Negar Amiri, Mojtaba Ahmadi, Yasser Vasseghian and Pegah Amiri
Chemical Engineering Department, Faculty of Engineering, Razi University, Kermanshah, Iran

Meghdad Pirsaheb
Department of Environmental Health Engineering-Kermanshah Health Research Center (KHRC), Kermanshah University of Medical Sciences, Iran

Nader Bahramifar and Habibollah Younesi
Department of Environmental Science, Faculty of Natural Resources, Tarbiat Modares University, Imam Reza Street, Noor, Iran

Maryam Tavasolli
Department of chemistry, Payame Noor University (PNU), Tehran, Iran

Kanakkan Ananthakumar
Department of Chemistry, Kamarajar Govt. Arts College, Surandai – 627 859. Tamil Nadu, India

Zahra Rahimi, Ali Akbar Zinatizadeh and Sirus Zinadini
Water and Wastewater Research Center (WWRC), Department of Applied Chemistry, Faculty of Chemistry, Razi University, Kermanshah, Iran

Mehraban Sadeghi, Akram Najafi Chaleshtori and Neda Masoudipour
Department of Environmental Health Engineering, School of Public Health, Shahrekord University of Medical Sciences, Shahrekord, Iran

Behnam Zamanzad
Department of Microbiology, School of Health, Shahrekord University of Medical Sciences, Shahrekord, Iran

Lee Fergusson
Principal Consultant, Prana World Consulting, Oxenford, Queensland 4210, Australia

Golshan Moradi and Laleh Rajabi
Polymer Research Center, Department of Chemical Engineering, College of Engineering, Razi University, Kermanshah, Iran

Farzad Dabirian
Department of Mechanical Engineering, Razi University, Kermanshah, Iran

Laleh Rajabi
Department of Chemistry, University of Victoria, 2329 West Mall, Vancouver, BC V6T 1Z4, Canada

Ali Ashraf Derakhshan
Environmental Research Center, Faculty of Chemistry, Razi University, Kermanshah, Iran

Parviz Mohammadi and Shaliza Ibrahim
Department of Civil Engineering, Faculty of Engineering, University of Malaya, 50603 Kuala Lumpur, Malaysia

Parviz Mohammadi
Department of Environmental Health Engineering, Public Health Faculty, Kermanshah University of Medical Science, Kermanshah, Iran

Mohamad Suffian Mohamad Annuar
Institute of Biological Sciences, Faculty of Science, University of Malaya, 50603 Kuala Lumpur, Malaysia

Masoud Shariati-Rad, Mohsen Irandoust and Somayyeh Amri
Department of Analytical Chemistry, Faculty of Chemistry, Razi University, Kermanshah, Iran

Mostafa Feyzi and Fattaneh Ja'fari
Department of Physical Chemistry, Faculty of Chemistry, Razi University, Kermanshah, Iran

Mohammad Eisapour Chanani, Nader Bahramifar and Habibollah Younesi
Department of Environmental Sciences, Faculty of Natural Resources and Marin Sciences, Tarbiat Modares University, Noor, Iran

Masoud Hatami, Habibollah Younesi and Nader Bahramifar
Department of Environmental Science, Faculty of Natural Resources, Tarbiat Modares University, Tehran, Iran

Peyman Mahmoodi, Mehrdad Farhadian, Ali Reza Solaimany Nazar and Amin Noroozi
Department of Chemical Engineering, University of Isfahan, Isfahan, Iran

Adeleh Afroozan, Ali Mohammad-khah and Farhad Shirini
Department of Chemistry, Faculty of Sciences, University of Guilan, Rasht, Iran

Godfred Owusu-Boateng and Victoria Adjei
Faculty of Renewable and Natural Resources, Kwame Nkrumah University of Science and Technology, Kumasi, Ghana

R.Kannan Seenivasan
Department of Chemistry, Government Arts College, Tamil Nadu, India

Veerasamy Maheshkumar and Palanikumar Selvapandian
Post Graduate and Research Department of Chemistry, Raja Doraisingam Government Arts College,Tamil Nadu, India

Negin Shaabani, Sirus Zinadini and Ali Akbar Zinatizadeh
Environmental Research Center, Department of Applied Chemistry, Razi University, Kermanshah, Iran

Vitthal L. Gole and Apurva Alhat
Department of Chemical Engineering, AISSMS College of Engineering, Kennedy Road, MS, India

Batool Shahroie and Laleh Rajabi
Polymer Research Center, Department of Chemical Engineering, Razi University, Kermanshah, Iran

Ali Ashraf Derakhshan
Environmental Research Center, Faculty of Chemistry, Razi University, Kermanshah, Iran

Ali Roholamin Kasmaei
Member of Water and Sewer Company of Gilan, Rasht, Iran

Mehdi Nezhad Naderi and Zaynab Bahrami
Department of Civil Engineering, Tonekabon Branch, Islamic Azad University, Tonekabon, Iran

Sirus Zinadini and Foad Gholami
Environmental research center, Department of Applied Chemistry, Faculty of Chemistry, Razi University, Kermanshah, Iran

Arezoo Fereidonian Dashti, Mohd Nordin Adlan, Hamidi Abdul Aziz and Ali Huddin Ibrahim
School of Civil Engineering, Engineering Campus, Universiti Sains Malaysia, Nibong Tebal, Penang, Malaysia

Maryam Habibi and Ali Akbar Zinatizadeh
Water and Wastewater Research Center (WWRC), Department of Applied Chemistry, Faculty of Chemistry, Razi University, Kermanshah, Iran

Mandana Akia
Department of Mechanical Engineering, University of Texas Rio Grande Valley, Edinburg, USA

Behnaz Jalili and Seyed Mehdi Borghei
Department of Chemical and Petroleum Engineering, Sharif University of Technology,Tehran, Iran

Vahid Vatanpour
Department of Applied Chemistry, Faculty of Chemistry, Kharazmi University, Tehran, Iran

Christopher Sarkizi
Chemical Engineering Department, Tarbiat Modares University, Tehran, Iran

Saeed Aghel, Nader Bahramifar and Habibollah Younesi
Department of Environmental pollution, Faculty of Natural Resources & Marine Science, Tarbiat Modares University, Tehran, Iran

Zahra Shaykhi Mehrabadi
Young Researches Club, Kermanshah Islamic Azad University (IAUKSH)

Soraya Mohajeri and Hamidi Abdul Aziz
School of Civil Engineering, Universiti Sains Malaysia, Nibong Tebal, Pinang, Malaysia

Mohamed Hasnain Isa
Civil Engineering Department, Universiti Teknologi PETRONAS, Perak, Malaysia

Mohammad Ali Zahed
Faculty of biological sciences, Kharazmi University, Tehran, Iran

Mojtaba Ahmadi and Pegah Amiri
Chemical Engineering Department, Faculty of Engineering, Razi University, Kermanshah, Iran

Ali Ahmadpour
Department of Chemical Engineering, Faculty of Engineering, Ferdowsi University of Mashhad, Mashhad, Iran

Mohammad Zabihi
Chemical Engineering Faculty, Sahand University of Technology, Sahand New Town, Tabriz, Iran

Tahereh Rohani Bastami and Ali Ayati
Department of Chemical Engineering, Quchan University of Advanced Technology, Quchan, Iran

Masoomeh Tahmasbi
Department of Chemistry, Ferdowsi University of Mashhad, Mashhad, Iran

Mazen Hamada
Department of Chemistry, Faculty of Science, Al Azhar University, Gaza Strip, Palestine

Hossam Adel Zaqoot
Environment Quality Authority, Gaza Strip, Palestine

Ahmed Abu Jreiban
Institute of Water and Environment, Al Azhar University, Gaza Strip, Palestine

Palanikumar Selvapandian and Arulsamy Cyril
Department of Chemistry, Raja Doraisingam Govt. Arts College, Sivagangai – 630 561. Tamil Nadu, India

Index

www.ingramcontent.com/pod-product-compliance
Lightning Source LLC
Chambersburg PA
CBHW080259230326
41458CB00097B/5202